China's Space Program – From Conception to Manned Spaceflight

Springer
London
Berlin
Heidelberg
New York
Hong Kong
Milan
Paris
Tokyo

Brian Harvey

China's Space Program – From Conception to Manned Spaceflight

 Springer

Published in association with
Praxis Publishing
Chichester, UK

Brian Harvey, M.A., H.D.E., F.B.I.S.
2 Rathdown Crescent
Terenure
Dublin 6W
Ireland

SPRINGER–PRAXIS BOOKS IN ASTRONOMY AND SPACE SCIENCES
SUBJECT *ADVISORY EDITOR*: John Mason B.Sc., M.Sc., Ph.D.

ISBN 1-85233-566-1 Springer-Verlag Berlin Heidelberg New York

Springer-Verlag is a part of Springer Science + Business Media (*springeronline.com*)

British Library Cataloguing-in-Publication Data
Harvey, Brian, 1953–
 China's space program: from conception to manned
 spaceflight. – (Springer-Praxis books in astronomy and
 space sciences)
 1. Astronautics – China
 I. Title
 387.8′0951

ISBN 1-85233-566-1

Library of Congress Cataloging-in-Publication Data
Harvey, Brian, 1953–
 China's space program: from conception to manned spaceflight / Brian Harvey.
  p. cm.
 Includes bibliographical references and index.
 ISBN 1-85233-566-1 (alk. paper)
 1. Astronautics–China–History. I. Title.
TL789.8.C55H36 2004
629.4′0951–dc22
                                                                                2004041731

Apart from any fair dealing for the purposes of research or private study, or criticism or review, as permitted under the Copyright, Designs and Patents Act 1988, this publication may only be reproduced, stored or transmitted, in any form or by any means, with the prior permission in writing of the publishers, or in the case of reprographic reproduction in accordance with the terms of licences issued by the Copyright Licensing Agency. Enquiries concerning reproduction outside those terms should be sent to the publishers.

© Copyright, 2004 Praxis Publishing Ltd.

The use of general descriptive names, registered names, trademarks, etc. in this publication does not imply, even in the absence of a specific statement, that such names are exempt from the relevant protective laws and regulations and therefore free for general use.

Cover design: Jim Wilkie
Typesetting: BookEns Ltd, Royston, Herts., UK

Printed in the United States of America on acid-free paper

Table of Contents

Preface xi
Acknowledgements xv
List of illustrations xvii

1 THE FLIGHT OF YANG LIWEI 1

2 ORIGINS: THE FIERY DRAGON 15
    Tsien Hsue Shen 17
    On the trail of the German rocketeers 19
    Return to China 21
    Scientific construction in China 22
    Foundation: 8 October 1956 22
    The Fifth Academy 22
    Chinese–Soviet cooperation 24
    Project 1059: the challenge of copying a Russian rocket 24
    The idea of a Chinese Earth satellite 25
    Project 581 27
    Earth satellite shelved; rocket programme continues 27
    Development of a sounding rocket 29
    Launch site in the Gobi desert 31
    The great split 34
    Assessment and conclusions 35
    References 35

3 DONG FANG HONG – THE EAST IS RED 37
    Military imperatives: Dong Feng 2 37
    China a nuclear power 38
    Dong Feng 3 and Dong Feng 4 39
    Dong Feng 5: China's long-range rocket 40
    A submarine-launched missile 42

China's rocket strike force 42
Sounding rockets make progress 43
Biological sounding rockets 44
Dogs fly into the atmosphere: the missions of Xiao Bao and Shan Shan . . . 44
Resumption of the Earth satellite project 46
Project 651. 47
Reorganization and dispersal 48
A 64-day design conference 49
The cultural revolution engulfs China 50
A rocket for China's first Earth satellite. 52
Design challenges of the Long March 1 52
The problem of the third stage of the Long March 1 53
Long March 1: the final laps 55
The moment of truth 56
'We did it through our own efforts' – Zhou Enlai 58
Dong Fang Hong: the aftermath 60
A second satellite: Shi Jian 1 60
Epilogue: Tsien Hsue Shen. 62
Assessment and conclusions 64
References 65

4 EXPANDING THE SPACE PROGRAMME 67
'No space race' 67
Four modernizations 68
Project 701: the Ji Shu Shiyan Weixing series (1973–76) 70
The Feng Bao rocket 71
Shanghai mobilizes. 72
First hot test 72
In orbit at last 73
The recoverable satellite programme 75
The challenges 76
Mission profile 79
A launcher for the FSW: the Long March 2 79
A broken wire 82
Coasting to orbit 83
Helicopters scrambled 84
A crisis-ridden second mission 85
An improved version 86
Introduction of microbiology experiments 87
Shi Jian 2. 87
New cargo 88
First 3-in-1 launch attempt. 89
Shi Jian 4 and Shi Jian 5 91
Assessment and conclusions 92
References 93

5   COMMUNICATIONS AND CONSPIRACIES ... 95
    Communications satellites: Project 331 ... 95
    Mao's last decision ... 96
    The rocket ... 97
    Configuring the launcher ... 98
    A new engine: introducing the YF-73 ... 98
    Arriving on station: the apogee motor ... 99
    Troubles, leaks, explosion ... 100
    A new launch site needed ... 101
    The communications satellite ... 102
    Preparing for the first launch ... 103
    Beating the summer thunder ... 104
    Zhang Aiping's telephone call ... 105
    New languages from space ... 106
    Dong Fang Hong 2A: 3,000 telephone calls at a time ... 107
    The Dong Fang Hong 3 series: 8,000 telephone calls at a time ... 109
    Benefits of comsats ... 113
    Commercialization of the Chinese space programme ... 115
    Restrictions, prices and quotas ... 115
    Export licence crux ... 116
    First commercial mission ... 117
    Fire on the pad! ... 119
    Débris, recrimination ... 120
    Long March crashes in flames ... 121
    Saint Valentine's day massacre ... 122
    Loss of confidence ... 122
    The Cox report ... 124
    Iridium ... 131
    Assessment and conclusions ... 133
    References ... 134

6   APPLYING THE SPACE PROGRAMME ... 135
    Meteorological satellites (Feng Yun) ... 135
    First polar, Sun-synchronous, Taiyuan, Long March 4 ... 137
    Feng Yun 1-2 carries balloons ... 138
    Haiyang – new ocean satellite ... 140
    A geosynchronous meteorological satellite ... 141
    A fuelling disaster in the processing hall ... 142
    Recoverable satellites: the new FSW 1 series ... 143
    China flies animal passengers into orbit ... 144
    Rogue satellite on the loose ... 144
    FSW 2 series: manoeuvrable, heavier, 18-day profile ... 145
    Results from the FSW series ... 147
    CBERS and Zi Yuan Earth resources programmes ... 153
    Navigation satellites: stars of the heavens ... 157

viii **Table of Contents**

      Microsatellites: smaller and smaller 160
      Assessment and conclusions 164
      References 164

7 BEHIND THE SCENES 167
      Designers and engineers: the problem of the intellectual and 167
        the revolution
      The first group of chief designers 169
      Design bureaux and organizations 169
      Chinese National Space Administration (CNSA) 170
      China Aerospace Corporation 171
      The testing infrastructure 176
      Institute for Aviation and Space Medicine 180
      Chinese Academy of Sciences 181
      Mission control, Xian 182
      Comships 185
      The space programme and the Chinese economy 187
      International contact 189
      Assessment and conclusions 192
      References 193

8 LAUNCH CENTRES, ROCKETS AND ENGINES 195
      Launch sites 195
      Jiuquan 196
      Xi Chang 204
      Taiyuan 206
      Hainan 208
      Launchers 209
      Sounding rockets 228
      Rocket engines 229
      Reliability 233
      Assessment and conclusions 236
      References 237

9 SHUGUANG'S FALSE DAWN 239
      Origins of a Chinese manned space programme 239
      First astronauts 241
      Disbanded 243
      Rumours and refutations 244
      Fresh plans 246
      Decision 247
      Renewal of contact with Moscow 248
      Mission control 252
      Project 921 in the open 253
      Assessment and conclusions 253

References 254

10 THE SHENZHOU MISSIONS 257
Roll-out 257
Shenzhou 1 flies! 259
Aftermath 261
Shenzhou 2 266
Shenzhou 2 postscript 270
Shenzhou 3: the long wait 273
Jiang, yuhangyuan at launch site 274
Mongolian landfall 276
Last trial: Shenzhou 4 278
What next? 282
Assessment and conclusions 287
References 288

11 THE FUTURE OF CHINA IN SPACE 291
The Chinese space programme in international perspective 291
China's space budget 293
The rhythm of China's space programme 294
Official policy: the white paper 296
A Chinese space shuttle 299
On to the Long March 5 301
Small launchers 305
New spacecraft programmes 306
Assessment and conclusions 315
References 317

List of Chinese satellite launchings 319
Principal milestones in the development of the Chinese space programme 323
Glossaries 329
Bibliography 334
Index 339

Preface

On 15th October 2003, China became only the third country in the world to put its own astronaut into orbit, following the Soviet Union's Yuri Gagarin in 1961 and John Glenn of the United States in 1962. China's term for a space flier was 'yuhangyuan', joining the now familiar vocabulary of 'cosmonaut' and 'astronaut'. The world realized that China had become a world space superpower and Asia's leading spacefaring nation.

The emergence of China as a spacefaring nation should, over the long course of history, be no surprise. Way back in what were sometimes called the dark ages in Europe, the rocket had been invented in China. In the 20th century, many of the engineering calculations necessary for rocket flight were done by one of the world's great space designers, Tsien Hsue Shen. The Chinese space programme was founded on 8th October 1956, a year before the first Sputnik was launched. On that day, China's political leadership decreed the foundation of the Fifth Academy to spearhead China's space effort and requisitioned two abandoned sanatoria to be its first laboratories. Had it not been for subsequent political upheaval – the great leap forward and the cultural revolution – China might have achieved much more, much sooner.

As it was, China's first satellite in orbit was the biggest of the superpowers. China was the third space power to recover its own satellites, put animals in orbit and develop hydrogen-fuelled upper stages. By the end of the century, China has launched over 60 satellites, including spacecraft for Earth observations, navigation, communications, weather forecasting and materials processing.

The Chinese space programme has sometimes been called the last of the secret space programmes. In its earlier days, China reminded one of what the Soviet space programme used to be like before the torrent of *glasnost* swept away its culture of secrecy. Although the Chinese space programme is far less secretive than it is often portrayed, there are still some blank pages to be filled in. A programme of military satellites was flown in the 1970s, the Ji Shu Shiyan Weixing, about which we still know virtually nothing. Details about many key points of development are few and far between and the diaries and memories of the key personalities of the period have yet to be written and published. The relationship between the designers, the academies and the political leadership remains *terra incognita*. Writing about the early Chinese space programme is like trying to assemble a jigsaw where some of the pieces are not coloured in and others are missing altogether. From the late 1980s, secrecy began to relax and by the time of Yang Liwei's flight, many parts of the programme were publicized in some detail. The Chinese became more and more comfortable about declaring their future intentions.

A complicating feature – one familiar to students of the Soviet space programme – is the use of different designators for the same satellites. In the west, Chinese satellites were named China 1,2, 3 and so on, also PRC-1, PRC-2 (People's Republic of China) and even Mao 1,2 and 3. At the time, the Chinese simply referred to these missions by their date of launch. Eventually, the Chinese introduced a set of designators and applied them retrospectively. That should have been an end to the problem, but the Chinese then revised some of these designators several times over – and then changed them again! This is not just a historical problem, but one that applies to some present programmes. Not only have satellites had their designators changed, but many facilities have changed names and identity. As if this were not complicated enough, inconsistent translations mean that many institutes, bodies and organizations acquire, over time, slightly different names. Sometimes similar sounding names turn out to be the same thing – but sometimes not. The Chinese also applied a series of numerical codes to their various space projects. Some were based on dates, others not. All this must be carefully disentangled.

Penetrating the fog enveloping some aspects of the Chinese space programme is one problem. The level of western misunderstanding of the programme is a challenge of similar magnitude. With some honourable exceptions, many in the western media who ought to know better responded to Chinese space developments with a mixture of puzzlement, patronizing downputting and dismissal. Chinese capabilities are often played down on the basis that their equipment is alternately primitive or imitative. If it works, the presumption is that it must have been stolen. When the first pictures of the Shenzhou's launcher, the Long March 2F, were published, they were quickly rubbished as a computer-designed forgery. When the rocket actually flew, the Chinese were then accused of robbing the design from elsewhere. Some seemed to deny the Chinese the credit of having created, designed and built their own equipment. This is a problem not peculiar to the space programme, for the west often forgets how China pioneered so many things – from medicine to mathematics, the idea of the suspension bridge, paper-making, the compass, chemistry, printing. The observations by the ancient Chinese astronomers are now recognised for their accuracy.

This book is the history of the Chinese space programme from its earliest times to the historic breakthrough of manned flight. The book was originally published in its first edition as *The Chinese space programme – from conception to future capabilities* by Praxis/Wiley in 1998. Much has happened since then, of which the first manned flight is the best-known feature. China has surged ahead with its unmanned programmes, developed a line of minisatellites and planned new launchers. Exciting projects are in the pipeline, like the planned lunar probe.

Here, chapter 1 tells the story of the flight of the Yang Liwei. Chapter 2, *The fiery*

Preface xiii

dragon, tells of the historical origins of the Chinese space programme. Chapter 3, *Dong Fang Hong*, is named after the first Chinese satellite to orbit the Earth, in 1970 and gives an account of the events leading up to China's arrival as a space power. In chapter 4, the story of the expansion of the space programme in the 1970s is told. Chapter 5 relates how China developed communications satellites and new, powerful rockets in the 1980s, only to see its efforts to commercialize its launchers caught up in the difficult international diplomatic environment of the 1990s. Chapter 6, *Applying the space programme*, tells how China applied its space programme to develop its current range of navigation, observation, weather forecasting and applications satellites. Chapter 7 goes behind the scenes to describe China's space infrastructure – launch sites, tracking facilities, industrial plant and organization while chapter 8 looks at the hardware which makes it all possible – its launchers and engines. Chapter 9 begins the story of Chinese manned spaceflight and how China planned to put its first yuhangyuan into space in the early 1970s in the Shuguang (Dawn) project – but it was before its time and turned out to be a false dawn. Chapter 10, *The Shenzhou missions* recounts how the Chinese resurrected the manned spaceflight programme, project 921. Finally, chapter 11 takes a glance at China's space plans and assesses its future capabilities.

Acknowledgements

Many people provided generously of their time and energies so that this book and its predecessor could be written. I especially wish to thank Rex Hall and Phil Clark who shared their knowledge, files and information over many years. Many others kindly provided reports, information and advice, such as Dave Shayler, Anders Hansson, Dwayne Day, David Harland, Paolo Ulivi and Jim Oberg. Neil da Costa provided photographs of Chinese astronaut instructors in training while He Ying of *Aerospace China* gave permission for the use of photographs; as did Audrey Nice of Surrey Satellite Technologies Ltd (SSTL). Permission to use photographs was given by Deng Ningfeng and Hou Mingliang of the China Astronautic Publishing House and I am also grateful to Lu Danlei for assistance. James Harford in the United States gave me his ground-breaking report on the first western visit to Chinese space facilities. Mark Wade kindly gave permission for the extensive use of graphic material. A number of valuable and important photographs were kindly supplied by Sven Grahn of Sweden and reproduced with his permission. Other photographs are reproduced by permission of Getty Images, Reuters and the Smithsonian Institution. From China I received the assistance of Dr Zhu Yilin of the Chinese Academy of Space Technology. Many individuals, institutes and companies in the Chinese space programme assisted with information, photographs and in other ways and these included Winnie Pang, Corporate Affairs Manager, Asiasat in Hong Kong; China National Space Administration; China Aerospace Corporation and the Great Wall Industry Corporation in Beijing. I am grateful to them all. Other photos come from the first edition and the author's collection.

List of illustrations

1 The flight of Yang Liwei
   Yang ready to go.. 2
   The launch of Shenzhou 5 7
   Yang in the the cabin... 9
   Yang back on Earth 12
   Shenzhou 5 back on Earth...................................... 12
   Yang's triumphant return...................................... 13

2 Origins: the fiery dragon
   Ancient Chinese astronomical instruments 15
   Fiery dragon 17
   Tsien's graduation 18
   Tsien's early spaceplane 20
   Zhou Enlai greets Tsien.. 21
   Huang Weilu 23
   Tracking early satellites 26
   Mao Zedong 26
   Deng Xiao Ping... 28
   Tsien and a sounding rocket 30
   Mao Zedong examines a rocket 31
   Early days in Jiuquan 32
   Construction in Jiuquan.. 33

3 Dong Fang Hong – the East is Red
   Tu Shoue... 41
   Xiao Bao 45
   Chinese space mice... 46
   Tsien during the 1960s... 47
   Zhou Enlai... 51
   Dong Fang Hong... 53
   The Long March 1 launch pad.................................... 55
   Long March 1 on the launch pad................................. 56

xviii List of illustrations

Dong Fang Hong 1 57
Dong Fang Hong 1, close up 59
Dong Fang Hong on show 61
Shi Jian 1 62

4 **Expanding the space programme**
Feng Bao 71
The FSW design 75
Yang Yiachi 77
Long March 2 ready for launch 80
FSW, prepared for launch 82
FSW reentry test 84
FSW recovery 85
Shi Jian 2 in assembly 90
Shi Jian 2 90
Shi Jian 5 92

5 **Communications and conspiracies**
Engines undergo tests 99
An apogee motor 100
The liquid hydrogen engine 101
Early building at Xi Chang 102
Long March 3 launch 104
Zhang Aiping's telephone call to Weng Enmiao 106
Dong Fang Hong 2 comsat 107
A satellite dish in a rural area 108
A satellite TV receiver 110
Dong Fang Hong 3 comsat 111
Satellites to computers 112
Satellites in rural areas 113
A future comsat 114
The Asiasat footprint 118
The launch of Freja 119
Long March 2E launch 120
Asiasat 2 121
Long March 3B on the pad at Xi Chang 123
Long March 3A night launch 126
Hardest hit: the Long March 2E 128
Ground comsat terminal 129
Luan Enjie 131

6 **Applying the space programme**
Feng Yun 1 series 136
Feng Yun 1 in preparation 136
The Long March 4 pad at Taiyuan 137

Feng Yun 1 panel test 138
Feng Yun 1 139
Haiyang 1 140
Feng Yun 2 141
FSW readied for the mission 146
An experiment with silkworms. 147
FSW in orbit. 148
Caterpillars – before and after 149
FSW imaging from orbit 150
A Feng Yun 1 image 151
Long March 2 carried FSW series 152
CBERS 1. 154
Beidou ready for launch. 158
Preparing Beidou 159
Beidou control. 159
The Tsinghua team 161
Tsinghua rendezvous 162
Tsinghua in orbit. 162
An environmental microsatellite. 163

7 **Behind the scenes**
Vibration test tower. 172
Vacuum testing. 173
Testing a satellite. 177
Rocket engine test stand 178
Thermal vacuum testing. 179
Large vacuum test facility 180
Mission control, Xian 183
A tracking station 184
A *Yuan Wang* tracking ship. 186
Cooperation with Italy. 190
Cooperation with Brazil. 190

8 **Launch centres, rockets and engines**
Jiuquan: an overview. 197
Rail transport, Jiuquan 198
The Swedish team with Freja. 199
Convoy on way to the pad. 200
The satellite assembly building. 200
The tower at Jiuquan. 201
Jiuquan: the new pads 203
Xi Chang: an overview. 204
The clean room 205
Taiyuan: an overview. 207
Taiyuan launch centre 207

xx List of illustrations

Long March evolution 210
Long March 2C launch 212
Long March 2 launch 215
Long March 2EA design 217
Long March 2F 219
Long March 3A launch 221
Long March 3B 223
Long March 4 night launch 225
He Ping 2 228
The YF-2A engine 230
The Long March family 234
Long March 4B leaving Taiyuan 236

9 **Shuguang's false dawn**
Shuguang 240
Possible Shuguang design 244
Project 921: the original design 247
Li Tsinlong 249
Wu Tse 250
Training in Moscow 251

10 **The Shenzhou missions**
Long March 2F 258
Shenzhou night launch 260
Shenzhou: the spaceship 263
The escape system 265
Shenzhou: a side view 266
Shenzhou deploying its panels 267
Shenzhou: a view from beneath 268
Shenzhou 2's landing track 270
Shenzhou's descent 271
Shenzhou reaches orbit 272
Long March 2F in the vehicle processing building 273
Shenzhou enters orbit 275
A dummy yuhangyuan 276
Shenzhou with sensors 278
Shenzhou 4: final tests 279
Shenzhou heads for orbit 280
Tracking Shenzhou in orbit 281
A Chinese space station 283
A Mir-class space station 285
A future orbital station 286
Robotics under test 286

11 The future of China in space

Communications – a priority 297
Quality control – an over-riding priority 298
A model of a Chinese space shuttle 300
Long March 5 design. 302
Long March 5 climbs to orbit 303
Long March 5 304
The Long March 5 family 305
Doublestar. 307
Doublestar's mission 308
The space telescope 309
Moonward bound?. 311
A Chinese lunar robot 312
A Chinese lunar base. 314
Shenzhou orbiting the Moon 315

1

The flight of Yang Liwei

Autumn had already begun to give way to winter in Jiuquan in the Chinese Mongolian desert. Heralded by a sudden blast of Siberian air, temperatures fell by 8° C to 12° C. Cold rain and snow were forecast. Launch towers, assembly buildings, gantries, machinery stood against the harsh desert landscape, with the light and dark browns of the low surrounding mountains as backdrop.

China's journey into the cosmos had begun here almost fifty years earlier when the army engineers of the 20th corps first came out here in the 1950s. Seeking a place where China could conduct missile tests in secret, they had come to this bleak spot. Jiuquan was well inland, 1,600 km north west of Beijing, about as far off the beaten track as one could find even in a country as large as China. Conditions were harsh then and until the first proper buildings were erected the engineers, soldiers, officers and rocketeers lived in tents, where they endured the bitter cold of winter, the baking heat of summer and the winds that blew across a landscape of brush and thorn trees. Although Jiuquan town was itself a small oasis ('Jiuquan' means 'liquor springs'), water was first brought in by truck until more wells were dug and, eventually, an artificial lake constructed. But the engineers had chosen well for Jiuquan offered not only secrecy but, being 1,000 m high in a mountain desert, clear skies in which to track ascending missiles and rockets.

Now in October 2003, Jiuquan was a different place. The old rocket towers and the crude but useful cement launch pads built by the pioneers of the 20th corps are still there, but Jiuquan space city had now built up to a true town in the desert, with up to 20,000 people during launch campaigns, housing blocks, three parks, restaurants, kindergartens, schools, railway station, airport and recreation areas. Windbreaks broke up the howling winds and the city streets were planted with poplars and willow trees providing colour and shade in summer. Dominating the skyline was a huge, white, block shaped, reinforced concrete windowless vehicle assembly building, with two enormous vertical doors 8 m wide and, along a tarmacadamed roadway, new launchpads, with, dotted in the surrounding landscape, small tracking buildings. This was China's new journey to the cosmos. White and yellow lights gleamed in the desert night around the vehicle assembly building, the roadway, the launch pad and what was new Jiuquan.

2 The flight of Yang Liwei

Early morning, 15th October 2003, still in the darkness. 5 am. Motorcycle engines roared into life. In a few moments, five motorcycle engines were throbbing, churning fumes into the air until the engines settled down and purred a steady beat. Behind the cyclists was a small bus, ready to drive the well-rehearsed short journey down to the launch pad. The bus engine was already on. Everyone was ready to go. They waited close to the single-floor building outside the astronauts' apartments where they were kitted up for space missions. Hundreds of journalists had gathered there. Although the mission was taking place under a certain amount of controlled access to information, the wall of secrecy had broken down several days earlier and no less than 50 buses had been hired by media organizations to make a mad dash thousands of kilometres to the launch site. The Chinese military had formally checked their papers at roadblocks, but they had all been waved through in the end.

Now stepped forward a short, 38-year-old spacesuited man, Yang Liwei. His identity was no surprise at this stage, for his picture had already been published by the ever-indiscrete *Wen Wei Po* newspaper in Hong Kong. The suit was white, with blue seams, the red flag of China stitched onto his left shoulder. He carried what looked like a workman's toolbox – but in reality the all important control system for his spacesuit until he was plugged into his cabin. Photographers were ready for him and their cameras flashed the moment he emerged. He raised his white-gloved right hand to acknowledge them, smiling in his black and white communications soft hat, his visor pushed back behind his neck. Waiting for him close to the bus was none other than the President of China, Hu Jintao. The president greeted the astronaut, making a few remarks on the importance of the mission for China to the assembled press and mission officials. Yang Liwei told him: 'I will not disappoint our motherland'. This was a formal occasion, for only thirty minutes earlier Hu

Yang ready to go. (Courtesy Reuters.)

Jintao had spoken to a just-suited Yang Liwei who had conversed with him behind a glass bubble designed to protect him from infections. Now, the brief ceremony done, Yang Liwei stepped forward to board his bus down to the pad. Right behind him were his two backup pilots, Nie Haisheng and Zhai Zhigang. They were there to take his place if for some reason something went wrong, but they must have known that the chances of Yang Liwei changing his mind at this stage were as close to zero as made no difference. Hu Jintao had flown in there only the day before, coming with his senior political colleagues straight from a communist party meeting. They had toured Jiuquan and put flowers on the grave of an old military and party comrade, Marshal Nie Rongzhen, a hero of the early Chinese space programme.

5.30 am The motorcycle engines now roared and the escort moved ahead as the bus drove down to the pad. Walking out to the bus was a small matter compared to the scene down at the pad. As he climbed down the steps of the bus, Yang Liwei could see that there were now hundreds of people assembled behind an orderly line on either side. He stepped forward, two senior military men on either side, Nie Haisheng and Zhai Zhigang in blue flight suits and black military boots taking up the rear. The spacesuit is slightly awkward for walking – it is designed for sitting in a spaceship after all – and the wearer is just a little hunched, making him look shorter than his 168 cms and it took him a couple of minutes to reach a preset standing microphone at the foot of the launch tower. But the hundreds of well-wishers applauded together as he took his final walk down to the pad in grey soft boots. He smiled nervously but behind him Nie Haisheng and Zhai Zhigang had broader grins – maybe they could afford to be more relaxed. The crowd comprised men, women, schoolchildren, many in thick coats and scarves as protection against the cold but calm early morning air. Some local people were there in brightly coloured traditional dress too. They clapped and cheered, waving bouquets of flowers to wish Yang Liwei a *bon voyage*. Yang Liwei reached the microphone, saluted the commanding military officer and in a few short words briefly reported that he was ready to fly and carry out his mission. Close to hand and supervising was manned launch centre director Zhou Jianping.

Now he climbed into the elevator at the foot of the white Long March 2F rocket which rapidly whisked him up nine floors to the top of the gantry. It was now three hours to liftoff, set for 0900. Climbing into the cabin of his spacecraft, the Shenzhou, was a complicated matter. First, grasping a rail above him, he slid on a white mattress over the sill of the spacecraft, the soft padding being necessary to make sure he did not tear his suit. By now he was in the top module of the Shenzhou. An orange-suited technician was there to pull in his legs and bring him in. Now Yang Liwei had to gently lower himself down the tunnel into the descent module below, again being careful not to rush and settle into the cabin where he would spend the mission.

The descent module of Shenzhou is acorn shaped, with a porthole at either side, tunnel in front and instrument panels around the tunnel. The couch is individually contoured and set on springs so as to absorb a bumpy landing. Around the walls is soft padding, both to protect the astronaut in the event of bumps but also to avoid hard surfaces that might tear or damage suits. Yang Liwei had plenty of space for himself. Shenzhou is designed to carry three astronauts (some say it could take four at a pinch). Whilst it is tight for three suited men (Russia's smaller Soyuz can seem claustrophobic), Yang Liwei had plenty of space for himself as he settled down to what would be his home for the next day. Once in his cabin, he closed the tunnel (the lever turns like an interconnecting door on a submarine). The technician closed up and evacuated the orbital module above him. He and his colleagues ran a series of tests to check both modules for airtightness. It was 6.15 am. He was over 50 m above the ground and the rocket could just be felt rocking in the light wind.

Now it was 7 am and getting light in Jiuquan. Although there was still two hours to go before launch, there was still much to do to prepare the rocket for flight. With Shenzhou now closed out, the gantry moved back from the top of the cabin. A rescue team of fourteen people stood ready in case they were needed. If they were, the gantry could be rolled back up close to the cabin and the astronaut quickly evacuated. For this there was an explosion-proof elevator or an escape slide and bomb-proof bunker. If the worst came to the worst and the rocket was in danger of exploding, the escape tower could be fired (the system was armed 15min before take-off). The escape tower was a pencil shaped set of rockets on the top of Shenzhou, 58.3 m above the ground. They were small but powerful solid fuel rockets that would burn for a few seconds, lifting the entire cabin free of the rocket. The Shenzhou cabin would then be dropped out and there would be sufficient time for the parachute to fill with air before it reached the ground probably only a couple of thousand metres from the launch pad. A similar launch escape system had once been used when a Soviet rocket exploded on the pad in September 1983. Cosmonauts Vladimir Titov and Gennadiy Strekhalov were grateful when it did indeed work as advertised. They had a bumpy landing but were very much alive – but it was the end of the mission. This Long March 2F had been the most carefully prepared of them all and the builders had known that this was the most important one ever sent out from the factory.

Over the next two hours, the gantry was pulled back and all the systems on the rocket were carefully checked. All the electric and electronic circuits were in order. The Long March 2F does not give any of the tell-tale signs of an impending launch, like the American Shuttle or the Russian Soyuz, where viewers can see cold liquid oxygen boiling off around the ready rocket. The Long March 2F uses a fuel with the unwieldy name of unsymmetrical dimethyl methyl hydrazine mixed with nitrogen tetroxide. They are kept at room temperature (and can be so for long periods) and do not need to be cooled. The rocket had been brought to Jiuquan on the rail network as long ago as the end of August. It had been stacked in the vehicle assembly building, tested and prodded to make sure everything was alright and had been fuelled up in a 7 hr operation on Monday 13th October. Chief designer of the Shenzhou, Qi Faren, had been there and this was a moment of truth for him too.

Out at sea, the Chinese tracking fleet had now taken up position. These were big ships, commissioned many years ago to follow the Shenzhou when it was far away from the Chinese landmass. They were called *Yuang Wang*, which means 'Long View' in Chinese. They had sailed from their home port in Shanghai in September to take up position at strategic points along the route of the Shenzhou: the mid-Pacific, the southern ocean, the Atlantic and the Indian Ocean. They had been on station since the end of the first week of October. Thirteen tracking stations were on call to track the Shenzhou: the four shipborne stations, six land stations in China and three overseas stations in Namibia, Malindi (Kenya) and Karachi. Of the land stations, the pre-eminent ones were Jiuquan (for the launch and landing), Xian (national tracking centre) and Beijing (manned mission control). The others were Weinan (Shaanxi Province), Qingdao (Shandong Province), Xiamen (Fujian Province) and Karshi (Xinjiang Uygur Autonomous Region). Between them all, their role was to track the spaceship as it passed overhead and regularly update Shenzhou's computers to carry out all the key milestone events of the mission.

We don't know what Yang Liwei thought as he lay in his cabin as the clock counted down. He had passed every possible psychological test to ensure he could get through all the difficult hours that lay ahead. From Suizhong in Liaoning, born 21st June 1965 the son of an economist father and a teaching mother, he excelled in maths at school. He had joined the People's Liberation Army and then the air force from whose aviation college he had graduated in 1987 and where he had accumulated 1,350 hours flying experience. Since then he had been through endless theoretical and practical training, as well as survival training in the event that he came down far off course. Like Yuri Gagarin before him, he had made no secret of his desire to be first. Although he had set some time aside for ping pong and basketball, he had rarely left the astronaut training centre during the previous few years. He had never had the opportunity to bring his son Yang Ningkang to school and knew of his son's teachers only by name, never personally. He rarely went to bed before midnight. He had only two weeks' holiday a year, spent with his parents.

By 9 am though, the appointed moment, the countdown had been through all its procedures. The process was supervised in a large control hall in Beijing, the manned mission control centre. Here, ninety engineers, most of them young and in their twenties, checked item after item from behind their consoles. On the huge screen in front, they could see the launch pad in Jiuquan 1,600 km distant. It was now properly light and the rocket stood out along against the desert and the mountains. With dawn, temperatures had begun to rise and were climbing up to 10°C. The escape system was armed, the electrics were live, the Long March 2F was operating on internal power, the fuel tanks were pressurized, the communications line to the cabin was clear.

0900. A dull thud beneath the rocket. The Long March 2F began to shake and the engines belched out the characteristic telltale orange and brown plume of the nitrogen fuels. Power built and built and in seconds, the Long March 2F was rising slowly, ever so slowly, up its launch tower. Its rise seemed agonisingly slow as it stood out against the lightly blue sky. Safely some distance from the pad, the engineers and military watched, shielding their eyes, their hearts almost stopped as they watched the rocket rise. But as the flames beneath passed the bottom of the tower, the Long March gathered momentum and could be visibly seen to accelerate. 20sec after launch, sharding began to tumble from the rocket. No one had seen this on the four previous Long March 2F missions before, simply because they had all taken place at night, but the dropping of exterior shielding was a procedure familiar to the early European Ariane rockets and nothing to worry about. It was thermal weather protection blanketing to protect electronics on the interstage. Now the Long March 2F was rising ever faster, heading skyward and could be seen bending over in its climb toward the east. The rocket had soon pitched over, a long needle with its four liquid – fuel strap-on boosters on the bottom still burning brightly. Down below, the burning engines looked ever more like a bright pulsing star as the Long March headed ever higher into the atmosphere.

Then, after two minutes there was a sudden flash at the bottom of the rocket. The strap-on rockets had done their day's work, had burnt out and were now explosively separated from the main rocket. They tumbled back into the atmosphere, presumably to fall into uninhabited desert some place far downrange. The Long March 2F was soon 30 km high, outside the thickest part of the atmosphere. The escape system was soon jettisoned, its rockets firing it clear of the cabin. Shenzhou was now exposed in the airless open now. For Yang Liwei, natural light flooded in through the portholes. On the Soyuz, Russian cosmonauts bring a mirror with them so that they can see the Earth recede below them, but we don't know if he did the same. All this time, Yang Liwei will have felt the vibration and roaring of the ascending rocket. The ride became smoother as the strap-ons came off. In his hand was a clipboard pad and a pencil on a string and he noted each event in the launch sequence as it happened. Half a minute later, there was the next milestone as the main first stage fell away and the second stage took over. There was a brief moment of quiet while the second stage ignited, thrusting him back in his seat as the rocket sped for orbit. As the rocket climbed ever higher, 50 km, 80 km, 100 km, it pitched ever more over in its climb. The emphasis now was less and less on height, more on building up horizontal speed as it headed toward orbit, even though it did continue to climb. 15,000 km/hr, then 20,000 km/hr. Yang Liwei crossed over the coast of China. The sun was ever brighter above, for it was mid-day down below as he headed over the Pacific. In his cabin was a small bobble on a string, which fell down toward the vertical as his rocket climbed ever higher. Back on the ground, the rocket had long disappeared from sight, leaving only a smoky trail in the high atmosphere and on the ground a still steaming sizzling launch pad.

9.10 am. Suddenly, the engines of the Long March 2F died, their fuel exhausted, their job done. The second stage separated. It fell back into a lower orbit, where it began to tumble slowly. Amateur astronomers spotted it in the night sky over Matija Perne, Slovenia only three hours later. Yang Liwei heard and felt the clunking sound of the Shenzhou separating automatically, pushing it briefly forward of the cylindrical second stage. No longer was he thrust back in his seat. Now, encountering weightlessness for the first time, he could see his bobble begin to float and the pencil on his checklist began to wander across his cabin. Over the blue of the Pacific Ocean, Yang Liwei was now in orbit and China had become the third country in the world to send a man into space. It was 9.10 am Beijing time, 15th October 2003. Had the rocket failed to get Yang Liwei into orbit and had he splashed down in the Pacific, three ships had been on standby to pick him up. Now the ships – the *Beihai 102*, the *De Kun* and *De Hi* – were stood down and told they could return to port.

The launch of Shenzhou 5. (Courtesy Reuters.)

First, some important business. Although Shenzhou was in orbit, it was important to check was it the right orbit – was it high and stable enough for the mission to last? Second, would the spaceship's equipment deploy properly? Signals from the Shenzhou were calibrated with all the tracking data sent back to mission control. All was in order and Shenzhou was in the perfect, accurate planned orbit, 197 km by 328 km. It was the 30th straight launch success for China in a row, with a reliability achieved only after careful preparation, endless checking, fanatical quality control and, earlier, heartbreaking failures. Next, 22 min into the mission, Yang Liwei felt the vibration of the solar panels deploying on the propulsion module behind him and the orbital module in front. Signals in the cabin showed that Shenzhou could now take power direct from the solar panels and would no longer be dependant entirely on its batteries. The good news was relayed in telemetry down to the *Yuang Wang 2* tracking ship in the southern ocean. In turn, the message was relayed back to Beijing mission control, where operations were directed by Liu Chengjun, 31 and Hong Chunhui, 26.

Shenzhou had now deployed properly, was in the correct orbit and was stable in flight. This was as everyone had hoped, but people with long memories recalled the maiden flight of a Russian spaceship of this type, the Soyuz 1, where things had gone wrong from the start, dooming the single cosmonaut on board, Vladimir Komarov. Still, China waited for the absolute knowledge that all was well before telling the world. 0934 and Yang Liwei was over the *Yuang Wang 3* tracking ship in the South Atlantic. 'I'm feeling good', he relayed back by radio to the tracking ship as he passed overhead. Li Jinai, head of the manned spaceflight project, now declared that the first stage of the mission could be considered a success. At 0942, just half an hour after it entered orbit, the launch of Shenzhou 5 was announced by the official media in Beijing. Telecasts of the launch quickly followed, beginning a day of saturation coverage in the Chinese printed and electronic media. This was a conservative mission. Originally the Chinese had planned to put two astronauts in orbit for a week and this was the profile flown by Shenzhou 2,3 and 4. But caution prevailed and they decided to opt for a 21 hour, 14 orbit mission, like Shenzhou 1. This had happened before. Originally, the Russians had planned to send Yuri Gagarin into orbit for a day, but they had second thoughts and eventually went for a single orbit mission. The Chinese even agreed to a live broadcast of the launch, but lost their nerve at the last minute.

By 10.30 am Yang Liwei was back over China, having flown around the world in 90mins. Had anything been evidently amiss at this stage, Shenzhou 5 could have come down on either the third or fourth orbit. Commands to return could have been issued by tracking stations in Karshi, Xingjiang and Weinan, Shaanxi. But there was no need to. Yang Liwei called in to Beijing mission control (formally called Beijing Aerospace Command and Control Centre) and the ground told him he could now go ahead and take off his gloves. Things were going smoothly and he knew he could begin to relax. Have an early lunch and take a rest, they told him. The menu, the record shows, was sweet-and-sour shredded meat, pork and sliced chicken with eight-treasure rice including nuts, with a hot pickle from Sichuan called Zachai followed by herbal tea with some traditional medicine mixed in. The food was stickily coated to prevent crumbs from floating around the cabin. For the next two hours, he was officially resting, though like any previous space traveller he almost certainly spent the time looking out his two portholes at the blues, whites and browns of our planet by day, the sunrises and sunsets and by night the lights of Earth's cities and the spectacle of Earth's weather, lightning and storms. At 1.39 pm, during his fourth orbit, he took out his log book to write down his account of everything that had happened so far. Down below, his spaceship had been spotted by amateur observers crossing dark skies over Pennsylvania, Oklahoma and Washington DC. Nearly fifty years earlier, they had spotted Russia's Sputnik in an October sky; now it was China's Shenzhou.

The rest was important, for at 3.57 pm, over the Pacific, Shenzhou came to a key moment in the mission. This was the time for the propulsion system to fire to make an orbital change and get the spaceship in the right orbit for landing the following morning. On the fifth orbit, the propulsion system duly fired. The burn raised the perigee so that the spaceship was now in an almost circular orbit of 331–338 km, crossing the equator at 42.4°. Already, China's first manned flight had gone past Yuri Gagarin's single orbit in 1961 and John Glenn's three orbits the following year – and he had manoeuvred in orbit. Still, to demonstrate the caution of the mission, he did not enter the orbital module. For some of the telecasts, he even kept his visor down.

An hour later, at 5.05 pm, on his sixth orbit, Yang Liwei began a live telecast from the Shenzhou 5. Pictures showed him smiling and waving, his clipboard and pencil drifting in weightlessness in the cabin and clear pictures of the Earth. He unfurled miniature versions of the national flag of China and the United Nations. At 5.30 pm, there was an exchange with mission control, where he took congratulations from defense minister Cao Gangchuan. In a second telecast at 8 pm, he talked with his wife Zhang Yumei in the mission control centre. They had been married since 1990 and she was there with their eight year old son. She asked him about what he could see outside. The camera showed the blue Earth and one of Shenzhou's solar panels. His young son seemed most interested in the space food. The telecasts were relayed from the tracking ships by satellite through compressed digital video to mission control in Beijing. Other signals were relayed from the cabin through seven different wave bands – short wave, ultra short wave, S-band and C-band.

As he orbited the Earth, his control panels and three computers gave him up to date information on the progress of his mission. A world map displayed his position over the Earth's surface. Another read-out displayed altitude, speed, flight time, temperature, humidity and the status of all systems on the Shenzhou. Data were displayed in Chinese characters and alarms were read out by a pre-recorded voice, much like that of the automated voice in an airplane cockpit (pilots call her 'Bitchin' Betty').

11 pm. Now was the official sleep time. Yang Liwei had been up since early morning and needed sleep to prepare himself for the busy and dangerous reentry into the Earth's atmosphere the next day. Passing over the tracking ship on the 12th orbit at 00.18 am, telemetry showed that Yang Liwei was indeed asleep and that all was well in the quiet cabin. It was not a long sleep, for Yang was awake again by 2.52 am, but it was enough. At 4.34 am on the 13th orbit over China, ground control confirmed that they would go ahead with a landing on the next orbit. Yang Liwei acknowledged this and began preparations for coming down. In the landing area, the senior meteorologist had just issued his forecast. Wind speed would be a firm 5 m/sec, visibility over 10 km and temperatures –4°C to –8°C, all within acceptable limits. Landing site was the district of Dorbod Xi in Siziwang Si county, central inner Chinese Mongolia, 100 km north of Hohhot, 41.3°N, 111.4°E. In case things went

Yang in the the cabin. (Courtesy Getty Images.)

wrong, China had taken precautions in the event of a landing badly off course. Shortly before the mission, China made an agreement with the Australian government for a search-and-rescue mission to be mounted if the astronaut came down in the deserted outback.

5 am. Yang Liwei, now passing over South America, was approaching the moment of truth, for landing and take-off are the most dangerous moments of any spaceflight. The last American spacecraft to attempt a return to Earth, the shuttle *Columbia*, had broken up during its return to Earth only nine months earlier and this was still fresh in people's minds. Yang Liwei was in contact with the *Yuang Wang 3* again as he approached the Namibian coastline. Nearby, China had its main overseas ground tracking station in Swakopmund, Namibia. Between the two tracking systems, they were able to apply the maximum possible surveillance to this critical stage of the mission. Shenzhou 5 was now flying backwards, its retrorockets pointed in the direction of travel, the craft carefully aligned at the correct angle to the Earth's horizon. At 5.04 am, the stations confirmed that Shenzhou was in the correct position for reentry and issued the command to proceed. At 5.36 am, the orbital module was jettisoned to begin its period of flight as an independent space laboratory. The module carried a science payload and a CCD observation camera, mounted to the exterior of SZ-5 with a ground resolution of 1.6 m. It manoeuvred to its operating orbit six days later. The module carried out five manoeuvres to readjust its orbit between then and the end of January, concluding its scientific mission successfully in March 2004 after 152 days. The results were transmitted digitally to Earth, including detailed information monitoring the Sun's magnetic field.

5.38 am. The retrorocket system for Shenzhou 5 blasted for three minutes. It was soon clear that it had fired for the proper duration and thrust and at 5.44 am, Yang Liwei reported the burn had gone perfectly. The burn was sufficient to cut hundreds of kilometres an hour off Shenzhou's path, taking it out of orbit. Shenzhou swept in a vast arc over Africa, Arabia and west of the Himalayas. The shape of Shenzhou was such that it was possible to control the craft as it descended and generate a cushion of air underneath the cabin so as to steer it to a precise reentry point. Just a minute before reaching China's western border, at 5.59 am, the propulsion system was jettisoned, to burn up in the atmosphere later in a fireball over north China. The descent module was on its own now, its heat shield pointing in the direction of travel. By 6 am, the descent cabin was over western China. At 6.04 am, Shenzhou encountered the denser layers of the Earth's atmosphere. The heat shield began to turn redhot and then whitehot. An ion sheath of particles surrounded the cabin, cutting Yang Liwei off from contact with the ground for several minutes. Always a worrying period and it was during this time that *Columbia* had been lost.

Blackout did not last for long and soon after 6.06 am, signals were streaming back from Shenzhou. 'I'm just fine' said Yang Liwei and mission control got a good fix on his signals to predict the landing point, which was aimed at the district of Dorbod Xi. The area is in inner Mongolia, north of Hohhot, well to the east of Jiuquan, northwest of Beijing and west of the Beijing to Ulan Bator railway. Five recovery helicopters were already in the air, hovering like bees, ready to spot the descending cabin and rush to retrieve the astronaut. Yang Liwei was now falling through the atmosphere, the cabin cooling. He flicked the switch on to activate the parachute once the system sensed denser air.

6.11 am. At 15,000 m, out came the pilot parachute, quickly reported by Yang Liwei. 'I'm still fine', he reassured ground control on short wave radio, 'though it's warm in here', he added. At 6.14 am, he dropped the the heat shield, which was no longer needed. Without its weight, the cabin would now descend slower. Dropping the shield was also necessary to expose small rockets used to cushion the final descent. Now the drogue parachute was out and the cabin's speed had fallen from 201 m/sec to 80 m/sec. At 6.16 am the large, 1200 m^2 diameter main parachute came out. 'Deployed normally', reported Yang Liwei as Shenzhou 5 descended onto the dawn grasslands of Chinese Mongolia. Had the main parachute failed, a backup one would have popped out, but it was a third smaller and the landing would have been rougher. Soon he was spotted by one of the recovery helicopters. On the ground, a team of cross-country vehicle was parked in a line, ready to set off in pursuit. It was just getting daylight and they still had their headlights on. The landing was so accurate that an image of the descending Shenzhou was relayed by one of the camera crews with the landing team. 'Parachute deployed' had already been signalled around the world on the internet as space enthusiasts the world over stayed up during the night to follow the mission.

6.23 am: as Shenzhou 5 finally reached the ground, three small solid fuel rockets at the base of the cabin fired for a second to cushion the last moment of the descent, sending up plumes of dust engulfing the spaceship. But the dramatic finale made a big difference for Yang Liwei, for the Shenzhou cabin comes in at quite a pace and it is a bumpy landing without the final soft-landing rockets, as Russian cosmonauts have reported when they have not fired properly. The soft-landing rocket can also have the effect of tipping the cabin over on its side, exactly what happened this time to Yang Liwei. The parachute is then dropped, so as to prevent wind from catching it and dragging the cabin.

No one saw the precise moment of landing, but a helicopter crew radioed in at 6.24 am that it had spotted the parachute on the ground and gave coordinates. The news was relayed around the world immediately. Two minutes later, a team of cross-country vehicles was on its way, bumping across the grasslands. Helicopter #3 spotted the cabin first, estimating it was about 7 km away. The crew managed to contact Yang Liwei on short wave at 6.30 am and quickly hooked up a line between Yang Liwei and Chinese prime minister Wen Jiabao who rushed to congratulate him. We owe you all our gratitude, he said.

6.33 am and the first helicopter had touched down beside the cabin. Yang Liwei was still inside and their priority was to get him out safely. If he had to, he was well able to do so himself: the previous May, when three cosmonauts had landed in a Russian spaceship hundreds of kilometres off courses, they had got out by themselves and had simply waited in the steppe for rescuers to eventually arrive.

It took the orange-suited rescuers only five minutes to open the hatch at the top of the descent module, now lying on its side and get Yang Liwei out. When they opened the hatch, they could see him still strapped into his cabin, waving. They gradually lifted him out, letting him get his landlegs back again after a day's weightlessness. He

Yang back on Earth. (Courtesy Getty Images.)

slid out of the hatch onto the ground, and, looking just a little dazed from all the attention, waved to the rescuers, journalists and television crews. Yang Liwei had landed just before dawn, but there were good streams of early morning light now. He had travelled 600,000 km in 21 hours and circled the Earth 14 times. Yang Liwei took off his communications soft hat and put his gloves into knee pockets. Yang Liwei is recovered and well, the helicopter team formally reported at 6.38 am, but they didn't need to, for the television pictures told the full story. The landing point was Amugulang ranch, Dorbod Xi, Siziwang. Liang Qi, head of recovery operations,

Shenzhou 5 back on Earth. (Courtesy Reuters.)

explained that the site was chosen because of its flatness, the lack of power lines or inhabited buildings.

Yang Liwei sat down in a director's chair to talk to doctors, journalists and the rescue team. This done, he was put into blue astronaut coveralls, similar to those worn by Nie Haisheng and Zhi Zhigang at his launch walk-out. He was given a white scarf to keep him warm and asked to pose for photographs with his rescuers. Then he was brought away by helicopter first for a plane journey to Beijing. Again, more photographs. Soon he talked to his wife and son in mission control. There, prime minister Wen Jiaobao assembled his colleagues on a hastily made stage at the front of mission control to congratulate the team on its achievement.

There was warm enthusiasm for the flight across China. Whilst it may have lacked the fervid joy with which Soviet citizens greeted Yuri Gagarin in 1961, there was evidence of a great sense of satisfaction with China's achievement. *People's Daily* ran 100,000 extra copies which were quickly snapped up, as did other papers. There were demonstrations in some towns. School children drew pictures of spaceships and showed them to press and television. Wall posters appeared, combining twenty-first-century technology with more traditional styles of socialist realism. 10.2 m stamps were printed in Yang Liwei's honour. The *People's Liberation Army Daily*

Yang's triumphant return. (Courtesy Reuters.)

triumphed: For China this is the beginning and there will be no end. China was undoubtedly heartened by the comments and praise that flooded in from political and space agency leaders the world over, for it was universally warm and generous.

First, Yang Liwei was brought by plane to Beijing. As he came down the stairs, a band was ready to greet him. No sooner was he down the steps than he was presented with flowers and met by senior military and party officials and, once they had had their say, his relieved wife Zhang Yumei. Accompanied by her and by programme chief Li Jinai he was driven back to his apartment in the astronaut training centre.

Yang Liwei was then brought away for a week's debriefing and medical examinations. Where is he? the western media soon asked, quickly speculating that things were amiss. They weren't and after the debrief Yang Liwei, accompanied by his young son Ningkang, opened an exhibition in Beijing of his Shenzhou 5 cabin, spacesuit, parachute and model space food. The cabin then became part of the travelling show that went on to Hong Kong and Macau. When the cabin reached Shanghai, hundreds of thousands queued in freezing conditions to see Yang Liwei and his cabin on 24 hr exhibition.

For Chinese space officials, leaders, technicians and engineers, this was a time of relief, relaxation and achievement. The eleven-year project to put someone into space had reached its climax. Due diligence had been rewarded with its trouble-free outcome. As the official commentary put it, space travel had long been the dream of our ancient civilization, kept alive in stories and legends, like *Chang e*, the fairy who flew to the moon and Wan Hu who perished in a spaceship made of a kite, wickerwork chair and fireworks. Now China's space leaders could indulge their fancies and expand their ideas for space stations, flights to the moon, sending probes to the planets and building a base on Mars. Except that now, there was a real chance that they might actually come true one day. So where did China's long march to the cosmos actually begin?

2

Origins: the fiery dragon

China has a long history in astronomy, astronautics and rocketry. Some of the most ancient and reliable astronomical observations were made in China and they well stood the test of time. The rocket – the word means 'firing arrow' in Chinese – was invented in China. The ancient Chinese discovered the secret of gunpowder in the 3rd century – sometime between 220 and 265 AD: the formula took a thousand years to work its way westward, to reach England by 1248 where it was known to Roger Bacon (for the record, the formula is 50% nitre KNO_3, 25% sulphur, 25% carbon). In the 10th century, gunpowder was fitted to the heads of arrows to explode on hitting their target. Firecrackers were introduced for festivals at around this time.[1]

Ancient Chinese astronomical instruments.

16 Origins: the fiery dragon

The rocket – the use of gunpowder rather than the bow to propel the arrow – was invented by Feng Jishen in 970. Primitive rockets were used by the Song dynasty to fight Xia in 1083. Later, the Mongols learned to use these rockets and made them the basis of the expansion of their empire. When the Japanese invaded China in 1275, Kublai Khan fired rockets to drive them away. The Chinese began to put their rockets into launching tubes, rather than release them like conventional arrows. The first of these, *Flying fire spear*, used a paper container and was introduced in 1119.

During the Ming dynasty (1368–1644), these early Chinese rockets came into their own. They possessed the fundamental elements of modern rockets: a combustion chamber, firing system, explosive fuels and fin guidance systems (feathers). The Ming histories reported that over 39 types of rocket weapons were in use. At one stage, a group of rocket troops was formed. A squad of 64 soldiers was equipped with *Soaring flame bird*. This was an exceptionally intimidating instrument of war – a paper and bamboo contraption shaped like a bird which would be fired over the walls of a city. Soldiers would ignite and release the rockets under *Soaring flame bird* which would fly upward and dive terrifyingly on the enemy. Once it landed, it would explode, the gunpowder being mixed in with poison to do further harm. Similarly, *Burning crow* was a paper-and-bamboo contraption with four rockets, fired in a ballistic trajectory into enemy camps up to a range of 300 m. It was designed to explode a 250 g gunpowder warhead on landing, but the sight and sound of these large bird-shaped missiles zooming in must have been frightening.

Not only were basic rockets in use, but two-stage rockets were invented. The most famous one was *Fire dragon over water* rocket, used to destroy enemy ships as far as 3 km distant. This was a 150 cm long bamboo cylinder filled with gunpowder with a wooden dragon on its head and tail. Underneath were two more rockets. When the first set of rockets burned out, they dropped off and a fuse ignited the main part. It skimmed over the sea in a flat trajectory at a height of just over 1 m. *Fire dragon across water* was, in effect, the first anti-ship missile. A land-based version was called *Poison sand barrel* which was ingenious in another way. This was a boomerang type of rocket which dropped poison sand on the enemy before the second stage ignited to send the rocket back in the direction from whence it came, so the enemy never got a proper look at the device.

During land warfare, rockets such as *Swarm of bees* were often hidden under ground and released at short range against an unsuspecting advancing infantry and cavalry, generally with terrible effect. Next the Chinese began to group rockets together in multiple tubes, firing many at the same time. *Hundred arrows chest* comprised tube rockets 150 cm long able to hit targets 100 m away. *Seven arrows pipe* was carried by soldiers in a leather bag and could be ignited by an individual, mobile soldier, much like a modern bazooka (an upgraded version had nine arrows and was called *Nine dragon spear*). *Flock leopard* was carved in the face of a fierce leopard and comprised 40 arrows each 70 cm long with an iron head and fins, able to fly 120 m.

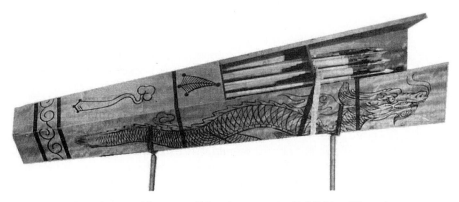

Fiery dragon. (Courtesy China Astronautics Publishing House.)

Names of early Chinese rockets

Flying fire spear	Magic flying crow
Soaring flame bird	Burning crow
Five tigers from arrow cave	Fire basket arrow (twenty tubes)
Nine dragon arrow	Swarm of bees (thirty-two tubes)
Hundred arrows chest	Hundred tiger fleeing rocket (two hundred tubes)
Seven arrows pipe	Fire dragon across water
Flock leopard	Poison sand barrel

Thunder cannon shocking heaven and and flying into the sky to strike the enemy

A 16th century inventor (some say he was an administrative official) called Wan Hu designed a rocket-propelled chair on which he planned to ascend into heaven. He built an open wickerwork cabin, to which he fitted 47 rockets underneath and above, with two kites to keep him aloft. Wan Hu disappeared in flame and smoke and was never seen again. A crater on the Moon is now named after him, so in one sense he made it to the heavens after all. This is the first recorded design of something approximating to a manned space rocket.

The use of rockets came to an end under the succeeding Manchu dynasty (1644 onward). The Manchus were much more interested in literature, social science, public administration and opening the country to western ideas. Interest in science, technology, chemistry, metallurgy and mechanics declined, as did the production of weaponry. Rocket development in China went was now interrupted for three hundred years.

TSIEN HSUE SHEN

Several names are irrevocably associated with the development of the world's great space programmes: in Russia, Tsiolkovsky, Korolev, Glushko, Chelomei; in the United States, Goddard, von Braun, Faget, Kelly. The person who contributed most to the development of modern space exploration in China was Tsien Hsue Shen.

18 Origins: the fiery dragon

Hardly known in the west, he was born in Hangzhou, Zhejiang in 1911, the only child of an educational official. His name – Tsien Hsue Shen, which means 'study to be wise' in Chinese – reflected his father's ambitions. Much of what we know about Tsien comes from a pioneering biography by Iris Chang, *The thread of the silkworm*.[2]

Tsien Hsue Shen attended a primary school for gifted children in Beijing. A model child with an outstanding school record, Tsien subsequently entered the Beijing Normal University High School. At 18, he applied to Jiatong university in Shanghai to study railway engineering, coming third in the nation in the entrance exam. A serious, aloof, immaculately dressed and perfectly behaved student, Tsien was a man who liked to study and work on his own, his only outside interest being classical music (he played the violin). His time in Shanghai was punctuated by typhus, which he only narrowly survived, social turbulence and the beginnings of the war with Japan.

Graduating as top student, with 89 points out of 100, he chose to pursue aeronautical engineering, competing for a scholarship in the United States on a scholarship in 1935. There he started at Massachusetts Institute of Technology (MIT), staying only a year before moving to the California Institute for Technology (popularly known as CalTech) in Pasadena, where he studied under the great

Tsien's graduation. (Courtesy Smithsonian Institution.)

Austro-Hungarian mathematician Theodore von Karman. Tsien graduated as doctor in 1939.

Here we begin to trace Tsien's interest in rocketry. Five fellow students and associates invited him to join a group interested in what would now be called amateur rocketry. They were a gang of experimenters buying up spare parts, assembling them and letting them off in the nearby desert. He was in effect the mathematics advisor to the group, in 1937 writing his first work on rocketry *The effect of angle of divergence of nozzle on the thrust of a rocket motor; ideal cycle of a rocket motor; ideal efficiency and ideal thrust; calculation of chamber temperature with disassociation*. Their first, often dangerous, experiments were presented to the Institute of Aeronautical Sciences and written about locally in the student press, where Tsien made some reckless comments about the possibility of eventually sending rockets 1,200 km into space. Rather like fellow rocketeers in Germany and the Soviet Union, their work soon became sponsored by the military who saw the potential for rockets both to make aircraft fly faster and to fly as ballistic missiles. Military funding rose from $1,000 to $650,000 in five years. By 1942, after the United Sates entered the war, Tsien was working on small solid rocket motors to help aircraft get airborne; shortly afterwards he helped to draw up plans for a missile programme.

Tsien became assistant professor of aeronautics in 1943. As well as doing research, Tsien taught students, though many found his manner intimidating, intolerant, arrogant, over-precise and unsympathetic. He was one of the co-founders of the famous Jet Propulsion Laboratory (JPL), whence American unmanned exploration of the Moon, the nearby planets and the outer solar system was to be subsequently guided. He was the first head of research analysis at JPL in 1944. By the following year, he was working for Karman in the Pentagon on advising the United States military on how to harness the latest discoveries in aeronautics and rocketry for the post-war defence forces (he later received a commendation from the Air Force for this work).

ON THE TRAIL OF THE GERMAN ROCKETEERS

In May 1945, having been given the temporary rank of colonel of the United States Air Force, Tsien arrived in Germany to survey the Nazi wartime achievements in rocketry, their rocket factories and secret test sites. On 5th May he met the leading German rocket engineer, Wernher von Braun, who had just surrendered to the Americans. Not long afterwards, the man who was to be his opposite number in the Soviet Union, chief designer Sergei Korolev, was scouring other nearby parts of Germany on an identical mission.

Returning to JPL, Tsien published his wartime technical writing in a book called *Jet propulsion*. After a return to MIT in 1946–8 and a brief visit to China in 1947 (where he married), he became in 1950 the Robert Goddard professor of jet propulsion at CalTech. He gave a presentation to the American Rocket Society in which he outlined the concept of a transcontinental rocketliner able to fly 400 km

20 Origins: the fiery dragon

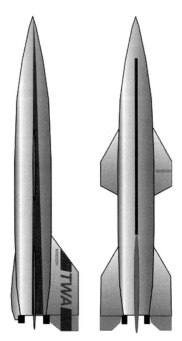

Tsien's early spaceplane. (Courtesy Mark Wade.)

above the Earth, its spacesuited passengers floating in its cabin as they briefly enjoyed weightlessness. His proposal was later covered in *Popular science*, *Flight* and the *New York Times*. His rocket plane looked like the German A-4 (more infamously known as the V-2) but with wings. The following year, he predicted that astronauts would travel to the Moon within 30 years. Some of his rocketplane ideas inspired the United States Air Force to develop its spaceplane project of the late 1950s, the Dynasoar (standing for 'dynamic soaring'), ultimately one of the ancestors of the space shuttle.[3]

The following year, 1951, at the height of the McCarthy witch-hunt in the United States, he was accused of being a communist. A period of confusion followed, in the course of which Tsien had his security clearance revoked. He was held alternately in jail, under a form of house arrest, under surveillance and unsure of his future. The various bureaucratic factions of the American government argued about whether he should be released, jailed or deported (the course of action favoured by the Immigration and Naturalization Service). As they did so, Tsien worked on the problems of rocket guidance and how computers could steer rockets in their ascent through the atmosphere. His papers were impounded. He was charged that one set comprised secret codes: closer inspection found that they were logarithmic tables. When he was put in San Pedro prison, the courageous president of CalTech, Lee Dubridge, flew to Washington to get him released.[4]

RETURN TO CHINA

In a September 1955 agreement between the American and Chinese governments, Tsien and 93 fellow scientists returned to communist China in exchange for 76 American prisoners of war taken in Korea. Reentering China through Hong Kong, then a British colony, Tsien and his family were warmly greeted in Shenzhen by the Chinese Academy of Sciences and welcomed in a series of homecoming celebrations that culminated in Beijing, just restored as China's capital city. A couple of months afterwards, he came to visit Harbin Military Engineering Academy (the circumstances that led him there are not known). The commandant of the Academy, General Chen Geng, asked him whether he thought China could make guided missiles. Why not? responded Tsien.

It is no coincidence that the beginning of China's modern missile programme may be dated to 1956, the year after Tsien's return. He brought with him – in his head, since he left his papers behind – the most up to date theory of rocketry from the United States. However, he had to start virtually from scratch. China in the early 1950s had barely emerged from a long period of great turbulence and destruction – social unrest, the war with Japan, then the civil war and finally the communist revolution of 1949. Making cars and trucks represented the limit of China's industrial and technical capacity – there were no aircraft factories, test sites, wind tunnels or the type of facilities he had begun to take for granted in California. However, reconstruction and modernization had begun, the first five year plan for socialist reconstruction having been adopted in 1953.

Zhou Enlai greets Tsien. (Courtesy China Astronautics Publishing House.)

SCIENTIFIC CONSTRUCTION IN CHINA

In January 1956, Chairman Mao Zedong proposed the rapid development of science and technology and that China should catch up quickly with the world's most advanced levels in economics and science. In response, the Supreme State Conference set up a Scientific Planning Commission under the leadership of prime minister Zhou Enlai. On 17th February 1956, Tsien presented a report to the Central Committee *Opinion on establishing China's national defence aeronautics industry*. The commission consulted widely with scientists and experts, after which it drew up *Long-range planning essentials for scientific and technological development, 1956-67*, which defined 57 priority tasks which would ensure China's independence in rocket and jet technology in 12 years. These included atomic energy, rockets, jet technology and computers.

In April 1956, the Zhou Enlai presided over a meeting of the Central Committee Military Commission which invited Tsien Hsue Shen to outline the potential of guided missiles and rockets. Within days, the government had appointed a State Aeronautics Industry Commission to develop the country's aviation and missile defences. Deputy premier Nie Rongzhen was made director and Tsien Hsue Shen was one of its members. On 10th May, Nie Rongzhen issued his first report *Preliminary views on establishing China's missile research*. On 26th May, the central military commission chaired by Zhou Enlai accepted the report and ordered the setting up of administrative machinery to get work under way.

FOUNDATION: 8 OCTOBER 1956

On 8th October 1956, the Central Committee of the Communist Party of China, presided by Mao Zedong, established the Fifth Research Academy of the Ministry of National Defence to develop the space effort. This is now officially marked in China as the birth day of the Chinese space programme. Within the academy, the first Rocket Research Institute was established under vice-premier Nie Rongzhen, with Tsien as its first director. The government took over two abandoned sanatoria and requisitioned a hundred university graduates to begin work there. Nie Rongzhen reported to the Central Committee on its establishment later in the month.

Leaders of the Fifth Research Academy of the Ministry of National Defence, Beijing, 1955

Tsien Hsue Shen
Guo Yonghuai
Xu Guozhi

THE FIFTH ACADEMY

Even though it had an enigmatic title anyway, the existence of the Fifth Academy was classified. This was a Russian trait too: the space industry there operated under

Huang Weilu. (Courtesy China Astronautics Publishing House.)

the sonorous title of the Ministry of General Machine Building. Likewise, the body responsible for nuclear development was similarly classified (the Ninth Academy; in Russia it was the Ministry for Large Machine Building).

The Fifth Academy's first task was not to experiment, as its leaders might have liked, but to build up a basic expertise among staff who were keen but had not had the opportunity to reach a third level education. Tsien was joined by a number of his contemporaries, some of whom had studied while he was in college and many of whom had, like him, worked in the United States. Typical of these were fellow rocket experts Liang Shoupan (MIT), Tu Shoue (also MIT) and guidance expert Huang Weilu (London University). The Fifth Academy set up ten research laboratories and two branch academies to deal with the many aspects of designing modern missiles and preparing them for flight.

156 university graduates were assigned to the Fifth Academy. None had even seen a missile before. Undeterred, Tsien drew up a curriculum *Introduction to guided missiles*, one that included instruction in aerodynamics, electronics, controls, computing, propellants, radio, physics, engines, assembly and body structures.

The leadership of China soon realized that the institute would still need outside help if it was to progress at all. No matter how good Tsien's theoretical knowledge and that of his colleagues, China simply did not have the ability to develop even basic rocket hardware from scratch. The leadership turned to China's communist ally, the Soviet Union.

CHINESE-SOVIET COOPERATION

Negotiations with the Soviet Union began in 1956. The Russians were then at a critical stage in the development of their own intercontinental ballistic missile and Earth satellite project. In September 1956, Nie Rongzhen led a delegation to visit Moscow. The Soviet Union agreed to sell China two R-1s, the reverse-engineered versions of the German A-4 wartime weapon (the V-2). These duly arrived in October 1956. However, they cannot have been much help to Tsien, for he had already seen the original A-4 during his visit to Germany eleven years earlier.

The Chinese reckoned that the Russian rocket did not represent the most up to date Soviet technology then available. Of course they were right: the Russians had moved well beyond the R-1, had now reached the R-5 and were already building the R-7 which was soon to launch Sputnik. In Dnepropetrovsk, the Mikhail Yangel Design Bureau (now known as NPO Yuzhnoye) had modified the R-1 and considerably improved it as the R-2. This missile had a range of 590 km.

China appealed to the Soviet Union to be more forthcoming, Nie Rongzhen leading another delegation to Moscow in July 1957. Under an agreement termed the *New Defence Technical Accord 1957-87*, signed 20 August 1957 but not ratified until 15 October, the Soviet Union now agreed to supply missile models, technical documents, designs and specialists. In January 1958, the first of several R-2s reached the Fifth Academy under cover of darkness. A hundred Soviet specialists arrived, bearing over ten thousand blueprints and technical documents. The first group of 50 Chinese graduates went to Moscow for study. While there, some zealous Chinese students at the Moscow Aviation Institute engaged in some freelance espionage on behalf of their country, copying and lifting restricted information. However, despite a much more intense level of cooperation, the Chinese felt all along that the Russians were being less than whole-hearted. Requested information was often not supplied or referred to higher authority, when the request disappeared into a bureaucratic black hole.

PROJECT 1059: THE CHALLENGE OF COPYING A RUSSIAN ROCKET

In January 1958, the Fifth Academy adopted the *Essentials of a ten year plan for jet and rocket technology, 1958-67*. With the Soviet Union making rapid strides in space exploration (a first probe was even sent to the Moon early in 1959), Chinese scientists were divided as to the wisdom of working on rockets as obviously out of date as the Russian R-2.

The Fifth Academy took the view that even to copy the R-2 would be a difficult enough challenge in itself. They decided to go ahead and code-named the enterprise 'project 1059', beginning a tradition of numerical allocations to keynote projects. Coding is an important aspect of the Chinese space programme and adjacent areas. In recent times, these codes have been related to the year when the project was approved, with the number indicating either the month or the number of the project for that year (e.g. 'project 863' for the third project or month of 1986). In earlier days, this was not always the case and the coding system remains obscure.

China lacked the ability to bring together the key material elements of any rocket, even the most basic ones. Throughout early 1959, there was continuous traffic between Russia and China as the academy tried to get the R-2 into production. Two Chinese delegations were in Moscow that August to ask about additional equipment. During the war, Germany's production of the A-4 had consumed a considerable part of the war effort, using scarce supplies of steel, aluminium and rubber at a crucial stage, one requiring huge numbers of hours of assembly. The Chinese were now learning the hard way that production of a rocket was a difficult, demanding and sophisticated task. In October 1958, under the Chinese-Soviet accord, another Academy of Sciences delegation had visited Moscow. Formally titled a High-altitude Atmospheric Physics Delegation, under the leadership of Zhao Jiuzhang, Wei Jiqing and Yang Jiachi, the tour became quickly aware of the level of complexity and organization required to run the rocket programme there.

The coded projects of the Chinese space programme

211	Moon probe (2003)
331	Communications satellite (1977)
581	Earth satellite project (1958)
651	Earth satellite project, as renewed (1965)
701	Ji Shu Shiyan Weixing (1970)
714	Shuguang manned spaceflight programme (1971)
761	Sounding rocket programme (1977)
863	Advanced technologies for the sciences (1986)
911	Recoverable satellite programme (1967)
921	Manned spaceflight programme (1992)
1059	Copy a Russian R-2 (1958)

THE IDEA OF A CHINESE EARTH SATELLITE

It is not known how prepared the Chinese were for the launch of Sputnik 1. The Soviet media had publicly announced its proposals to launch an Earth satellite several times since 1954 but whatever China's response, no-one in the west had paid much attention. When Sputnik 1 was eventually launched, on 4th October 1957, the Chinese Academy of Sciences set up an optical observation administration office and observation stations were built in Beijing, Nanjing, Guangzhou, Wuhan, Changchun, Yunan and Shaanxi. Their observations were coordinated in the Zijin Shan Purple Mountain observatory of Nanjing, one which dated to 104 BC and which now followed the Sputnik's path round the Earth and predicted its passes.

Tsien Hsue Shen had already proposed an Earth satellite during the period of the development of the Chinese R-2 rocket. At the second meeting of the VIII Communist Party Congress, Mao Zedong approved the idea. On 17th May 1958, apparently encouraged by the performance of the Soviet Union in orbiting a large, 1.5 tonne satellite two days earlier (Sputnik 3), he told his colleagues that China too must launch Earth satellites. His only rider was that a Chinese satellite should be

26 Origins: the fiery dragon

Tracking early satellites. (Courtesy China Astronautics Publishing House.)

Mao Zedong: 'We too must build artificial satellites' – 17 May 1958.

large and not the size of the small American satellites then being put into orbit (Vanguard 1 was a tiny 1.5 kg). In August 1958, the state Scientific Planning Commission endorsed Mao's proposal, arguing that a satellite would mobilize military rocket development, impress foreign powers and spur the development of other areas of science and technology. Nie Rongzhen, as deputy premier, called a special meeting of the Fifth Academy and asked Tsien to get together the best people to draw up a plan for the development of the satellite.

PROJECT 581

The Earth satellite was given a project code number (project 581) and made a priority national task. Here, the title '581' came clearly from the year ('58') and either the month ('1') or the number of the projects adopted that year (1). A '581 group' was brought together, soon to be divided into teams for engineering (institute 1001), control systems and research tasks. In October, the Academy of Sciences organized a scientific exhibition, which included models of rockets and satellites. Visitors included Mao Zedong and Zhou Enlai. In November 1958, the task of building the satellite was given to the Shanghai Institute of Machine and Electricity Design (SIMED), which was under the dual management of Shanghai municipality and the Chinese Academy of Sciences. The 1001 team was transferred to work there under the guidance of Prof. Yang Nansheng and Prof Wang Xiji. Young engineers were drafted in from Shanghai's industries and engineering centres, the most advanced in China.

There were long delays getting the project underway which have never been fully explained. Mao's support came in the middle of the great leap forward, when a period of rapid modernization in industry, agriculture and the economy had been announced, including the bizarre national campaigns to eliminate rats, flies, mosquitoes and sparrows. These campaigns – in which all citizens were expected to participate – certainly disrupted the space effort. Worse was to come in the autumn with the drive for steel production in which every home and factory was required to melt steel in order to increase national output. By the following year the economy was collapsing and millions were starving. Even the relatively pampered senior scientists around Tsien went hungry. At one stage, 70% were suffering from oedema due to malnutrition. Marshal Nie Rongzhen intervened, ordering they be supplied with rations from the navy.

EARTH SATELLITE SHELVED; ROCKET PROGRAMME CONTINUES

The idea of an Earth satellite was shelved temporarily in January 1959, the decision being taken to concentrate on sounding and military rockets, but to drop the satellite project. The view was taken that China did not have the resources to do both.[5] The Chinese lacked vital skills in the areas of machining, tooling, welding, pressing and punching. Facilities for rocket production were primitive and had to use converted

28 Origins: the fiery dragon

aeroplane hangars. The directive to cancel the project was issued by Deng Xiao Ping, then a medium-level but rising official. The three teams were broken up. Many years later, the decision was summarized as follows:

> In January 1959, Deng Xiao Ping issued an instruction to the effect that, in the current situation, when the national resources were not available for launching satellites, they should readjust their space technology research duties. The party group of the Chinese Academy of Sciences called a meeting to study the question of the fundamental policy for their work to develop artificial satellites and came to the conclusion that it was necessary to correct the erroneous tendency to be over-anxious to have artificial satellites in circumstances where the basic conditions for achieving this were not currently present.[6]

At one stage, there was considerable pressure to abandon the rocket programme (project 1059) as well, for some considered it also too advanced for China's state of development at the time. Things came to a head at the National Defence Industries Conference held in Beidaihe in July 1961. Mao Zedong and Zhou Enlai insisted that the rocket project press on.

Substantial numbers of personnel were requisitioned for project 1059. Beijing council assigned 6,000 of its workers to building work. At one stage, 1,400 institutes and enterprises and 14 factories were involved. By 1961, the workforce of project 1059 had swelled to 15,000 men working in sites adding up to an area of more than 1 m^2. Top graduates were assigned to the project as their first work and the army sent its best leadership cadres. Zhou Enlai insisted on regular progress reports. Some of

Deng Xiao Ping. (Courtesy China Astronautics Publishing House.)

the items necessary for the project, like aluminium alloy plate, seamless stainless-steel tubes and particular rubber items were simply not available in China. Up to 40% of the Chinese R-2 used substitute materials made in their stead. Some technical skills, especially in welding, were poorly developed and required intensive Soviet help. Those involved in the project worked in some very difficult conditions, in factories unheated in winter and unventilated in summer. Many slept in their workplaces by night and took their meals at their workbenches in order to save time. Although some had dormitory accommodation, others lived in tents. Some contracted dropsy from the poor conditions. The rocket engine was made in a shed and had to be hand-built by metal craftsmen. The project had one telephone line and to make sure it worked, militia were posted to guard every telephone pole.

A key personality in the development of Chinese rocketry in this period was Tu Shoue (1917–). Like Tsien, he had a background in the United States, having enrolled in MIT in 1941 and worked subsequently in the Curtis aircraft plant in Buffalo. Returning to China, he became professor in Qinghua University and subsequently Beijing. From 1957, he was drafted in as head of design in the Fifth Academy and led the development of the R-2 and military rockets in the late 1950s and early 1960s. One of his leading colleagues was Liang Shoupan (1916–), a 1939 graduate from MIT, professor of aviation engineering in China during the war and brought into the rocket programme in 1956.

DEVELOPMENT OF A SOUNDING ROCKET

Independently of the main effort with the R-2, the Shanghai Institute of Machine and Electricity Design began in 1959 to develop a sounding rocket called the T-5. Sounding rockets are small rockets used to probe the atmosphere, the most popular use being for weather forecasting. They are flown on up-and down missions of short duration and are not intended to reach orbit. Although the scale involved in designing and firing sounding rockets is much less than that of an orbital rocket, the basic principles of rocketry are the same. Granted the complete lack of experience of the institute, it was an ambitious project for its time, but one not out of keeping with the lofty idealism of the period of the great leap forward.

Designed to study the atmosphere and its geophysical phenomena, the T-5 was built and assembled in 1958. However, the young engineers were unable to get the materials necessary to construct an engine test bed, or, more seriously, obtain liquid oxygen. They had to abandon the project a year later. Instead, in October 1959, they decided to develop a smaller, simpler sounding rocket, one more tailored to the country's straightened economic circumstances. This was the T-7, but it was to be preceded by a development version, the T-7M. This was a small test rocket with a liquid-fuel stage and solid-fuel stage. Possibly because of the inability of Chinese industry to produce and store liquid oxygen, the T-7M used storable fuels (in this case, nitric acid as oxidizer and anilene and furfuralcohol as fuel). The T-7M weighed 190 kg, was 5.3 m tall, had a thrust of 226 kg and was designed to reach up

Origins: the fiery dragon

Tsien and a sounding rocket. (Courtesy China Astronautics Publishing House.)

to 10 km. Unlike the T-5, it was unguided, with four triangular tail fins. Chief designer was Yang Nansheng and chief engineer Wang Xiji. Their progress was inspected by President Liu Shaoqi and general secretary Deng Xiao Ping that December.

The T-7M's first launch took place at Laogang, outside Shanghai, on the East China Sea, on 19th February 1960. Launching facilities were crude. The team calculated trajectories with a hand computer, worked in an old airport shed and turned the tracking antenna by hand. A bicycle pump was used to pressurize the fuel tank! On its first mission, the T-7M headed into the clouds, reaching a height of 8 km. That May, the rocket was put on display at the Shanghai New Technology Exhibition and was inspected personally by Mao Zedong.

After the success of the T-7M, work began at once on the operational version, the much bigger T-7. The first hot engine run of its motor took place at a site set aside for them at Shanghai airfield on 18th April 1960, an event watched by Nie Rongzhen and Tsien Hsue Shen. In the meantime, it was decided to set up a new launch site for the programme. The Academy of Sciences, with Shanghai municipality, set up this new site at Shijiedu, inland in Anhui province in the hills of Guangde.

Shijiedu was ready for the operational version, the T-7, to make its maiden flight in September 1960. The T-7 weighed 1,138 kg on the pad, had a payload of 25 kg and reached an altitude of 60 km. It carried meteorological sounding equipment, designed to measure the atmosphere's pressure, density, temperature, wind speed and direction. The guidance engineer was Zhu Yilin.

Mao Zedong examines a rocket. (Courtesy China Astronautics Publishing House.)

LAUNCH SITE IN THE GOBI DESERT

Even as the Shanghai team made progress with sounding rockets, events were moving to a climax with the development of China's first 'real' rocket. By this stage, good progress had been made in building China's first launch site, also with Soviet advice. The Russians had already constructed three rocket launch sites – Kapustin Yar, on the Volga, where they had tested the German A-4s; Baikonour or Tyuratam, in Kazakhstan, where they had launched the first sputnik in 1957; and Plesetsk, northern Russia, which was their first operational intercontinental missile base. For China, one of the top priorities was to site the base away from the coastal regions that could be surveyed easiest and were most vulnerable in the event of conflict.

The Chinese launch site was settled northeast of Jiuquan oasis, 1,000 m high in the Gobi mountain desert, 1,600 km west of Beijing and on the ancient silk road. It is in Ejin Qi county, inner Mongolia, although often described (apparently wrongly) as part of Gansu province. Also known as Shuang Chengzi base or the east wind centre, like Baikonour it was selected on account of its low population density (in the event of rockets failing) and remoteness (for secrecy). A railway line was connected there from the Fifth Academy in Beijing via the Lanzhou-Urumqi railway to the launch site (this must have been a considerable journey for the rocketeers, for travel there takes five days). The 300 km branch line starts at Qingshui town, 40 km east of Jiuquan city and heads northward along the Ruoshui river, eventually ending at Saihan Toroi, inner Mongolia.

In April 1958, the 20th corps of the Chinese army was ordered to the area to construct the launch site. Sun Jixian was appointed first commander of Jiuquan,

32 Origins: the fiery dragon

with Li Zaishan its first commissar (his job was to ensure political discipline). Conditions must have been harsh then, bitterly cold in winter and baking hot in summer, exacerbated by strong wind. Water had to be brought in by truck while the first wells were drilled. Zhou Enlai came to investigate progress. Tsien Hsue Shen was out there early on, living in a tent like the others and sharing the very spartan grain-based diet.

It is possible that the Americans caught wind of these developments. In October 1958, the Central Intelligence Agency issued a report on possible rocket launching ranges in China. This was really a scoping exercise to calculate where the Chinese *might* locate a rocket launching centre, based on the known constraints of population density, downrange impact areas, terrain, climate, distance from hostile borders and communications links. The CIA identified five sites from whence the Chinese could launch rockets either westward or eastward. They were looking in the right place and from 1958 U-2 spyplanes were put up from airfields in Okinawa, Taiwan, Pakistan and Thailand, sometimes in American colours, sometimes in the colours of the Republic of China in Taiwan.

From 1960, the U-2s were joined by the Discoverer/Corona photo-reconnaissance satellites. The first satellite overflight may have been as early as December 1960. The two systems complemented one another, Corona providing safe, high, broad

Early days in Jiuquan. (Courtesy China Astronautics Publishing House.)

pictures, the U-2s detailed, closer images, but the attendant risk of the plane being shot down. We know that Discoverer/Corona satellites overflew and photographed northwestern China from as early as December 1960. As was the case with American reconnaissance of the Soviet Union, the primary objective was to watch nuclear and military programmes rather than the space programme. We can presume that the Americans had good on-going intelligence of the state of the Chinese space programme and related activities. In October 1962, Corona satellites found and U-2s confirmed missile installations in Lien Shan near the Gulf of Liaotung off the Yellow Sea. In August 1964, U-2 and space-based intelligence led the CIA to the conclusion that China would explode an atom bomb within a few months at its Lop Nor site, which it duly did.

By March 1962, the Americans had a detailed picture of Jiuquan launch centre. They had found three launch pads, in addition to two Surface to Air Missile (SAM) sites. They rightly surmised that the main role of Jiuquan was to test medium range missiles and impact them into the Taklamakan desert nearly 1,800 km due west. The three launch pads they identified as A (a double pad), B and C. The photography was so good that the National Photographic Interpretation Centre was able to pick out the housing and accommodation areas, servicing facilities, airfield and communications systems, making an extraordinarily detailed map. Even small garden plots were photographed.

The National Photographic Interpretation Centre was able to update the American intelligence community on the state of Jiuquan following three more overflights by August 1964. In 1965, KH-7 mission 4017 was sent to overfly Jiuquan as well as the Lop Nor nuclear facility. The new missions noticed increased numbers of vehicles at the site, fresh tents, a balloon site for weather tests, new barracks and at least ten new buildings at the main base. A KH-4 satellite overflew Jiuquan on 17th June 1967 and the CIA was able to give a detailed report on activities there, even though observing conditions were less than ideal. The American photographs were able to pick out the launch complexes, 45 vehicles, scarring of the ground where engines were fired, the construction of a causeway over a stream and 30 tents.[7]

Construction in Jiuquan. (Courtesy Sven Grahn.)

THE GREAT SPLIT

The 30-year unshakeable Chinese-Soviet accord ended in tears in August 1960. No single reason explained the breakdown in the relationship. The refusal of the Soviet Union to supply specifically requested nuclear technology exasperated Mao and appears to have been the main explanation. Khrushchev for his part became more and more convinced that China had every intention of using it in a nuclear war at the first available opportunity. Fourteen hundred Soviet technical advisors returned abruptly home on 12th August, bringing their blueprints with them and shredding anything they could not carry. Over two hundred joint projects were cancelled. The departure of the Soviets from the Fifth Academy was apparently good-natured and an occasion of genuine regret on both parts. Photographs taken at the time show the scientists bidding each other fond farewells.

The split between the two communist allies came at a crucial moment. In September 1960, the Chinese had made sufficient progress to launch a Soviet-supplied R-2 rocket, one using Chinese fuel. The real test was not far off. The construction and launch teams laboured in their factories and in the Gobi desert throughout October.

Director of the crucial upcoming test was Marshal of China Nie Rongzhen. Two months later, on 5th November 1960, China eventually launched its first real rocket actually made in China. Tsien himself was there to watch the rocket brought down to its pad, lifted into place, fuelled and fired. Nie Rongzhen presided. As the R-2 disappeared over the horizon, the celebrations began. Two more R-2s were successfully fired the following month. Although hardly anyone outside its scientific and political leadership knew it, China had joined the space race. The Chinese R-2 was named the Dong Feng 1, or 'east wind' 1. The success was publicly reported in December. A full scale silver grey exemplar with red Chinese markings on the side was put on display in the Beijing military museum many years later.

Leaders of the launch of the Chinese R-2, 5 November 1960

Responsible for the strategic rocket programme	Marshall Nie Rongzhen
Responsible for rocket development	Tsien Hsue Shen
Organiser of first test flight	Colonel-General Zhang Aiping
Responsible for control	Wang Zhen

For four nations, the A-4 rocket had marked the coming of age of their rocket programmes. The first A-4 had been fired into the atmosphere from Peenemünde over the Baltic by the Germans on 3rd October 1942. The Americans had copied the A-4 at White Sands in New Mexico in 1946, the Russians in Kapustin Yar the following year (the R-1). Now the Chinese, with Dong Feng 1, had emulated the German, Soviet and American achievements. Would China now be the world's next great space power?

ASSESSMENT AND CONCLUSIONS

China joined the space race on 8th October 1956, even before the first Earth satellite had been launched. The Chinese saw rocketry and space science as an important part of national defence, in leading industrial and technical development and in developing an economy in the full first flush of socialist construction. At an important conjunction of time, the modernizing Chinese leadership was facilitated by the return to China of the cream of China's scientists who had just been expelled from the United States. Many were leaders in their respective fields in the most advanced technological country in the world.

In the heady, early days of socialist construction, it seemed possible that China could soon build an intercontinental rocket *and* an Earth satellite, possibly joining the Soviet Union and United States in space as a close third. However, despite their great theoretical knowledge, the Chinese soon realized that their reach exceeded their grasp. The cruel realities of a war-blighted underdeveloped economy struck home and they soon realized that the construction of a basic rocket, even to the standards of what the Germans had achieved in the early 1940s, would take much time, effort and patience. In pragmatically turning to the Soviet Union, the Chinese were able to receive some grudging, then more forthcoming, help and assistance. It was only when Chinese scientists toured the Soviet Union that they realized the considerable industrial and technical base that was necessary to underpin the building of modern rockets and an Earth satellite.

Despite the enthusiastic start of projects to reverse-engineer a Russian rocket (project 1059) and build an Earth satellite (project 581), Chinese ambitions unravelled. The Earth satellite project fell victim to the great leap forward and the dawning, unpalatable realization that this project was one bridge too far. The Sino-Soviet split almost put paid to the project to fire the R-2. Thankfully for the Chinese, it was so advanced that they were able to complete the R-2 within months of the Soviet departure. But with the Russians gone and China embargoed by the western world, they must have now realized that they were on their own.

REFERENCES

1. For a description of early Chinese rocketry see Lai Chen Chien, Youn Chiung Luuo, Yu-Jen Su and Yu-Ting Ke: *Multi-tube and multi-stage rockets in ancient China*. Paper presented to the 50th International Astronautical Congress, Amsterdam, 4–8 October 1999.
2. Basic Books, New York, 1995. Note that this book uses the same form of Tsien's name as that used by Iris Chang. In several modernized versions of his name, he is also called Qian Quesen. Lest there be confusion, they are one and the same person.
3. See *Tsien spaceplane*. Mark Wade's Encyclopaedia Astronautica, www.astronautix.com
4. Reference: Frank Marble: Tsien revisited. *Caltech News*, volume 36, 2002.

5. Zhu Yilin: Development of Chinese satellites under Prof. Tsien. *Journal of the British Interplanetary Society*, 50, 185–188, 1997.
6. Zhang Yun (ed): *The Chinese space industry today*. China Social Sciences Publishing Co, Beijing, 1986.
7. Central Intelligence Agency: *Possible guided missile testing and training ranges in Communist China*. Geographic intelligence report, October 1958; National Photographic Interpretation Centre: *Shuang Cheng-tzu missile centre, China*. Photographic interpretation report, March 1962; National Photographic Interpretation Centre: *Shuang Cheng-tzu missile centre, China*. Photographic interpretation report, August 1962. I am grateful to Dwayne Day for making these papers available to me. See also: Kenneth Gatland: *Missiles and rockets*. Blandford, London, 1975, 216–222; Kevin C Ruffner (ed): *Corona America's first satellite programme*. Central Intelligence Agency, Washington DC, 1995; Central Intelligence Agency: *Search for uranium mining in the vicinity of Akosu, China*, August 1963; SNIE 13-4-74 *Chances of an imminent Communist Chinese nuclear explosion* (26th August 1964); *KH-4 mission 1042-1, 17–22 June 1967* [partly declassified reports]; Dwayne Day: Recon report (published on the internet).

3

Dong Fang Hong – the East is Red

China's initial success with the R-2, the Dong Feng 1 belied the difficulties which lay ahead. The development of the Dong Feng series of missiles in the 1960s took much more time and effort than was anticipated. It was some time before the idea of the Earth satellite was to be restored to the agenda: even when it was, progress was hesitant. Concentration on military needs came first.

MILITARY IMPERATIVES: DONG FENG 2

As the Earth satellite project faded into the background, military imperatives came to the fore. While western perceptions of China were of an aggressive, insurgent communist Asian troublemaker, China for its part felt isolated, friendless, surrounded by powerful enemies (principally the United States) and betrayed by its former friend, the Soviet Union. For these reasons, building up missile defences assumed the upper hand, rather than satellite projects.

The next rocket, the Dong Feng 2, would be China's first indigenously – designed and manufactured rocket, with a range of 1,500 km, capable of hitting the old wartime adversary, Japan. When seen in the Beijing military museum many years later, it looked like a long, tall, stretched V-2. It would match the Soviet ballistic missile capabilities of the mid-1950s. The Dong Feng 3 would reach the Philippines (10,000 km), the Dong Feng 4 the mid-Pacific and the Dong Feng 5 would be an intercontinental ballistic missile to match the American Atlas and the Soviet R-7 used to launch Sputnik. Further in the future, China planned to emulate the ability of the United States, the Soviet Union, France and Britain to launch missiles from submarines – which meant it could strike from anywhere underwater.

Denied Soviet assistance, the Chinese found the development of an indigenously-designed and built missile to be cruelly difficult. The first attempt to launch the Dong Feng 2 ended disastrously when it shook violently and crashed 69sec into its mission on 21st March 1962 at Jiuquan. The post-mortem found that the engine had not been structurally installed properly, the gyroscope was in the wrong position, the guidance system was defective and the rocket frame was unable to withstand elastic vibration. The project leaders were mortified with the failure, but Nie Rongzhen rallied them and told them to apply the lessons so that this problem never arose again.

Tsien set to work to redesign the guidance system. The redesign took over two years, involving 17 ground tests and did not achieve success until 29th June 1964. Further successful launches were carried out on 9th and 11th July. Although officially and politically this was a period of 'adjustment, consolidation, replenishment and improvement' after the great leap forwards, the rocket project continued to be a high priority. Resources streamed into rocketry and related technologies (e.g. computers). During 1962–4, work began on rocket engine test stands, a vibration testing tower, vacuum chamber, wind tunnels and solid-fuel motors. More constructively for the space industry, the role of the political commissars in the Fifth Academy was downgraded and the amount of all working time devoted to party activities reduced to less than 1/6th. Subsequent official histories refer to this as a period in which erroneous leftist methods of the great leap forward were rectified. In November 1962, on the proposal of President Liu Shaoqi, the rocket effort was brought directly under the Central Committee of the Community Party of China, reporting directly to one of its committees.

CHINA A NUCLEAR POWER

In 1964, Chinese nuclear science achieved its long-awaited breakthrough when in October its first nuclear weapon was exploded, the same week when, incidentally, the foe Khrushchev was deposed in the Kremlin. Tsien and his team modified the Dong Feng 2, so as to increase its range and give it an internal computer guidance system. Retitled the Dong Feng 2A, it was launched on 27th October 1966 on a 640 km westward path carrying a live 1,290 kg nuclear weapon which, on impact, duly exploded in the Xinjian desert, releasing a 12 kiloton nuclear explosion. Chang calls this test the most reckless nuclear experiment in history. She may well be right, though only a few years earlier the Soviets had likewise exploded warheads from rockets and the United States had let off nuclear bombs high in the atmosphere. The Chinese nuclear tests in the 1960s had the casual approach to safety not that dissimilar from the American tests in Nevada a decade earlier. Photographs show the scientists and workers gathering to watch the atomic blasts in ringside seats at what seems to be only a few miles from the blast, before eagerly rushing forward to admire its devastating effects. Radiation danger seemed but a minor consideration.

The Chinese nuclear test led the American media to rediscover Tsien. They ran a series of articles deploring the way in which he had been expelled from the United States and lamenting how he now headed up the rocket and nuclear programmes of one of America's political adversaries. Ten years earlier, the American press had passed over his deportation in silence. Tsien was not the only American-trained nuclear expert. There were others, like Zhu Guangya (1924–), who studied nuclear physics in Michigan from 1946 to 1950 and subsequently returned to head up the nuclear research programme; hydrogen bomb expert Deng Jiaxian (1924–86), who studied in Purdue university from 1948-50; and nuclear metal physicist Chen Nengkuan (1923–), a veteran of Yale, Johns Hopkins and the Westinghouse Institute. Several other leading atomic physicists trained in other western countries, like Wang Ganchang (1907–) who studied in Germany in the 1930s and Russia in the 1950s; and Cheng Kaijia (1918–), who trained in the University of Edinburgh, Scotland, from 1946–50.

DONG FENG 3, 4

The next project, the Dong Feng 3, had originally been for a 10,000 km range missile but had to be suspended due to a series of technical setbacks. The design was simply too ambitious for Chinese manufacturing capabilities at the time. Eventually, the project got under way in early 1965. There were two important innovations in the DF-3. First, the rocket clustered four engines together at take-off, not just the single engine of the DF-1 or DF-2. Such clustering required the four engines to achieve virtually exactly the same level of thrust simultaneously (otherwise the rocket would topple or go off course).

Second, the Chinese abandoned the fuels used on the DF-1 and DF-2 for storable propellants. Storable fuels had the handling advantage that they could be kept in the rocket for a long time before they were launched. The German A-4 and its successors used alcohol and liquid oxygen (lox): the liquid oxygen had to be kept cold and must be quickly drained if a launch did not take place, an event which happened frequently in a period of long and troublesome countdowns. Storable fuel systems did not have to be frozen and could be kept on board for long periods, at the ready, which was important for the military.

The principal disadvantage is that storable propellants use nitric acid or nitrogen tetroxide as oxidizer with UDMH (unsymmetrical dimethyl hydrazine). These fuels are highly toxic, corrosive and environmentally damaging. The dangers of corrosion required the engineers to use new metals in building the tanks, seals and motors. They were able to use new forms of high-strength aluminium alloys 50% stronger than those previously used, enabling a reduction of the rocket weight and an increase in the payload weight. The DF-3 was the first Chinese rocket subjected to intensive ground testing.

The four engines of the DF-3 were first tested together on a stand in July 1965. As rocket engines became more powerful, the engineers encountered new problems, such an unstable combustion caused by vibration. Zhou Enlai watched a countdown rehearsal during a visit to Jiuquan on 30th July 1966. The DF-3 eventually made its first flight there on 26th December 1966. The next batch of tests went well and the missile was considered operational in 1969 (a descendant was eventually developed and sold to Saudi Arabia).

The Dong Feng 4 was an improved version of the Dong Feng 3. It came to have a dual identity. With a small upper stage, the civilian version of the Dong Feng 4 became best known as the historic Long March 1 rocket which launched China's first satellite. It made its first flight on 30th January 1970 (this is now logged as the first test flight of the Long March 1) and was subsequently formed into a missile strike force near Harbin, capital of Manchuria, in the north-east.

DONG FENG 5: CHINA'S LONG-RANGE ROCKET

The Dong Feng 5 was important for the development of Chinese Earth satellites and for the Chinese military. Russia had built an Inter Continental Ballistic Missile (ICBM) first (the R-7, August 1957) and then turned it straight into a satellite launcher (Sputnik 1, October 1957). For their first successful satellite, the Americans uprated a medium-range rocket (the Jupiter), attached an upper stage and put Explorer 1 into orbit (January 1958). America's intercontinental ballistic missile came later in 1958 (the Atlas). In essence, the Chinese followed the American approach, using a medium-range missile (the DF-4) as the basis for their first satellite launcher (the Long March 1), then building a small intercontinental ballistic missile later (the Dong Feng 5). Chinese space histories give considerable attention to the Dong Feng 5, for with it, the Chinese achieved limited military rocket comparability with the United States and the Soviet Union. An ICBM could just now reach the United States, China's most likely superpower enemy.

The work done perfecting the DF-5 was passed on to two rocket teams – the Beijing team designing the Long March 1 to launch China's first Earth satellite and the Shanghai team designing a launcher for military satellites, the Feng Bao ('storm'). These three rockets have much in common. The DF-5 required the Chinese to make considerable progress in guidance systems, propulsion, rocket engines, steering devices and strong but light materials. Chief designer of the DF-5 was Tu Shoue. Later, it became the basis for the Long March 2 rocket.

Work on the DF-5 came to a virtual standstill for 1967–9 because of the cultural revolution and because of the switching of people and resources to the work of the Long March 1 rocket. DF-5 scientists and technicians were assigned for reeducation, land reclamation or manual labour. The first flight of the DF-5 eventually took place on 10th September 1971 from Jiuquan and it was described by the Chinese as 'essentially' successful (a term which implies that some aspects were not). A DF-5 base was later established at Harbin in Manchuria. Early test launches were made from there, to impact in Tibet and Taklimakan desert in Xinjiang.

At this stage, the Dong Feng 5 had been tested only on short to medium range missions, within China. Its long-range ability was unverified. A comprehensive round of tests was approved in 1977, with further flights from Jiuquan within China in 1979. Five took place: these were successful and it was decided to then fly the rocket to its full range, height and speed. This would require a trajectory far outside China's borders. The DF-5 would be launched to an altitude of 1,000 km, fly at 7 km/sec, impact in the southern Pacific half an hour later and reach a range 8,000 km distant. Its impact would be observed by China's ocean going communications ship (comship), the *Yuan Wang*. The main problem in preparing the launching was the unreliability of the electronic components, which repeatedly gave trouble. Deputy premier Zhang Aiping set up a trouble-shooting group

Dong Feng 5: China's long-range rocket

Tu Shoue. (Courtesy China Astronautics Publishing House.)

(prosaically called the Electronic Component Reliability Work Leadership Group) to try sort it out, with evident success.

On 26th April 1980, the new comship *Yuan Wang 1*, following inspection by Zhang Aiping and representatives of the state council and the Central Committee, was seen off from its port in Shanghai to monitor the planned impacts in the south Pacific. The fleet arrived on station two weeks later, close to the equator, about 1,000 km from the Solomon islands and Fiji. In the early dawn of the day of the first test, 18th May 1980, the wind had dropped, the waves lapped gently around the ships and numerous clouds drifted past.

Back in China, in the still nighttime morning, chief designer Tu Shoue, test team leader Zhang Lianfu and their colleagues gathered around the DF-5, bathed in floodlights. The atmosphere was tense as ten years of work reached the moment of truth. The DF-5 roared off into high clouds, bending over in its path over China toward the Pacific. Ten tracking stations reported its arc across the morning sky. At sea, sailors spotted a huge ball of fire, scattering into bright spots, as the DF-5 rocket crashed Earthward into the upper atmosphere. However, they could note one bright spot still falling and burning its way through reentry. With a sonic boom and a whistling noise, a parachute opened and the recovery cone made a tremendous 100 m high splash. A helicopter took off at once from one of the recovery ships, dropping diver Liu Zhiyou into the sea to attach a flotation system. The recovery cone was hoisted on board in 15 mins. Another DF-5 payload was observed impacting on 21st May by prying ships of the Australian navy, though it may not have flown the full distance intended. On 2nd June, the *Yang Wang* and its attendant ships returned to port in Shanghai in triumph. There was joy throughout the rocket community in China, Zhang Aiping even being moved to wrote a poem to mark the event. A week later, all those involved in the DF-5 tests were received by Deng Xiao Ping and Hu Yaobang in the Great Hall of the People in Beijing.

SUBMARINE-LAUNCHED MISSILE

Progress on the submarine-launched missile was slow. Because submarine-based missiles might spend long periods at time at sea before their launch, they had to be solid fuel rockets. The Chinese decided on a powerful, two-stage submarine-launched ballistic missile. The solid rocket motor nozzles were designed to be swivelled so as to ensure accurate guidance and that the rocket hit the right target. This proved to be a knotty problem which took years to resolve. The programme was also impeded by the cultural revolution. Worse, on 16th March 1974, technician Wang Lin, deputy head of the rocket workshop, was unable to rest that night as he contemplated new ways of mixing more effective propellant. He returned to the workshop. Later that night, the mixer exploded. Wang Lin was buried underneath the protective iron door of his laboratory and he died. The first and second stages underwent ten successful ground runs in the late 1970s and made three land-based launches in 1981–2. A further disaster, this time a natural one, intervened in August 1981, in the final stages of assembly of the rocket for the first underwater test. The factory which made the gyro platform found itself, ironically, under water itself when flash floods tore down from the mountains, destroying roads and parts of the factory. The staff disassembled the gyroplatforms, carried them in backpacks across mountains and delivered them to the nearest flood-free railway line.

Two years later, on 12th October 1982, a Chinese navy submarine left harbour and dived close off shore. After a few minutes, it launched its first submarine-fired ballistic missile, arcing out over the sea, coming down in the distant ocean. China thereby demonstrated that the country had a land and sea-based operational intercontinental ballistic missile system at least on a par with France and Britain. Designer of the submarine-launched ballistic missile was Huang Weilu (1916–), a 1940s graduate of the University of London who subsequently went on to be one of China's main pioneers of radio control and guidance systems and designer of solid rockets.

CHINA'S ROCKET STRIKE FORCE

Ultimately the Dong Feng 4 and 5 were developed as China's nuclear strike force. The DF-5 carried a hydrogen nuclear warhead weighing 3 tonnes. Come the late 1990s, China had up to 20 DF-4s (able to reach Moscow) and between seven and 15 DF-5s, just able to reach the United States, a small but potent arsenal.

Reports in the late 1990s indicated that the DF-4 and DF-5 would be phased out by the Dong Feng 31 (range 8,000 km, also submarine-launched) and Dong Feng 41 (12,000 km) respectively. These were solid-fuel road-based rocket missiles, broadly equivalent to the Russian SS-25 or the American Minuteman and were set for location in two secret silo-based centres. They could strike not only the American west coast but inland as far as the Rockies. The Dong Feng 31 was still being flight-tested in early 2002. The situation broadly parallels that in the Soviet Union many years ago, when the first generation of pad or silo intercontinental ballistic missiles was phased out for the second generation of lighter, more powerful, road-mobile solid fuel launchers.

American analysts were divided as to whether this signalled purposeful modernization of an obsolete military apparatus or a more aggressive foreign policy. Occasionally arguments flared as the Americans would accuse China of building up its nuclear arsenal, predictably followed by heated Chinese denials.[1]

China's missiles – the Dong Feng (DF) series

DF-1	5 Nov 1960	Reverse-engineered Russian R-2
DF-2	29 June 1964	Designed to reach Japan, 1500 km
DF-3	26 Dec 1966	Designed to reach Philippines, 10,000 km
DF-4	30 Jan 1970	Designed to reach mid-Pacific
		Became the Long March 1
DF-5	10 Sep 1971	First with intercontinental range
	18 May 1980 (operational)	Became the Long March 2
DF-31	1990s	Range: 8000 km, submarine-, land-launched
DF-41	1990s	Range: 12,000 km, land-launched
Submarine-launched ballistic missile	12 Oct 1982	

SOUNDING ROCKETS MAKE PROGRESS

Even as China was making progress with the development of its military missiles, the sounding rocket teams were paving the way for China's eventual entry to the space race. In 1962, the Academy of Sciences put forward the specification for a more advanced weather sounding rocket, better than the T-7. Designer Wang Xiji responded by improving the tail fins, reducing the weight of the engine, increasing the thrust, adjusting the upper stage to work at altitude and stretching the length of the tank. The new rocket was the T-7A. Weighing 1,145 kg, it stood 10.32 m tall. It was first launched in December 1963, carried 40 kg of equipment and reached an altitude of 115 km.

The T-7 and its improved version, T-7A, were used many times in the 1960s for meteorological, geophysical, technology development and biological missions, though, like similar Soviet missions, a full flight log does not appear to be available. The meteorological version first flew in December 1963. Scientific versions were used to survey the ionosphere, electron densities, cosmic rays and the Earth's magnetic field.

BIOLOGICAL SOUNDING ROCKETS

Under the management of the Biophysics Research Institute of the Chinese Academy of Sciences and the Shanghai Machinery and Electrical Equipment Design Academy, the T-7A was used for a series of biomedical missions in the course of 1964-66. The series, involving the redesign of the nose, was called the T-7AS-1. The nose of the T-7A was redesigned to take a sealed biological cabin with a standard cargo of four white rats, four white mice and 12 biological test tubes which in turn would hold fruit flies and other test items. The aim was to observe the behaviour of the animals – some restrained, some free-floating – in flight by camera. After the flight, some of the animals would be dissected to see whether there had been any effect on their biology; the others would be bred to watch for genetic change. The T-7AS had a weight of 1,165 kg and a height of 10.81 m.

On its first mission, the biological sounding rocket took off on 19 July 1964, reaching an altitude of 70 km. Film was taken of the reaction of the cargo to weightlessness. On 1st and 5th June, 1965, two further missions were carried out. The animals were recovered intact from 60 km to 70 km on all three occasions.

DOGS FLY INTO THE ATMOSPHERE: THE MISSIONS OF XIAO BAO AND SHAN SHAN

The Academy of Sciences decided at this stage to proceed with flights of dogs. Canines had been the main precursors of humans in the Soviet space programme. The T-7 was adapted a second time, with a larger sealed cabin, one able to take a dog, four white rats and 12 biological test tubes. The resulting sounding rocket, the T-7AS2, weighed 1,346 kg, of which the biological container came in at 170kg. It was 10 m tall, with a diameter of 0.45 m and was designed to reach between 60 km and 115 km. The purpose was to test the reaction of the dogs under weightlessness, ascent and descent and their reactions to noise and vibration. The level of radiation dosage was also to be tested. During the flight, the dog's heart-beat, temperature, respiration and breathing rates would be measured by a tape recorder. The carrying of a dog required a much more advanced life-support system, but as a precaution against a delayed recovery, arrangements were made for a pressure valve to be released during the descent to let in fresh air.

On 15th July 1966, the T-7AS2 made its first flight carrying China's first space dog, Xiao Bao ('little leopard' in Chinese). All went well. An Air Force helicopter spotted Xiao Bao's cabin drifting down, impacting at less than 10 m/sec. The helicopter touched down alongside. The dog's handler rushed forward to see Xiao Bao alive inside. The handler took him out, the dog wagging its tail and apparently glad to be back. He returned to kennels but the rats were dissected on the spot. The second dog flight took place on 28th July 1966, this time carrying a bitch, Shan Shan ('coral' in Chinese). It went equally well. Plans were under way to fly a monkey that September, but the cultural revolution intervened and the mission did not take place.

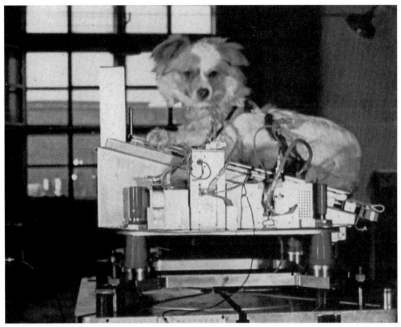

Xiao Bao (Little Leopard). (Courtesy China Astronautics Publishing House.)

The T-7 was then adapted to carry satellite test equipment – for example, satellite control systems – that would later be used in orbit on China's first Earth satellites. The designer was Lin Huabao. Jiuquan was used as a launching site, with a spiral cage launch-rail system devised by Jiangnan shipyard in Shanghai. Two launches were made – one in 1965, reaching 83 km, the other, in 1969, reaching 81 km. Instruments on board took pictures of the ground, photographs of the stars, obtained data on electrons in the ionosphere and tested out infrared scanners. The descent cabin used a mixture of airbrakes, braking parachute and main parachute to lower it to the ground at a speed of less than 6 m/sec.

The first mission nearly led to disaster. The cabin came down in the desert in Badain Jaran. The recovery team of 20 people, led by Lin Huabao, set off in trucks but there were sand dunes everywhere which the vehicles could not manage. Searching on foot, they did not find the cabin until after nearly a day's searching in extremely sandy desert. Only the tip of the cone was showing, the rest being buried in dunes. It took them three hours to remove the scientific and test equipment – whatever they could carry by hand. In their attempt to return to their trucks that night, they lost their way. Although they fired flares, nobody saw them. They had a terrible night, having run out of water and with their feet all blistered. Lin Huabao rallied the team and they managed to crawl out of the desert the following day.

Chinese space mice. (Courtesy China Astronautics Publishing House.)

RESUMPTION OF THE EARTH SATELLITE PROJECT

Project 581, the first Chinese Earth satellite, had been postponed indefinitely in 1959. The success of the R-2, the subsequent development of the Dong Feng military missiles and the successful flights of sounding rockets gave Tsien Hsue Shen the confidence to press for a resumption of work on plans for an Earth satellite. In the early 1960s, countries such as France and Britain began to develop launchers to put small Earth satellites into orbit and it was clear that space would not remain the preserve of the two super-powers indefinitely.

In 1962 Tsien recruited a satellite development team of four men from the Shanghai Institute of Machine and Electrical Design (SIMED). They spent a year with Tsien in Beijing working through possible designs and consulting the Russian and English-language literature. The four were Zhu Yilin, Kong, Li and Chu (full names are not available in the record). They seem to have had access to good information on a range of American space programmes, like Discoverer, Mercury and Ranger. The team met Tsien weekly to review progress. By the end of the year, they had drawn up a plan to cover the years 1964-73 for the development of Chinese Earth satellites. They also lectured students in the Chinese University of Science & Technology. While they were doing this, Tsien concluded his book *An introduction to interplanetary flight*, published in Beijing in 1963. This text brought together the sum of Tsien's knowledge of the subject to date, including the experience of developing the R-2 and also served as a basis for the lectures of the students.

Inspired by the flight of Yuri Gagarin around the Earth in April 1961, Tsien Hsue Shen convened a series of 12 symposia on space flight. The first was held in June 1961. The meetings continued for three years, in the course of which progress in spaceflight world-wide was reported and noted. They made studies of American application satellites, working through the open literature. The four designers acted as its secretariat. In 1963, the Academy of Sciences formed a Commission on Interplanetary Flight. Two papers were published by the University of Nanjing about a simple probe able to hit the moon, like Russia's Luna 2 in 1959.

In January 1963, the Shanghai Institute of Machine and Electricity Design was made organizationally, on the order of the State Council, part of the Fifth Academy. When the four designers returned to Shanghai, they formed a satellite research division. Fifty technicians and experts were drafted into the process. They divided into five groups – for overall design, structure and thermal control, attitude control, electrical power and tracking, telemetry and command. They were permitted to cooperate with other institutes in Shanghai, but not further afield.

A further spur to the development of the satellite came when the Academy of Sciences in Beijing established a committee to investigate the desirability of a man-made satellite. Its feasibility study began in May 1964 and concluded in July 1965. However, further authorization for the project to continue was held up by the State Council. This caused much frustration in Shanghai, Tsien writing them letters of encouragement as he tried to grapple with the decision-makers above him.

PROJECT 651

Tsien sought authorization for the resumption of work on an artificial Earth satellite from the Central Committee of the Communist Party of China in January 1965. He circulated a proposal, pointing out that an artificial Earth satellite had been on the agenda for seven years (project 581). With the development of the DF-3 rocket

Tsien during the 1960s. (Courtesy China Astronautics Publishing House.)

progressing well, it should be possible to launch an Earth satellite. Nothing was to be gained from waiting further.

Four months later, the party's Defence, Science and Technical Commission supported the project, recommending a satellite launch in the 1970-1 time period. The Commission proposed that the Academy of Sciences develop the satellite and the Seventh Ministry the rocket. The Academy responded to the invitation by presenting, in July 1965, *A proposal on the plan and programme of development work of our artificial satellites*. The academy proposed an evolutionary long-term development plan for China, starting with simple, experimental satellites and then progressing to more difficult tasks, applications programmes and the recovery of satellites. Prime Minister Zhou Enlai and the Central Committee approved the plan on 10th August 1965, stipulating that the satellite be visible from the ground and that its signals could be heard all over the world (similar considerations governed the approval of Sputnik 1). Decided in 1965, it was coded project 651.

REORGANIZATION AND DISPERSAL

The decision to approve construction of the satellite laid down the lines of responsibilities between the different organizations and bureaux and also commissioned the construction of naval tracking and communications ships (comships). Zhou Enlai took the opportunity to make further organizational changes to streamline and rationalize the space industry, concluding a process which had been under way for some years. The key elements of these moves are simplified here, though they were infinitely more tortuous and tangled than this abridged description indicates.

On 23rd November 1964, the Central Committee of the Communist Party of China and the State Council issued a joint resolution to restructure the space industry. Their main purpose was to bring together the many different elements working on space activities in different parts of the country, in a dispersed and, they felt, uncoordinated way. All of them, including the Fifth Academy, were now transferred to what was called the Seventh Machine Industry Ministry. It was subdivided into four academies: long-distance rockets (1), ground to air (2), cruise (3) and solid (4), with sounding rockets and the bureau of tactical missiles located in Shanghai.

Whilst technically a change in name, in practice the change also meant a sharp change in status for the scientists and engineers. They lost military wages and were transferred to local wages which were generally lower. Their food rations were reduced. In January 1965, Liu Shaoqi appointed Wang Bingzhang the first head of ministry, with Tsien as one of his six deputies. The Seventh Ministry quickly developed an eight year plan for the development of Chinese rocket technology, to cover the years 1965–1972. Over 3,000 people in the ministry's design, production and user departments put forward their views. The new plan was adopted by the Central Committee in March 1965. In May 1965 the new Seventh Ministry received a high-powered official visit from Liu Shaoqi, Zhou Enlai, Marshal Nie Rongzhen and Jian Qing (Mao Zedong's wife).

Responsibility for the satellite, which had been designed in Shanghai, was now handed over a special bureau of the Chinese Academy of Sciences called institute 651. This paralleled Soviet practices whereby the various design bureaux were given code numbers (Korolev's was OKB-1, Chelomei's was OKB-586, and so on). Not only that, but the Shanghai Institute was moved to Beijing, where it was renamed the Seventh Research Division of the Eighth Institute of Design in the Seventh Ministry of Engineering Industry, but was also known as the Beijing Institute of Spacecraft Systems Engineering (BISSE).

Whilst the transfer of the satellite project must have galled the Shanghai team, it nonetheless cooperated with the Academy of Sciences and joint working groups were formed. The former Shanghai team and the Academy of Sciences institute 651 team formed a one joint working group, the Chinese Academy of Space Technology (CAST), on 20th February 1968. Tsien Hsue Shen was appointed the first president. All the existing institutes and bureaux working on satellite programmes were brought under the academy on the orders of Zhou Enlai. In a swop, several Beijing institutes were moved wholesale to Shanghai to form what subsequently became called the Shanghai base of the space industry (the Shanghai Academy of Space Technology (SAST), also known as Research Institute #805). The direct descendants of both bodies – CAST and SAST – survive to this day.

There was a further organizational change in November 1967 with the establishment of the Beijing Wan Yuang Industrial Corporation, later renamed the Chinese Academy of Launcher Technology (CALT), its current identification (although some Chinese literature still uses the older terminology to the present day). Thus by the end of 1967, China had a unified academy of launcher technology and another academy of satellite technology, both persisting to the present.

A complicating theme of all of these moves was dispersal. Regarding the military threat to China at this time as very serious, the reorganization specified in August 1965 that as many production units as possible should be shifted away from the coast, the south-east and the Sino Soviet border, to as far inland as possible (called 'thirdline regions').

64-DAY DESIGN CONFERENCE

The Chinese Academy of Sciences set up its own Artificial Satellite Design Academy in September 1965. A historic conference was held later that year, one bringing together all the groups concerned with the satellite project. The conference lasted 64 days – from 20th October to 2nd December 1965 – and reviewed all the progress made so far. This was called the '651 conference' after the project code and the number of the institute building the satellite. The meeting was of the view that, although China had made a late start in the development of satellites, its first sputnik should be much more advanced than either the first satellites of the Soviet Union or the United States. It should be visible and audible to the Chinese people. The satellite would have a weight of at least 100 kg, fly out of its launch site in an easterly direction and enter orbit at 42°. Sun Jiadong (1929–) was appointed chief designer.

Unlike most of his American-trained colleagues, he was Soviet-trained, having spent the 1950s in the premier Zhukovsky Air Force Engineering College.

A further conference was held for the development of the space programme in March 1966. 120 delegates attended. The purpose of the conference was to hammer out a 10-year plan for space development. The first objective was the launch of an artificial satellite, then a recoverable satellite and, as we found out many years later, a manned spaceflight. Later, the programme would develop a complete system of application satellites for communications, weather, nuclear detonation detection, early warning of missile attack and navigation.

In May 1966, the first satellite was designated the Dong Fang Hong-1 ('the east is red'). It was decided to increase the weight of the satellite to 170 kg and that it would fly to a much higher inclination, 70°, which would mean that it could be seen and tracked over a much greater proportion of the Earth's ground mass. The decision specified that the visibility of the rocket be that of a fourth magnitude star (it was fitted with an extension to help) and the satellite that of a fifth magnitude star. The solar cells and scientific instruments were eliminated. Instead, a tape recorder would play stirring revolutionary tunes.

Dong Fang Hong 1 was a spherical polyhedron 1 m in diameter and made of fibreglass wound on aluminium alloy. The surface was plated with insulation material designed to protect the instruments from the extreme heat and cold of Earth orbit. A sealed instrument module contained five silver zinc batteries able to provide power for 20 days. An electronic circuit was used to generate the tune *The east is red*. On the outside, four 3 m short-wave antennae, four transmitting antennae and four transmitting beacons were fitted. Two sensors were installed – an infrared horizon sensor and a solar angle gauge. Five mockups of the satellite were produced and these were subject to tests designed to measure the design's ability to withstand heat and cold, to transmit, to resist radiation and humidity. These tests were completed in October 1969.

At the same time, work began on the national tracking and control system. Originally, it had been intended to have a national control centre in Xian, Shaanxi with seven observation posts scattered round the country. Xian was not completed on time and the task was undertaken by the launching centre in Jiuquan instead.

THE CULTURAL REVOLUTION ENGULFS CHINA

However, again there were delays and once again these were for political reasons. In March 1966, Mao Zedong launched the cultural revolution. By summer, the Seventh Ministry had come to a standstill. The Academy of Sciences, developing the satellite, was overrun by seething political factions. The Director of the Institute of Geophysics, Zhao Jiuzhang, was killed. The ministry divided into two rival political factions – one favouring the red guards, the other the Liu and Deng Xiao Ping cliques. There were even armed clashes among the units responsible for building ground stations. Tsien Hsue Shen was deposed in early 1967 and reduced to the status of an ordinary employee. He had to sign confessions (his wife, Ying, was a

The cultural revolution engulfs China 51

Zhou Enlai.

teacher of western vocal music in Beijing conservatory and had to hide out for nearly a year). Tsien's colleague Chien Wi Zhang was sent by the Red Guards to work in the country first and then to stoke a steel furnace. Later, Tsien called the cultural revolution 'the ten lost years'.

Scientific approaches were ridiculed as revisionism, academics as reactionary and leading scientists as bourgeois intellectuals. Subsequent histories recorded that the revolutionaries spread the slogan that 'when the satellite goes up, the red flag goes down' and urged concentration on political rather than scientific tasks. Titles in the workplace were abolished, chief designers being given no more authority over rocket design than the lowliest technician or support staff. Some scientists were locked up in cattle sheds, others banned from research and many sent to be reeducated in the countryside. Those who resisted were killed. The rest bided their time, waiting for the bad times to pass. Some kept up their reading and scientific study clandestinely.

Zhou Enlai and Nie Rongzhen attempted to protect the satellite project from the ravages of the cultural revolution. On 17th March 1967, Zhou Enlai persuaded the Central Committee and the State Council that the space and defence ministries and their associated projects should be brought under military control. The principal feature of military control was that such controlled areas were exempt from what the cultural revolutionaries called the big four activities (free expression, free assembly, big posters and the holding of great debates). Zhou designated 15 scientists as being 'under state protection' but the rest had to look after themselves. Zhou Enlai issued instructions that the former leadership was to be reinstated, the missile programme was to remain a national priority and its leading scientists were to receive state protection. In April, the army moved in to enforce the decision. The new China

Academy for Space Technology (CAST) was first put under military law, so as to be an exempt body from big four activities. It did not escape the revolution entirely and its early work was impeded by the shortage of materials throughout the country. Units which had been transferred inland to third-line regions suffered acutely from revolutionary activity. But according some subsequent Chinese histories, the space industry suffered less under the cultural revolution than many other parts of the Chinese economy. The rest must have suffered grievously indeed.[2]

ROCKET FOR THE FIRST EARTH SATELLITE

A new rocket was developed to support the project, called the Chang Zheng 1 (CZ-1), or the Long March 1, named after the formative moment in Chinese communist history when Mao Zedong led his armies 8,000 km from the clutch of the nationalists into the distant but safe province of Shaanxi. Essentially, the Long March 1 was the civilian version of the DF-4 military missile, which in turn was an improved version of the DF-3. Design of the Long March 1 began in the Seventh Ministry but during the reorganization in November 1967 was switched to the Chinese Academy for Launcher Technology (CALT).

The specification was for a liquid propellant rocket able to send 200 kg into low Earth orbit. Thirty leading engineers and technicians were assigned to the project. Ren Xinmin (1915–) was chief manager of the Long March 1 project, sorting out the principal problems of unstable engine combustion. From Ningguo, Anhui, he studied ordnance in college, before going to the United States in 1945 to obtain a doctorate in Michigan. Returning in 1949, he developed solid-fuel rockets for the People's Liberation Army before becoming director of the Liquid Propellant Engine Design Division of the Seventh Ministry of Machine Building. Outline design of the Long March 1 was completed within a month.

DESIGN CHALLENGES OF THE LONG MARCH 1

The two stages used nitric acid as oxidizer and UDMH as propellant. Designers of the first and second stage were Ren Xinmin, Ma Zuozin and Zhang Guitian. The Long March 1 design required the use of new alloys, such as titanium, glass-fibre reinforced plastic and new forms of high-strength steel. New forms of advanced welding were used to complete the components to a high degree of perfection. For the first time, the first stage used a shared tank, rather than two individual ones. Whilst saving weight, this approach required high standards of plumbing, for any leakage would result in an immediate explosion. The engines for the Long March 1, whose design was also led by Ren Xinmin, involved new challenges. Four YF-2 engines were grouped together for the first stage. The engine for the Long March 1 proved to be extremely difficult, for it failed 17 running tests, mostly due to unstable combustion. The second stage engine had to be adapted so that it could light at altitude (60 km) in low pressure and air density: accordingly, a special air

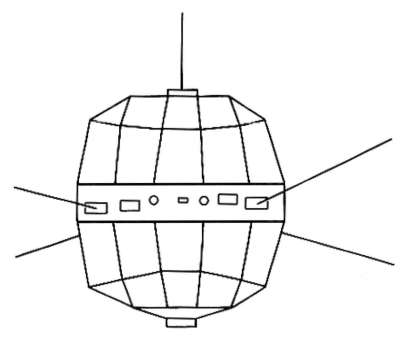

Dong Fang Hong.

evacuation chamber was built to simulate high altitude conditions and tested in November 1966.

Under the leadership of Huang Weilu, an inertial guidance system was developed for the Long March 1, a system much more advanced than the early Soviet rockets and up the American standards of the mid-1960s. For electrical systems, the rocket marked the end of valves and the introduction of transistors, though this was far from trouble-free, owing to the poor production quality of Chinese transistors at the time. Two telemetry systems were fitted to the Long March 1 – one on the second stage, the other on the third – and were designed to transmit back several hundred different parameters as the mission progressed. Finally, if the worst should come to the worst, a destruct system was fitted. This could be activated either from the ground or by sensors on the rocket itself should it veer off course.

THE PROBLEM OF THE THIRD STAGE OF THE LONG MARCH 1

The DF-4 had insufficient thrust to get a satellite into orbit and for this reason the Chinese developed a small third stage with a solid rocket motor to get the satellite on the final leg of its journey into orbit. Designer of the third stage was Yang Nansheng. This was the first time that the Chinese had built a solid fuel rocket. They had to start from scratch, for they had received no assistance from the Soviet Union in this area and other countries were embargoed by the United States from assisting China.

The chemical industry had no experience of producing the kind of fuel which powers solid rocket motors (the Americans use a form of perchlorate, which is poured into the rocket casing like a dark grey sludge).

The Chinese had recognized the need for solid rocket motor research when the space programme had begun in 1956. The Fifth Academy had included a solid propellant research group – but it comprised only three new graduates. By 1960, 70 people were working there, led by Li Naiji. They went to some lengths to try learn about solid rocket technology in other countries, picking up any engineering and industrial information they could. The Changchun Applied Chemistry Research Institute of the Chinese Academy of Sciences was commissioned to produce fuel – liquid polysulphide rubber; and the Harbin Military Engineering Academy produced the first batch of fuel – potassium perchlorate. It was two years before a compound was produced of sufficient quality to burn. In 1959, the Fifth Academy teamed up with ordnance ministry factory #845 in Xian to produce the first significant quantities of pourable material. To take the process a step further forward, a national propellant conference was convened in January 1961 and in March the following year the group of engineers became the Solid Fuel Rocket Engine Research Institute in the Fifth Academy. Xiao Gan was the first head. Within a short period, the institute had developed a fuel pouring mixer and production line and was making engine tests.

Then disaster struck. On 6 December 1962, 200 kg of solid fuel exploded in the mixer, killing two women technicians instantly. Two others died later of their injuries. Much more stringent safety systems were then introduced. It was the first of eight accidents in the Chinese space programme. A second national propellant conference was convened in August 1964 to review the – clearly slow – progress of the previous three years. Additional resources were devoted to attack the key problems which the institute was experiencing, such as the solid fuel cracking, the nozzle overheating and unstable combustion. There was progress when in summer 1965 six engines were fired successfully.

Disasters and accidents in the Chinese space programme

6 Dec 1962	Explosion of solid rocket mixture in preparation, killing 4 technicians
26 Jan 1968	GF-02 solid rocket motor explodes during tests
16 Mar 1974	Solid fuel mixer explodes, killing workshop head Wang Lin
Jan 1978	Explosion in LM-3 engine, unspecified number of fatalities
22 Mar 1992	Launch crews injured in on-pad abort, Xi Chang
2 Apr 1994	Feng Yun 1 metsat exploded in fuelling: 1 technician dead, 31 hurt
26 Jan 1995	6 villagers die, 23 injured when LM-2E explodes at 70 sec, Xi Chang
14 Feb 1996	80 injured, unspecified fatalities, LM-3B crash on first flight, Xi Chang

The specification for the Long March 1 required a third, upper, solid stage 4 m long weighing 1.8 tonnes. Its role was to fire at 600 km – when the second stage had completed its work – to put the Earth satellite in orbit. Preliminary design work was completed in April 1967, the institute opting for polysulphide rubber propellant,

The Long March 1 launch pad. (Courtesy Sven Grahn.)

high strength steel casing made by Anshan Steel Works and a graphite nozzle throat liner. Their efforts at this crucial stage were complicated by the relocation of the academy to north-west China, a third-line region, where accommodation was very limited; and the cultural revolution. The upper stage was declared a State Priority Crash Construction Programme on more than one occasion in 1969 and 1970 by Zhou Enlai in order to have work completed on time.

The first test run of the third stage, on 26th January 1968, was a failure, the engine exploding after 30sec due to a failure in the adhesive. However, by 1970, 19 tests had been carried out, five being high-altitude simulations. The one test that really mattered lay ahead. The T-7 sounding rocket was used to test the solid rocket engine of the upper stage of the Long March 1 rocket, the carrier of the upcoming Chinese Earth satellite. The GF-01A solid rocket engine was fitted to the top of the sounding rocket in place of the biological or scientific payload. Two tests of the GF 01A were carried out – on 8th and 20th August 1968. The powerful third stage managed to boost the payload to an altitude of 311 km, which must have been the Chinese altitude record at the time.

LONG MARCH 1: THE FINAL LAPS

A priority area of work in the satellite project was the building of a rocket engine testing site. This was declared to be a priority national project in 1965. Two years later, near Beijing, the first rocket engine test stand had been constructed on the edge of a gully so that the flames could be deflected down the cliff face. The Long March engines underwent all-up testing there in June 1969. Afraid that this crucial task would be interrupted by the revolutionaries, Zhou Enlai issued orders that the tests were a matter of national honour and interfering with them would be regarded as unpatriotic. So protected, the workers toiled for eight days without interruption to complete the tests.

56 Dong Fang Hong

At Jiuquan launch site, construction of a new pad able to take the larger Long March 1 rocket and its new upper stage began in 1965 and was completed two years later. This was called 'area 2' and was close to the pad which had first fired the R-2 missile in 1960 (area 3).

Final testing of the Long March and its upper stage took place in the course of 1968–9. The various parts were tested to exhaustion, the final rocket assembly being put together to test whether they would fit and function together (this is called the science of 'integration' in rocketry). The new rocket was successfully fired on 30th January 1970, though without the third stage (also logged as a Dong Feng 4 launch).

MOMENT OF TRUTH

All the different components of the Long March 1 were brought together for the first time in July 1969, even as the Moon race of the two superpowers reached its final, frenzied climax. A fully-integrated mockup left the works on 5th February 1970 and was shipped to Jiuquan, the real one following on 26th March, Tsien attending the launch preparations. Two versions of the satellite had been built, should a problem occur with one of them. They arrived on 1st April. The next day, Zhou Enlai called a meeting in the Great Hall of the People to take progress reports on the preparations. The satellite was installed on the rocket and a series of inspections and tests followed. Several days later, Tsien Hsue Shen, Ren Xinmin and their most senior colleagues

Long March 1 on the launch pad. (Courtesy China Astronautics Publishing House.)

Dong Fang Hong 1.

reported to Zhou Enlai in Beijing that the satellite was ready for launch. The meeting went on till late. Eventually satisfied, Zhou Enlai approved the launch, sent the engineers back across China to the launch site and asked for daily reports from their arrival. Mao Zedong gave the go-ahead some days later. The final assembly was erected on the pad on 17th April and cleared for flight on the 24th. The tower was moved back.

That morning, the weather was warm and sunny with a spring breeze. The forecast that evening was fine: some clouds at 7,000 m and wind speeds of less than 4 m/sec. Fuelling began. Zhou Enlai was on the phone at 3.50 pm with a final good luck message. The day clouded over. 20 mins before the launch time, 9.35 pm, floodlights were switched on, to bathe the gantry and rocket in milky light. Just then, the clouds parted, stars beginning to wink through the darkness of night.

At 9.35 pm that evening, the historic *Ignition!* command was given. Red and orange flames streaked out of the bottom of the Long March 1. China's first Earth satellite lifted off from its pad in Jiuquan and headed for orbit. Flames streaked across the night sky, at one stage making a tail 500 m long. At 60 sec, the second stage began firing, even as the first stage was completing its burn. The first stage dropped off, to fall over Gansu. Telemetry signals were presented on coloured recorder pens in the control centre, marking the rocket's ascent to orbit. Each key stage was reported by loudspeaker, to be greeted with cheering. The second stage eventually tumbled into the South China Sea. The third stage, the capsule atop, glided upward. By the time the third stage ignited, 600 km high and 200 sec later, the climbing rocket was over Guangzi. Qi Faren, then a 37-year-old engineer, recalls dashing out of the bunker to watch it climb into the night sky – the happiest moment of his life, he recalled later.

Everything went perfectly. At 9.48 pm, launch control announced 'the satellite and rocket have separated and the satellite has entered orbit'. Only two minutes later, the anthem *The east is red* was picked up, loud and clear on 20.009 MHZ. Tsien Hue Shen and his colleagues gathered on the still-hot launch pad, some cheering, some dancing, some even crying. Tsien made an impromptu speech. His life's dream had at last come true in the deserts of north-west China.

'WE DID IT THROUGH OUR OWN EFFORTS' – ZHOU ENLAI

China had become the fifth country to send a spacecraft into orbit. Strong signals were at once picked up by the American space command on 20,000 megacycles. It was the heaviest first satellite launched by any country.

Zhou Enlai was telephoned with the good news at exactly 10 pm. He told them he would pass on the word to Chairman Mao Zedong at once and that a celebration was in order. Later that night, he boarded a plane for a conference between China, Vietnam, Laos and Cambodia. He was able to tell them the news of China's achievement as soon as he arrived.

The official announcement was made the following morning. Zhou Enlai personally insisted that a small note be added to the press communiqué: 'We did it through our own unaided efforts'. The following evening, after the announcement had been made in Beijing, parades were held all over the cities, towns and villages of China. People vied with each other to be first to see the satellite in the spring night skies. In Beijing, fireworks were set off, bands played, coloured banners were unfurled. At 10.29 pm that night, Dong Fang Hong 1 passed over Beijing. Three nights later, it was spotted passing over Hong Kong, then a British colony.

In order to make the flight more visible to ground observers, a shiny metallic ring was added to the bottom of the final stage. This increased its brightness to magnitude +3.3, comparable to the stars of the Little Dipper. The satellite itself was magnitude +5 to +8, only just visible to a naked eye observer with acute eyesight in extremely dark skies. It is still up there, object 4392 in the log of US Space Command.

Key steps toward a Chinese Earth satellite

January 1958	Proposal by Tsien Hsue Shen for Earth satellite (project 581)
May 1958	Approval of project by Mao Zedong
January 1959	Project shelved indefinitely
1962	Recruitment of satellite team in Shanghai
January 1965	Proposal by Tsien to Central Committee for satellite launch
April 1965	Support from the party Defence, Science and Technical Commission
10 August 1965	Proposal approved by prime minister Zhou Enlai
May 1966	Design bureau 651 made responsible for satellite construction
24 April 1970	Launch of Dong Fang Hong 1.

The Dong Fang Hong went into an orbit of 441 by 2,386 km, inclined at 68.55° to the equator and circling the Earth every 114.09 mins. Though its silver zinc batteries were designed to work for only 15 days, its transmitter continued to function 28 days until the end of May. As hoped for, its signals could be picked up over great distances. Air density will gradually drag the satellite downwards and it is expected to burn up in the atmosphere before its centenary in 2070. The carrier rocket, which was the object that most ground observers probably actually saw, burned up in the upper atmosphere on 29th December 2000 at 5.44 am. As the new millennium dawned, the satellite was in a higher, safer orbit at 431-2,124 km.[3]

'We did it through our own efforts' – Zhou Enlai

Press communiqué

Our great leader Chairman Mao has stated 'We too should produce man-made satellites'. In the midst of the triumphant march of the people throughout the country to hail the 1970s, we are happy to announce that this great call issued by China has come true. China successfully launched its first man-made satellite on 24th April 1970.

The launching was hailed in the Chinese media as one of the great events of the century. Granted the heightened level of political tension in China at the time, it is no surprise that the media announcements dwelled more on the politico-revolutionary portentousness of the event than its scientific import. Pictures of Dong Fang Hong were not released until ten years later. Chinese radio repeatedly rebroadcast the 'east is red' theme from the satellite for days. Details were given of when the satellite would pass over China and where. Photographs duly appeared of skywatchers scanning the heavens for a sight of the satellite (or, more likely, its larger and more visible rocket). China received an avalanche of congratulatory messages from all over the world – not something China had been used to for some time.

In fact, the satellite's main role appears to have been to broadcast 'the east is red'. Of each minute's broadcast, the first 20 sec comprised 'the east is red'. This was followed by a second rendition of 'the east is red'. After a 5sec gap, telemetry was transmitted for 10sec. After a further 5 sec, the whole programme was repeated.

On the Mayday parade in Tienanmen square a week later, Tsien Hsue Shen and Ren Xinmin stood on the podium along with other Chinese leaders as the band

Dong Fang Hong 1, close up. (Courtesy: Sven Grahn.)

played the same tune that the first satellite was broadcasting all over the world. Mao Zedong commended Tsien Hsue Shen and Ren Xinmin in his Mayday address to the nation.

First Earth satellites in orbit

1957 October	Soviet Union	Sputnik	84 kg
1958 January	United States	Explorer 1	14 kg
1965 November	France	Astérix	40 kg
1970 February	Japan	Osumi	38 kg
1970 April	*China*	*Dong Fang Hong 1*	*173 kg*
1971 October	Britain	Prospero	66 kg
1980 July	India	Rohini 1B	35 kg
1988 September	Israel	Offeq 1	156 kg

DONG FANG HONG AFTERMATH

In August 1970, after the Dong Fang Hong launch, the Lin Biao revolutionaries were dominant in the Military Commission that directed much of China's planning. The Military Commission persuaded the government and party to adopt a new five year plan (1971–6) which had the slogan 'three years catching up, two years overtaking', committing the country to developing eight new launch vehicles and 14 new satellites in five years (other reports speak of an average of nine satellites a year). Many of these projects, which most scientists considered to be unnecessary and unrealistic, got under way though few saw the light of day. This was disruptive to existing projects and saw the commencement of several projects which later had to be abandoned.

The very survival of the space programme in this time of turmoil may in many ways be attributed to Zhou Enlai. He went to considerable efforts and probably some political risks to put the space industry off-limits to the revolutionaries. He drew up lists of the scientists involved in the Earth satellite project, putting them under protection from vilification. When Zhou Enlai died, personnel in the space industry turned out in vast numbers to mourn him, placing wreaths and poems in his honour in Tienanmen square, despite large-scale intimidation by the red guards.

SECOND SATELLITE: SHI JIAN 1

China's second satellite followed Dong Fang Hong into orbit nearly a year later. Now that the propaganda value of launching a first satellite had been demonstrated, China's second satellite could concentrate on scientific tasks. It is possible that this second satellite achieved what the first one had been intended to do, had not political and propaganda imperatives risen to the fore in 1966, during the early design stage.

The tasks of the second satellite were agreed at a conference held in the Chinese

Dong Fang Hong on show. (Courtesy Mark Wade.)

Academy for Space Technology in Beijing in May 1970. The second satellite received a new designation, Shi Jian (meaning 'practice' in Chinese). Sun Jiadong was in charge of the overall project. The first mockup was assembled in 1969. Full testing was carried out on three engineering models throughout 1970. A sophisticated manufacturing process was involved, for it had 72 sides. There were delays in May 1970 when staff in the Shanghai Scientific Instruments Plant were reassigned by the red guards to work in the countryside. Some scientists were put under political investigation and others were put under house arrest. Nor were they helped by Lin Biao who in October 1970 dispersed the Space Physics Institute from Beijing.

Slightly heavier at 221 kg, Shi Jian eventually entered orbit on the evening of 3rd March 1971. Signals were transmitted on 20,009 MHz and 19.995 MHz. Although the explosive bolts separating the satellite from the third stage had fired, the satellite did not separate from its carrier rocket. Enveloped within the third stage, the signals received were weak, only about 1% of what had been hoped for. The designers were, as one might imagine, perplexed and worried. On the eighth day, the signals suddenly came through loud and clear. Ground observations confirmed that the satellite had now separated from the launcher. The reason for the failure to separate, followed by its sudden and unexpected release, were never understood.

The satellite's orbit was similar to Dong Fang Hong – 267 by 1,830 km, 69.9°, period 106.15 mins. Beijing did not announce the launch until 16th March, presumably when separation had been confirmed and stronger signals had been received.

Shi Jian appears to have been a modified version of Dong Fang Hong – it was a 72-side polyhedron. The chemical battery system was taken out and replaced by a nickel-cadmium system rechargeable by solar cells. In place of the tape recorder that

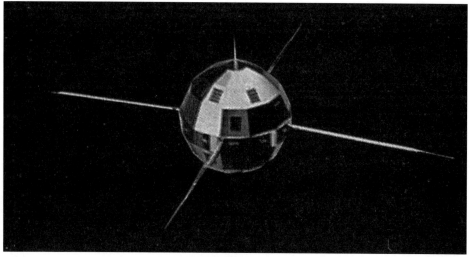

Shi Jian 1.

played 'the east is red', Shi Jian carried three scientific instruments – a 11 mm cosmic ray detector, a 3 mm X-ray detector and magnetometer. A heat flow meter was carried. A hundred automatic thermal shutters closed as the spacecraft entered darkness, opening again as it entered light (a similar system was carried by Sputnik 3).

Using four short-wave antennae, the radio transmitter emitted a stream of scientific data on 16 channels – this time there were no periods of silence, nor music – and could be picked up 3,000 km away. The system's power needs were under 2 watts. Shi Jian continued to transmit scientific data until it burned up in the upper atmosphere on 17th June 1979. The battery and telemetry systems showed no evidence of deterioration and maintained the same high level of performance throughout the mission, despite 10,000 charging and recharging cycles (one for each orbit as the satellite went into and came out of darkness). The passive solar louvres, likewise, opened and closed a similar number of times. The design teams rightly received commendations for these achievements in 1978. The 3,028-day mission appears to have been completely successful.

EPILOGUE: TSIEN HSUE SHEN

Tsien Hsue Shen's ambitions of the 1940s and 1950s of launching a satellite were fulfilled. He must rate as China's greatest scientist of the 20th century.

Tsien Hsue Shen is now in his 90s. He is bed-ridden and grants interviews only very occasionally. He gave permission to his secretary to write a biography, but he is only permitted to start after his death. He never publicly returned to the United States (though there is an unconfirmed report he once made a private, family visit in 1972). Sadly, the Cox investigation by the Congress in the late 1990s resurrected the spying charges, accusing him of making off with the Titan rocket design (which would have been remarkable, as it had not even been commissioned then). The irony was that when Tsien left America, he took the deliberate decision to leave all his notes behind. Tsien was very hurt about the misery he endured there in the 1950s, the failure of the government of the United States to apologize for its wrong-doing and resolved not to have any further formal dealings with the US government. He did correspond with some of his colleagues in CalTech until the cultural revolution but then the trail went cold. He only left China once and that was for the Soviet Union, but he didn't like it.

One American institution kept faith with Tsien, CalTech. The institute acclaimed him a Distinguished Alumini in 1979. He did not come to collect, but did send a gracious acknowledgement. Not until 1981 did one of his old friends from Caltech, Frank Marble, get to meet him again when he was giving guest lectures at the Academy of Sciences Graduate School of Science & Technology in Beijing. Following their friendly reunion, Marble tried to persuade him to take his old papers back. No, said Tsien, your American students need them more than my Chinese ones!

But he changed his mind. The Tsien papers did return to China in 2001. Some went to the Institute of Mechanics which he founded, but most went to the Tsien library set up by the government at Jiatong University in Xian. Extracts from the papers were published in a commemorative *Manuscripts of HS Tsien 1938-55*, ready in time for a commemorative symposium held in December 2001 in honour of his 90th birthday. He also eventually picked up his award from CalTech. Too ill to travel, his friend Frank Marble brought it to his home in a ceremony which received widespread coverage in the Chinese media. His American colleagues reminisced with him about CalTech and the days of Theodore von Karman. His American-born son became a graduate of CalTech, his daughter a doctor with a successful practice.[4]

When the Shenzhou returned to Earth in November 1999, President Jian Zemin went to visit him to tell him about the successful mission. They had a 'lively chat', recorded the *People's Daily*.[5] In 2001, in his 90th year, when several journalists visited him, they reported that 'in the home of the venerable Tsien, his bed faced the most conspicuous part of his bookshelf on which is a model of the spaceship Shenzhou'.[6] Two years later he was bed-bound when he received the great news that China had sent a man into space and successfully brought him back to Earth.

Chang's view is that Tsien made four main contributions to the resurrection of Chinese science in the 1950s. First, he gave the Chinese leadership, Mao Zedong and Zhou Enlai, the confidence that in investing in rocketry the money would be well spent and there would be positive outcomes. Second, Tsien was able to bring discipline and coherence to the engineers and scientists charged with modernizing China's rocket forces. Third, he helped to build up the intellectual infrastructure for the development of Chinese science. In Russia at this time, scientists were also obliged to build up a totally indigenous space industry, though in practice a very small group had secret access to western materials.[7] Tsien insisted his scientists and engineers build up a proper system of reference books and materials, not just in Chinese, but in Russian and English. Fourth, Tsien built up the organization necessary to ensure that China develop a proper space programme. He established

China's institute for missile design and all the other important elements that are essential for a national, coordinated space effort.

Iris Chang's biography of Tsien Hsue Shen comes to the conclusion that Tsien was not an original rocket theoretician. Had he died in 1955 or never gone to China he would have merited only a footnote in the history of science. The return to China was the crucial moment, for he was able to apply his American knowledge to China's growing military ambition. His real skills were in leadership.

Although deported from the United States in 1955 for being a communist (which he almost certainly was not), he joined the Communist Party of China in China the following year and from then on followed the party line, even issuing statements to denounce colleagues who stepped out of line and supporting the great leap forward. Maybe he should have stuck to the certainties of rocket science, for politics was to get him into trouble, even danger.

In late 1975, Tsien made the mistake of attacking his superior, Zhang Aiping, accusing him of national chauvinism in a poster which was widely circulated at the time. Zhang Aiping was then trying to sort out the problems around the failed first launch of the Long March 2 rocket. Zhang Aiping was not anybody but had been a senior revolutionary and military leader in the 1960s and 1970s. He was then Minister of the Commission of Science and Technology for the National Defence. The Gang of Four led the chorus of attacking not only Zhang Aiping but Deng Xiao Ping, then chairman of the military committee of the chiefs of staff of the People's Liberation Army. Tsien accused Deng of being 'the sworn enemy of all scientific workers who take the revolutionary road' (a bad mistake, for Deng was soon to become China's undisputed leader).

Tsien was now publicly associated with the Gang of Four, who ruled China following the death of Mao Zedong September 1976. Their ascendancy was short-lived. They were deposed in a military coup the following month and were sent to cool off in prison. Deng Xiao Ping became, by early 1978, chairman of the communist party, a position he held till his death in 1997. Tsien's old boss, Zhang Aiping, was reinstalled. One of Tsien's last career achievements was founding, in May 1978, the Institute for Military Operations Research and Analysis.

Unsurprisingly, Tsien became increasingly marginalized in China's space leadership. He tried to recant by writing newspaper articles explaining how it was really, after all, the Gang of Four who had retarded China's scientific progress. He managed to get back on side in 1989 when he publicly rushed to support the action of the leadership of Deng Xiao Ping and Li Peng in their bloody suppression of the demonstration on Tienanmen square. They obviously appreciated it, for in 1991, the government bestowed on the 80-year old the award of State scientist of outstanding contribution. The *People's Daily* published some of his reminiscences in a slavishly pro-party piece in June 2001 ('All achievements belong to the party').

ASSESSMENT AND CONCLUSIONS

This concludes the first stage of the Chinese space programme. The return of Tsien

Hsue Shen to China coincided with a period of military and scientific expansion. Concepts of Earth satellites were first explored in China in 1956, just as designs for early Earth satellites were being hardened up in the United States and the Soviet Union. The prospects that China might be the third space power, or enter the space race not far behind the Soviet Union and the United States, were dashed because of the political upheavals of the great leap forward (1958-9) and then the cultural revolution (1966–76).

Instead, efforts by China to build an Earth satellite were peristaltic, with periods of rapid progress coming to nothing because of political upheaval, the inability of the political leadership to make or execute decisions and endemic issues about who should lead the Earth satellite programme, leading to many complicated organizational changes. As a result, China became a space power much later than it could have. Even then, the first satellite, sent up in the course of the political turbulence, was used for political purposes. Not until the following year, 1971, did China put a proper scientific satellite into orbit. Order eventually prevailed over chaos. Shi Jian 1, while little publicized and hardly remembered compared to the Dong Fang Hong, was an outstanding success, returning scientific data from three instruments throughout the 1970s, a vindication of the engineers who designed and built the spacecraft.

REFERENCES

1. Joseph Anselmo: US eyes China missile threat. *Aviation Week & Space Technology*, 21 October 1996; China said to be developing missiles that could hit Rockies. *International Herald Tribune*, 24th May 1997; Joseph Anselmo: China's military seeks great leap forward. *Aviation Week & Space Technology*, 12th May 1997; Bill Gertz & Rowan Scarborough: Inside the ring. *Washington Times*, 4th January 2002.
2. For an account of this period, see Zhang Yun (Ed): *The Chinese space industry today*. China Social Sciences Publishing Co, Beijing, 1986; and Chen Hyi: *Into outer space*. China Pictorial Publishing Co, 1989, 156.
3. Phillip S Clark: *Chinese space activity, 1996–2000*. Molniya Space Consultancy, 2001; First satellite reportedly still in orbit. BBC monitoring reports, 3rd May 2000.
4. Frank Marble: Tsien revisited. *Caltech News*, volume 36, 2002 .
5. Chinese President visits Tsien Hsue Shen. *Go Taikonauts!* 9th December 1999. http://www.geocities.com/CapeCanaveral/launchpad/1921/news).
6. Xinhua news agency, 18th December 2001.
7. James Harford: *Korolev*. Wiley, New York, 1997.

4

Expanding the space programme

Developing new confidence from the launch of Dong Fang Hong and Shi Jian 1, the 1970s were to see a considerable expansion of the Chinese space programme. The first decade saw the Chinese space programme recover spacecraft from Earth orbit, build two new launchers and put three scientific satellites into orbit. The main line of the development was the recoverable satellite programme, of which three had been launched by the end of the decade and nine by the end of the first series. In operating a recoverable satellite programme, China became the third country, after the United States and Soviet Union, to bring satellites back from Earth orbit. In addition to these achievements, China put in orbit three mysterious satellites in the Ji Shu Shiyan Weixing series. Almost 30 years later, their purpose has never been explained.

' NO SPACE RACE'

Like the 1960s, the decade was affected by political turmoil, though this time it was less destructive than the first phase of the cultural revolution. There were fresh political interruptions after the dramatic events of September 1971 when Lin Biao, one of Mao Zedong's most trusted lieutenants, fled China for the Soviet Union: *en route* his plane was shot down by Chinese fighters and it crashed in flames. The consequent purges affected the Seventh Ministry, its head, Wang Bingzhang being jailed for many years. Paranoia was rife. Builders of a astronaut training device were implicated, even accused of preparing equipment to cure Lin Biao's insomnia. There were further purges in 1976 of those scientists thought to be sympathetic to Den Xiao Ping, who had been sacked by Mao Zedong. Order did not return until after the death of Mao in September 1976 and the overthrow by the military of the Gang of Four the following month. Once this took place, a rectification campaign was organized. This restored discipline in the space industry and reinstated rank to scientists who had been unjustly accused by the revolutionaries. Pre-cultural revolution working procedures were put back in place.

Key events in modern Chinese political history

1949	Chinese revolution
1958	Great leap forward
1966	Cultural revolution
1976	Ascendancy of Gang of Four
	Death of Mao Zedong
	Jailing of Gang of Four
1977	Ascendancy of Deng Xiao Ping
1978	The four modernizations; period of rectification
1989	Crushing of democracy movement
1997	Death of Deng Xiao Ping

The 1971–6 plan, which had involved – at least on paper – a furious expansion of the space programme – was scrapped and a new, more realistic plan was drawn up by Zhang Aiping. He defined the key tasks for the space programme in the 1980s as being to put the DF-5 into operation as an intercontinental ballistic missile; to launch a geostationary communications satellite; and to develop a submarine launched missile. This less ambitious, but more defined and practical plan was promptly approved by the Central Committee. The new leadership under Hua Guofeng and Den Xiao Ping encouraged newer, younger and more pragmatic engineers and managers to come forward in industry, concentrating on modernization rather than ideological struggle. At the same time, it was decided to reduce defence spending from 12% of national income to 5%; demobilize a million soldiers; and convert some military facilities to civilian use.

In August 1978, Deng Xiao Ping articulated what he believed China's space policy should be. Receiving a report from the Seventh Ministry, he told them that China was a developing country. 'As far as space technology is concerned, we are not taking part in the space race. There is no need for us to go to the Moon and we should concentrate our resources on urgently needed and functional practical satellites'.[1] The space budget was trimmed to meet its new, more modest ambitions and fell to 0.035% of Gross National Product, trailing not only the big space powers but Japan (0.04%) and India (0.14%).

FOUR MODERNIZATIONS

In October 1978, Deng Xiao Ping announced the 'four modernizations' for China in the post-Mao epoch: science & military technology, agriculture, education and industry (dissidents cheekily added a fifth modernization, that of democracy). Hand in hand with the four modernizations went an opening up of the economy and science. Foreign investment was welcomed, significant areas of the economy (for example retailing and services) were privatized, special economic zones were identified and developed and cooperation in technology was promoted. The four modernizations were approved by the 3rd plenary session of the 11th party central committee in December 1978. In a pointed reference to the space industry, it reaffirmed that quality control was the chief task of scientific production, rather than the intensification of the class struggle.

However, the effects of the cultural revolution could not simply be undone by party resolution. China's technical schools and universities had been effectively closed for ten years. Virtually no new graduates had come into the space industry and there was an acute shortage of scientific labour.

One of the features of rectification was the restoration of the chief designer system. This system was introduced in China in May 1964 and was a Soviet-originated one (*glavnykonstruktor*, in Russian; the first Soviet chief designer appointments dated to 1946). In China, each chief designer was assisted by a technical designer and an administrative designer. Each chief designer had a general department in his bureau which was responsible for leading and coordinating the project in question. The following appointments were made:

Chief designers, 1970s

Tu Shoue	DF-5 missile
Huang Weilu	Submarine-launched missile
Ren Xinmin	Satellite communications project
Sun Jiadong	Dong Fang Dong 2 comsat
Xie Gyuangxuan	Long March 3 rocket

Openness and purposeful modernization, with science at the top of the list, provided a much more promising environment in which a space programme might flourish. In 1980, China joined the International Astronautical Federation, though on condition that it displace nationalist Taiwan. The Chinese membership body was the Chinese Society of Astronautics, established the previous year with Tsien Hsue Shen as its honorary president and Ren Xinmin its president. Forty scientists subsequently became academicians, with Ren Xinmin becoming a member of the board of directors. In 1983, Yang Jiachi was elected executive committee vice-president. Tu Shoue was made chairman of the educational commission and Min Guirong joined the publications committee. China also joined the International Telecommunications Union and the UN committee on the peaceful uses of outer space.

In effect, China's twenty years of complete isolation from the world space community was quickly brought to an end. In 1977, Chinese space experts visited France and the following year Japan. In 1979, China received visits from the European Space Agency, France, Japan and the United States. An American delegation came to China. The first of many regional and international conferences on space in China were held in 1985. In due course of time, the Xi Chang space centre even came to feature on tourist itineraries.

While awaiting the development of their own Earth resources satellites, China negotiated with the United States for the use of Landsat data. A ground station was purchased from American industry and operated by the Academy of Sciences, China paying an annual fee of $200,000. It was operational from 1986. In 1988, China sent its most promising engineering graduates to courses in the Massachusetts Institute of Technology – the first time they had studied there since their predecessors had been driven out in the 1950s.

The Chinese space programme opened up within China itself. Workers in the space industry had been prohibited, on pain of extreme penalties, from telling their families where they worked. In a practice borrowed from the Soviet Union, they were assigned to mail box numbers, their institutes never being geographically identified. The greatest challenge faced by new graduates assigned to the space industry was to actually find their future place of work, since so virtually no one was allowed to tell them where it was! Likewise, the railway line from Qingshui to Jiuquan was not to be found on any map. The former policy not only concealed the space programme from foreigners and frustrated graduates but also inhibited cooperation between scientists within China itself. From now on, most space organizations were publicly named, identified and listed.

PROJECT 701: THE JI SHU SHIYAN WEIXING SERIES (1973–6)

There was a gap of over four years between the launch of Shi Jian 1 (1971) and the next Chinese satellite (1975). In the event, the next series of satellites, which took place before the period of openness and modernizations, raised more questions than it answered. The series comprised three successful launches and three failures during the period 1973-6. The series has been mentioned but never described in the Chinese literature. In China, it was codenamed project 701. Construction of the JSSW satellite had begun in early 1970 (hence '70' and '1'), although we know virtually nothing of its development or history.

Ji Shu Shiyan Weixing stands literally for 'technical experimental satellite'. The term *Chang Kong*, or 'Long Sky' has also been applied to the series and in the long tradition of Chinese renaming the subsequent satellites may eventually be formally renamed Chang Kong 1, 2 and 3. Because so little information has been made available about the programme, it is assumed that it may be military. JSSW may have been an attempt to develop a satellite for electronic intelligence gathering, a dominant theme in the military satellite programmes of the Soviet Union and the United States. Less likely, it could have been photographic in purpose, though it is unclear how the images were returned. The series took place at the same time as the development of the Chinese recoverable satellite programme and in the absence of information either from China, the two series were several times confused (indeed, their orbital paths were not that different). When the first launching took place, the official announcement appeared to confirm the military thrust of the programme, stating that the satellite was part of 'preparations for war'. The subsequent official history refers to the importance of the satellite entering a very precise orbit and that small errors in perigee were simply not acceptable.[2] Intriguingly, this was a familiar characteristic of some Soviet electronic ocean intelligence satellites so it is possible the Chinese series had a similar purpose.

Feng Bao. (Courtesy China Astronautics Publishing House.)

THE FENG BAO ROCKET

Project 701 used a new launcher, the Feng Bao, made in Shanghai (the classification '701' has also been applied to the rocket). The chief designer was Shi Jinmiao. Feng Bao was based loosely on the Dong Feng 5 missile, data from which was passed to the design team in Shanghai. In effect, the Feng Bao was a parallel, Shanghai programme to the Beijing Long March 2 which in turn built on the experience of the Long March 1. Responsibility for the Feng Bao was given to the Shanghai #2 Bureau of Machinery and Electrical Equipment and it was built in the Xinzhong Hua plant. The subsequent official histories of the Chinese space programme looked askance at this decision:

> The Shanghai region had never previously undertaken the research and development of a carrier rocket and was ill-equipped to do so in every respect,

be it technological resources, production capability or testing facilities to research and develop a large rocket.[3]

There appear to be two reasons for the decision to build the new rocket in Shanghai. One was probably political – it was Mao Zedong's power base and he probably liked to allocate pet projects there (this was not just an eastern habit, for Lyndon Johnson, after all, moved mission control from Cape Canaveral to his home state of Texas, 1,500 km distant). Second, there may have been an intention to build up a space industrial base outside the national capital. Shanghai was the most advanced industrial city in the country and was the best candidate. The bureau received considerable help from the Chinese Academy of Launcher Technology in Beijing which received study teams and sent them specialized personnel and the blueprints of the DF-5.

SHANGHAI MOBILIZES

The Shanghai team mobilized to the full the industrial and technological resources of the city and the region, quickly drawing in all the leading state and municipal companies. A local computer company, the Hua Dong Computer Technology Research Institute, designed the on-board computer, apparently from scratch. The rocket's aluminium copper alloy tanks were welded by the Shanghai Jiangnan shipyard. These team members were a resourceful lot: when the local factory was just too small to carry out necessary welding work for the Feng Bao, they jacked up the roof of the building by 1.7 m to gain the extra space.

Design of the Feng Bao was completed in December 1969. They used the DF-5 design where they could, making improvements as necessary, for example to the pressure system, valves and guidance. The Feng Bao was a 192 tonne rocket, almost 33 m tall, able to put 1,500 kg into a 190 km orbit at 69°.[4]

Unlike Beijing, the Shanghai region had no testing equipment. Undeterred, the designers managed to adapt existing factories for their purposes. Within a few months, they had built a final assembly building, shake table, engine test stand and materials strength-testing unit. Despite its inexperience, the Feng Bao team managed to build and test the test version of the rocket in only ten months. This was no mean achievement, for its performance was somewhere in between that of the Long March 1 and Long March 2, where equivalent work had taken their designers four years. The Beijing team, whilst annoyed at scarce resources being used elsewhere to match their efforts, grudgingly came to admire the speed and talent of the Shanghai designers.

FIRST HOT TESTS

A non-flight test version of the Feng Bao left for Jiuquan in November 1970, just over a year after the team had started work. In March-April, 1971, the engines were hot-tested in Jiuquan and worked perfectly. The test revealed a number of problems, such as the computer and poor valves for the first stage engine, all the result of rushed work and lack of rigourous quality control. After a campaign of improvement and renewed quality control, the first flight test rocket was completed exactly a year later when it was sent by train to Jiuquan in April 1972. Not until 6th August did the design team feel able to tell Zhou Enlai that the rocket was ready.

The first test of the Feng Bao took place on 10th August 1972. It was a suborbital mission. Although a success in some respects, the attempt revealed areas where further improvement was required in order to carry a heavier payload. Fuel tanks were recast with thinner walls, fuel flow to the engines was improved and it was decided to run the all engines to exhaustion in the ascent to orbit. The second stage manoeuvring engines, or vernier engines, would be used to get the rocket into orbit. Cables were re-wired, redundant equipment removed, miniaturized systems introduced. As a result of these measures, the payload increased by 50%. Some of the equipment was tested against leaks for as long as a year at a time; other equipment was alternately baked or cold soaked for weeks on end.

Chinese engineers seemed to have encountered considerable difficulty with the Feng Bao design. The rocket was substantially more demanding than the Long March 1, being required to lift payloads of 1.9 tonnes compared to the 300 kg of the first Long March. Official histories have blamed the difficult history of the Feng Bao on the interference of the cultural revolution, with its negative effects on quality control during the crucial phases of design.[5] The revolutionaries had an especially strong following in Shanghai, where the Feng Bao was built. Scientists who tried to confront technical problems were accused of sabotage and acting on behalf of class enemies – always a hard charge to refute.

IN ORBIT AT LAST

Their best efforts were to no avail, for the first orbital attempt failed on 18th September 1973, the second stage vernier engine being the culprit. When the Feng Bao was launched again, on 14th July 1974, thrust in the second stage vernier manoeuvring collapsed and the satellite failed to reach orbit. In later accounts, the Gang of Four was blamed, but the technical decision was taken to replace the second stage engine with engines taken from the Long March 2.

Eventually, patience was rewarded on the third attempt. On 26th July 1975, the Ji Shu Shiyan Weixing 1 entered orbit. JSSW-1 orbited at 183 – 460 km, 69°, 91 mins. The launch announcement gave the barest details about the satellite (only orbital parameters), instead offering readers a weighty political commentary on the current state of development of the proletarian revolutionary line. JSSW-1 decayed after 50 days in orbit, crashing into the atmosphere over the Pacific ocean on 14th September 1975.

JSSW-2 entered orbit on 16th December 1975. This time, the launch announcement did not even given the orbital parameters, instead providing more appropriate information on the struggle against Lin Bao and Confucius. JSSW-2 flew 70 km lower than JSSW-1 (183–387 km, 69°, 90.2 mins), burning up in the atmosphere after only 42 days.

JSSW-3 came nine months later, on 30th August 1976. It had markedly different orbital characteristics, flying much further out than its predecessors (198 – 2,100 km, 69.2°). Like its two predecessors, it weighed 1,110 kg. The launch announcement gave even fewer details about the satellite (only the date), paying more attention to its political significance (this satellite marked the struggle against Deng Xiao Ping and the right deviationists). JSSW-3 decayed in 817 days. None of the three satellites manoeuvred in orbit. Nor were signals picked up in the west: presumably they were sent only over China itself.

The final JSSW launch took place on 10th November 1976. The satellite was slightly heavier at 1,210 kg, but the second stage vernier engine let the launch teams down again and it never reached orbit. The JSSW programme then closed. This may have been because it did not achieve the intended results. Officially, they were technology test satellites, but it is not clear what technology was tested nor how it was subsequently applied.

There was a footnote to the JSSW story. American aerospace experts visiting China in 1979 were shown what they were told was a spare military spacecraft in the Shanghai Huayin Machinery Plant – a domed cylinder 2.5 m tall, 1.7 m diameter, weighing 1.2 tonnes, with 1cm by 2cm solar cells. They were told that China had launched three of them, each with 10-day missions, which certainly fits the JSSW profile, though they were given no further information about the purpose of the missions. The official history, many years later, makes reference to costly projects hastily entered into without proper discussion during the period of the cultural revolution.[6] This could be an oblique reference to this programme. In 1992, giving a lecture in Stockholm, Long Lehao, the head of operations at Jiuquan, showed a tantalizing slide of the JSSW satellite, giving it an external appearance not unlike the FSW cabin.[7] Many years later, the series remains obscure. Unlike the case of the Soviet Union, where hitherto obscure missions have come out into the open through the histories of the design bureaux, this has not been the case in China. JSSW must have been important, for six were launched, even though only three reached orbit.

Ji Shi Shiyan Weixing series, 1973–76

Ji Shu Shiyang Weixing	18 Sep 1973	Launch failure
Ji Shu Shiyang Weixing	12 Jul 1974	Launch failure
Ji Shu Shiyang Weixing 1	26 Jul 1975	Orbited, decayed after 50 days
Ji Shu Shiyang Weixing 2	16 Dec 1975	Orbited, decayed after 42 days
Ji Shu Shiyang Weixing 3	30 Aug 1976	Orbited, decayed after 817 days
Ji Shu Shiyang Weixing	10 Nov 1976	Launch failure

THE RECOVERABLE SATELLITE PROGRAMME (*FANHUI SHI WEIXING*): PROJECT 911

China was the third country to recover a satellite from Earth orbit.[8] The idea of a recoverable Earth satellite in China went back to 1964 and the work of the Shanghai design team. They had been inspired by what they read of the American Discoverer series of recoverable satellites in the late 1950s and early 1960s – indeed the eventual design of the return capsule bears clear similarities with Discoverer. Whether they realized that the Discoverer programme was a military enterprise (it was declassified in 1995 and revealed as the Corona programme), designed to film Soviet missile bases and return a film-bearing capsule to Earth, is not known.

The idea of a recoverable Chinese satellite was first formally proposed in the Chinese Academy of Sciences *Proposal on plan and programme of development work of our artificial satellites* (spring 1965). The project was approved in August 1965. The Shanghai team applied for responsibility for this task, which was awarded them in spring 1966. Conceptual studies were carried out from June 1966 by Wang Xiji. Two scientists, Zhao Jiuzhang and Qian Ji visited enterprises to determine what experiments could usefully be carried out on recoverable satellites.

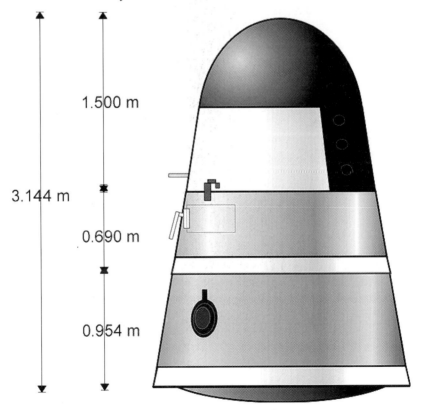

The FSW design. (Courtesy Mark Wade.)

76   **Expanding the space programme**

Detailed design work was carried out the following year. His team settled on a satellite weight of 1,800 kg, with a typical orbit of 173–493 km, 91 mins, 59.5°. A three-day conference was held on the progress of the project from 11–13 September 1967 and the programme was given a code, project 911 (a code out of sync with date sequences). It was given the name in Chinese Fanhui Shi Weixing (FSW), or recoverable experimental satellite. We later know that the FSW was also designed with a view to proceeding to a manned flight later on. Design for the recoverable satellite was frozen on 11th September, 1967. The orbital paths to be flown were settled by the Institute of Mechanics in the Academy of Sciences in March 1968.[9]

The precise purpose of the programme has never been entirely clear. 30 years later, it was given the general rubric of 'Earth observations'. If the programme was modelled on Discoverer, as seems likely, it was probably military Earth observations, or photoreconnaissance. The code 'Jian Bing', or 'pathfinder' or 'vanguard' was later known to have been assigned to at least part of the programme and this was associated with military photographic intelligence (the designator Jian Bing was, much later, applied to the mission launched in November 2003).

Whatever it original purpose, later versions of the FSW were also used to conduct a range of microgravity experiments in orbit. The cameras were put to civilian use. Whether this was because of an improvement in the international climate, or the limited military reconnaissance value achieved from the FSW missions, or a form of diversification, is a matter of guesswork. The military intelligence benefit from orbiting a recoverable cabin for a week every year is probably quite limited. Ultimately, the lasting achievements of the programme were in civilian applications.

The task of developing the project fell to the China Academy for Space Technology (CAST) which by the early 1970s had grown to 81 units spread over 16 different provinces. A new rocket, the Long March 2, was developed by the China Academy for Launcher Technology (CALT). Progress was held up by the later phases of the cultural revolution, but for which the first launch might have appeared several years earlier than it eventually did.

THE CHALLENGES

Recovering satellites poses several difficult engineering challenges: devising a protective heat shield to ensure the capsule survives reentry temperatures of 1,200° C, the development of retro rockets, a very precise attitude control system, quality ground tracking to prepare the cabin for the precise moment of reentry and search and recovery systems. These engineering challenges required the development of ever more sophisticated ground testing equipment. A thermal vacuum chamber called the KM-3 was constructed by the Institute of Environment Test Engineering and the Lanzhou Institute of Physics in 1970. The level of vacuum achieved was 10^{-9} torr (1 torr = 1/760 atmospheres). A tracking centre was built for the programme, the Xian Satellite Surveying and Control Centre. In the course of 1968-9, FSW prototype models were manufactured and put through a series of engineering, ground and vacuum tests. Others were given to the recovery teams to practice finding returning cabins.

The Chinese had no previous experience of making heat shields. They did not wish to use ablative heat shields of the type developed by the United States and Soviet Union in the 1960s: these were heavy shields, in which the material progressively burned off during the descent, enough remaining for the cabin to survive. Equally, they knew they did not have the capacity to go straight to low-density foam-type shielding, of the type subsequently used by the American and Soviet shuttles (tiling). They eventually found a non-ablative material whose qualities lay somewhere in between – a carbon composite material called XF, able to withstand reentry temperatures of 2,000°C. The main aerodynamics and ablation expert involved in the reentry problem was Zhuang Fenggan. Like Tsien a graduate of Jiatong University, he graduated from the University of California College of Science in 1950. He was responsible for the construction of wind tunnels to test the spacecraft.

The recoverable series required a relatively advanced level of automation. The person responsible for the automation in Chinese satellites was Yang Jiachi (1920–), like Tsien a graduate of Jiatong University. He received his doctorate in Harvard where he worked on early computers. After his return to China in 1956, he concentrated on automatic control systems in space. A new three-axis attitude control system was developed by Zhang Goufo of the Beijing Control Engineering Research Institute, using an infrared horizon scanner and a gyrocompass. The scanner was tested out on two T-7A sounding rocket tests at Jiuquan in June and

Yang Yiachi. (Courtesy China Astronautics Publishing House.)

July 1969. The system was finally settled in 1970 by the Institute of Automation in the Chinese Academy of Sciences. The whole system comprised a three-axis control system, analogue computers, sun and earth orientation sensors for orientation, inertial measurement unit and a cold gas thruster system to orient the spacecraft.

A camera system was developed by the Changchun Institute of Optics and Fine Systems which had some previous experience of aerial cameras, but none of space or stellar imaging (more sophisticated charged couple device (CCD) cameras were still some years away). The early series carried cameras for ground photography and side-pointing cameras for stellar photography (so as to work out the precise position of the spacecraft in orbit). The man credited with the design of the ground imaging was Wang Daheng, an optics specialist whose previous work involved the design of rocket and satellite tracking devices. Designer of the stellar camera was Wang Jintang. Tao Hong was responsible for cameras and film. They had to start from scratch and trawl foreign and domestic literature on advanced cameras as their starting point. In the midst of their efforts, the turmoil of the cultural revolution forced them to leave Changchun and move operations to Beijing. They produced the first cameras in 1969, testing them on aircraft first, having to overcome a series of problems such as films jamming.

One of the most crucial links in the chain was the retrorocket required to fire the FSW out of orbit and begin the final descent. The solid fuel rocket, developed by the Fourth Academy, would be required to fire in a vacuum after orbiting the Earth for several days in weightlessness. This was one of the slowest components to master. Ten tests were carried out in the course of 1971 to 1974 in the course of which the problems associated with the new engine were resolved.

The parachute system proved problematical. The importance of a safe system was underlined in April 1967 when Soviet cosmonaut Vladimir Komarov died when his parachutes snarled during an already-troublesome return to Earth. In July 1970, the Chinese air-dropped two FSW capsules to test the parachute system. Both failed and the cabins were destroyed. A second set of tests in October had an identical, dismal outcome. Chief designer Wang Xiji was jailed by the red guards and accused of sabotage. He was later able to prove that mechanical causes were to blame. With the help of the air-force, drop tests went on and by 1974 they were confident that a reliable system had now been devised.

The FSW satellites were a quantum leap in size and scale beyond the first two satellites. The FSW satellites were specified to weigh in the order of 1,800 kg, the actual payload being in the order of 150 kg and involve both a service module and recoverable capsule. The cabin itself was beehive-shaped, 3.1 m tall and ranging from 1.4 m in diameter at the forward end to 2.25 m in diameter at the large end. The normal procedure was for the 800 kg service module to remain in orbit for about 20 days.

Leaders of the Chinese recoverable satellite programme (project 911)

Zhu Yilin
Wang Xiji
Sun Jiadong

MISSION PROFILE

The FSW satellites consisted of a blunt cone capsule placed on a service module. During the mission, the nose is pointed in the direction of travel. At the end of its mission, when it reaches Chinese territory, the spacecraft is swivelled through 100°, pointed directly toward the Earth and the solid retro-rocket is fired. It descends almost vertically from orbit. This is a crude means of returning to Earth, one which uses up a substantial amount of fuel, but which has the advantage of ensuring that retrofire can be commanded over China and recovery can take place in China. By contrast, Russian spacecraft returning to Earth make a more gentle and economical braking manoeuvre in the direction of travel over the south Atlantic, the spacecraft descending in an arc that brings them in a high, gentle curve over the Mediterranean, eventually to land in southern Russia or Kazakhstan. The angle of retrofire must be very accurate, for each degree out means a 300 km difference in the landing spot. At 16 km, the FSW drops its heat shield and retrorockets, a parachute opens and the cabin comes down at 14 m/sec in Sichuan province in southern China. The Chinese landing technique, guaranteeing landing in China, is important if sensitive film is on board (the Russians fitted self-destruct devices to their spacecraft to stop their falling into unfriendly hands). On the other hand, the Chinese reentry manoeuvre requires a big velocity change, 650 m/sec, much more than the standard Russian or American reentry profiles (about 175 m/sec).

Sichuan province in the south-west of the country was chosen as the recovery zone, although it was hilly and often subject to thick clouds and mists. Photographs from the recovery area have frequently shown Mil-type recovery helicopters hovering against a background of mountains, following the descent craft down and then lifting it away for post-flight examination. The scene is one of the space cabin lying on the hillside, its red and white parachute streamed out alongside; the recovery teams saving and checking the cabin; and rural workers gathering on the nearby hills to watch the excitement.

LAUNCHER FOR THE FSW: THE LONG MARCH 2

The much greater weight of the payload compared to the first missions required more lifting capacity. Accordingly, the Long March 2 was developed by the Chinese Academy of Launcher Technology (CALT) in Beijing. CALT used as its starting point, the design of the Deng Feng 5 ballistic missile. The chief designer of the Long March 2 was Tu Shoue. At the same time, the Shanghai Academy of Space Technology was tasked with building a similar rocket which became the Feng Bao.

The Long March 2 new rocket had two stages, was 32 m tall, used nitrogen tetroxide as oxidizer and unsymetrical dimethyl hydrazine (UDMH) as fuel, weighed 190 tonnes and had a liftoff thrust of 280 tonnes. Made of high-strength aluminium cooper alloy, it was the first Chinese launcher to use full computer guidance and gimballed engines.

The initial design of the Long March 2 took four years, from 1965 to 1969. The payload requirement of almost two tonnes was substantially greater than the requirement for the Long March 1, then also in design but as yet unproven. The main technological challenges were to build an inertial guidance platform to guide the rocket during its ascent (essential for placing the satellite in the precisely required orbit), the development of an on-board computer (necessary for accurate recoveries) and the use of lighter metals (aluminium – copper). In reality, the main challenges were political – getting on with the work despite the interruptions of the cultural revolution.

New engines were designed for the Long March 2. Called the YF-20, four were used together on the first stage (when so clustered, they were called the YF-21). A version of the YF-20 was used for the second stage (the YF-22). Li Boyong led the design of the new engine, a process which involved many heartbreaks, with recurrent problems of vibration and turbopump failures. The four engines were first tested all together on 14th June 1969. Crowds gathered on the hills adjacent to the test stand. They then cheered to the roar of the engines whose thunderous rumble was matched by the hissing of 7.9 tonnes a second of water being doused from 30,000 nozzles attempting to keep the test stand cool. The second stage was given a similar test in November 1970.

Long March 2 ready for launch.

A key innovation with the Long March 2 was that the engines could be swivelled about their axis to steer the rocket. In rocket science, this is called gimballing. Hitherto, Chinese rockets had used vanes, or fins for steering, a technique introduced by the Germans on the A-4. Gimballing saves the weight of vanes and is more precise, but requires a much more sophisticated engine design. To steer the second stage, a complex of four small vernier engines was used (the YF-24). The first gimballing tests were made on a test stand in December 1969. The second stage engine was tested out at the rocket engine testing station in late 1970. The engines required for the Long March 2 represented a massive jump in performance – from 28 tonnes of thrust to 73 tonnes.

A further innovation was the use of self-pressurization: instead of the oxidizer and fuel tanks being pressurized by separate tanks of pressurizing gas, a vaporizer was used, effectively cutting out the need for separate gas bottles. The propellant tanks were made of aluminium alloy, giving a weight reduction of 30% on the aluminium-magnesium alloy of the Long March 1, though the material posed formidable welding problems.

Computerized guidance systems for the Long March 2 were designed by Liang Sili and the Micro-Electronics Research Institute, established in 1965. Its lightweight medium-speed, small-capacity digital computer was the first of its kind in China. The Chinese found the development of advanced guidance and computer systems to be a problem. China lagged far behind the west and the Soviets had been of little help in the 1950s (in Stalin's time, computers had been condemned as a 'false bourgeois science' and the Russians had tried desperately to keep up ever since). Two Chinese scientists were called in to resolve these problems – mathematics experts Liang Sili and Zheng Yuanxi of the Inertial Devices Research Institute. Significant improvements to space electronics at the time may be attributed to Wang Zheng (1890–1978), veteran of Mao's army in the 1930s, the man who modernized telecommunications, computers and electronics in China.

Another innovation, developed by designer Yu Menglun, was the interstage glide. Once the second stage engine had completed its burn, the manoeuvring vernier engines would continue to fire as a main engine. Also called the low-thrust orbital entry technique, this approach enabled an extra 500 kg of payload to be carried, though it meant that the ascent to orbit took several more minutes. This technique was also a feature of the Feng Bao, but with less success. Having done all this, the designers then organized a campaign to reduce the weight of the rocket still further, each kilo saved meaning more payload. By refining the equipment still further and by using lighter alloys, they made gains of 700 kg, quite an achievement.

The quantum leap forward of the Long March 2, compared to the Long March 1, even before the latter had flown successfully, required that verifying procedures should be more exacting than ever. The entire assembly was tested in the vibration testing tower and shaken time after time against the various vibration loads that affect rockets from side to side and end to end. After these tests, the rocket would be stripped down and every part examined to determine the effects of stress.

Premier Zhou Enlai inspected the Long March 2 in September 1972. It was one of 30 occasions on which he had met with scientists, engineers and technicians involved

82 Expanding the space programme

in the rocket programme. Encouraging them in their efforts, he reminded them of the absolute importance of quality control, safety and reliability and of not being blown off course by the cultural revolutionaries (or left deviationists as they were termed pejoratively at the time).

Designers of the Long March 2 (CZ-2)

Tu Shoue
Wang Yongzhi
Wang Dechen
Li Zhankui
Yu Menglun

BROKEN WIRE

The first attempt to launch a recoverable Earth satellite on the Long March 2 took place on 5th November 1974. It was a disaster. The rocket had barely lifted off

FSW, prepared for launch. (Courtesy Sven Grahn.)

before it began to sway from side to side and had to be destroyed in a fireball by the range safety officer. There was an intensive post-mortem which involved analysing telemetry, the wreckage and ground simulations. Due to vibration, the wire from the gyro to the control system had fractured – so it was later determined – and the control system had no basis for stabilizing the rocket. It begun to veer out of control, so the range safety officer set off an explosive device in the rocket, destroying the whole assembly a mere 20 sec into its first mission. All because a single wire broke.

This disaster – China's first – led to a high-level political intervention. Zhang Aiping, who held the lofty title of Minister of the Commission of Science and Technology in the National Defence, descended on the rocket teams to carry out inspections. There was an all-out drive for quality control. All wires were strengthened or double wires put in. There was a ten-month campaign of more vibration testing of key components to make sure nothing like this would ever happen again. The final assembly of the second rocket began in July 1975, Zhang Aiping inspecting the assembly workshop in person. Not until 20th August 1975 did Zhang Aiping give permission for a fresh attempt to be made. Soon after, the next FSW cabin left its factory for its long rail journey to Jiuquan. The improvements on the rocket were significant enough for it to be given the designation of the Long March 2C (the failed mission was later also given the designation 'Long March 2A').

COASTING TO ORBIT

The second attempt was made in November 1975. FSW 0 was launched into orbit on 26th November 1975 from Jiuquan. Seven seconds after liftoff, the rocket turned toward the south east. After 130 sec, the first stage engine shut down. The verniers on the second stage ignited, explosive bolts fired to separate the two stages and the first stage fell to the ground over uninhabited parts of Gansu. The second stage lit up, burning for 112 seconds. The small verniers continued to fire a further 64 seconds as the rocket coasted on toward orbit. The orbital insertion point was 176 km altitude, 1,800 km downrange, over Hunan. Back at the launch site, there was the ever-present memory of the previous year's broken wire. When the report was given that the rocket had entered orbit, several of the engineers sobbed quietly with delight. As the FSW entered orbit, in Sichuan, representatives of the Satellite Test and Control Centre arrived from Xian to install radio masts on a number of hills in the region to receive the signals of the cabin which was due to return three days later – if all went well. Local people were told of the impending arrival from the sky and the people's militias were mobilized.

No sooner was it in orbit than there was a loss of pressure of the gas orientation system. Tsien Hsue Shen believed the chance of a successful recovery was now nil. However Yang Jiachi calculated that the pressure loss was simply due to heating of the spacecraft during ascent followed by the cooling in space. He argued that a full-duration mission was still possible. A compromise was reached about what to do and the decision was taken to bring down the spacecraft slightly early, after three days aloft, on 29th November.

84 Expanding the space programme

The remote ground control station, on the western edge of Chinese territory, sent out the command to the FSW to take up orientation for reentry and then to fire the retrorockets. However, the tracking system broke down not long before the moment of retrofire. Ground tracking lost its lock on the spacecraft. Against all safety rules, four brave radio technicians climbed up onto the roof and manipulated the tracking system manually, subjecting themselves to high-frequency radiation in the process. They were not the only people following the satellite carefully. So too was the ageing Chairman Mao who constantly asked for progress reports.

HELICOPTERS SCRAMBLED

As retrofire approached on the 47th orbit, helicopters were scrambled to watch the cabin come in. The return to Earth was problematical, the cabin being badly burned and approaching far from the originally intended spot.

Observers were deployed to the tops of high mountains to spot the descent of the capsule in Sichuan. Nobody saw a thing. Eventually ground control calculated that the cabin might have touched down in Guizhou. There, four coal miners having their lunch in the canteen were startled to spot a red-hot ball falling from the sky and crash into trees. They found a blackened hulk in a crater. One of them threw a stone at the smouldering object and it bounced off with a metallic clang. The miners called the authorities. FSW was way off course and the cabin was very badly charred, indications of a less than perfect reentry. China had succeeded in recovering a capsule at the first attempt, like the Soviet Union many years before (the United States experienced a dozen failures).

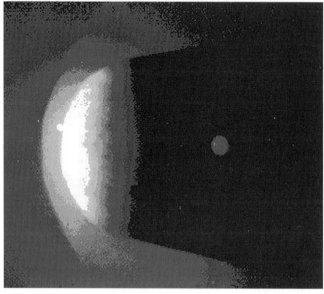

FSW reentry test

The Chinese designated the next series of FSW missions the FSW 1 series, so this series was retrospectively but oddly named the FSW 0 programme, the individual missions being numbered 0-1, 0-2, 0-3 and so on (thankfully, it did not have a precursor set, which would, to be logical, have to receive a minus designation, FSW -1!). Later, the Jian Bing designator was introduced, being applied to the first mission after the long gap in the programme between 1996 and 2003.

First recovery of satellites from Earth orbit

August 1960   United States (Discoverer 13)
August 1960   Soviet Union (Korabl Sputnik 2)
November 1975   China (FSW 0-1)

CRISIS-RIDDEN SECOND MISSION

Following the reentry problems experienced with FSW 0-1, the cabin was redesigned. The heat shielding material XF was extended to those parts of the shielding that had been badly burnt on the first mission. After several months of intensive work, the new spacecraft was ready by October 1976 and its mission timed, according to the Beijing media, to mark the crushing of the Gang of Four.

Launch was set for 7th December 1976 but was delayed when, two minutes before the scheduled liftoff, the swing arm of the rocket gantry failed to retract. Soldiers rushed forward – bravely or recklessly, depending on one's perspective – climbed the tower to the 30 m level, pushed the arm manually back from the fully-fuelled rocket

FSW recovery. (Courtesy China Astronautics Publishing House.)

and ran back into the bunker, all in five minutes. Launch director Zheng Songhui approved the launch, which now went ahead smoothly. Although there were problems during the mission – the attitude control system nearly exhausted its fuel – retrofire took place smoothly.

Hopeful of a more accurate landing this time, the recovery team mobilized. Four helicopters were scrambled. At headquarters, a plotting map marked the projected descent point while loudspeakers relayed the latest reports. At noon, a sonic boom from the returning cabin rumbled through the valleys of Sichuan. Sharp skywatchers noticed a black dot hurtle in from the north-west, splitting in two. One of them was the discarded heat shield, which was eventually found beside a road. The other was the cabin. Once the timer activated the parachute, the cabin could be seen gently descending, ending up in a vegetable garden on the side of a hill. One of the four helicopters found a flat spot 100 m away. The crew jumped out, mounted guard, began inspection and removed the precious film.

The third mission of the recoverable FSW satellite took place in January 1978 and was also successful. The post-flight announcement confirmed that remote sensing tests had been carried out. The following month, the deputy head of the Space Technology Research Committee, Sun Jiadong, reported on the conclusion of the first phase of the programme. He told his superiors that a new version of the FSW was in design.

IMPROVED VERSION

There was a gap of over four years before the next mission appeared. Not only were the spacecraft systems reviewed and improved, but the on-orbit lifetime was extended to five days. A new pointing system enabled the retro-rockets to be fired more accurately. New charge-couple device cameras were mounted to test the possibilities of transmitting data in real time. A radar transponder was added to facilitate recovery. More stringent quality control measures were introduced and five air drops were made from 10,000 m, each successful.

In the improved series, the FSW 0-4 appeared in September 1982 (the FSW 0 designator was unchanged). Further missions followed in August 1983, September 1984, October 1985, October 1986 and August 1987 (FSW 0-9). The charge-couple device transmissions were declared to be successful. The October 1985 mission took part in a general territorial survey of the land mass of China. By this stage, the missions had become routine. FSW 0-8 was distinguished by coming down in a small inland lake, thus making it the first splashdown in the Chinese space programme, although the lake concerned seems to have been thankfully quite shallow. The 1984-6 missions were land surveys taking more than 3,000 pictures using wide-angle cameras.

INTRODUCTION OF MICROBIOLOGY EXPERIMENTS

FSW 0-9, the last of the early series, broke new ground. It was the first mission to fly microgravity experiments and biology tests. FSW 0-9 was also the first to carry a western commercial payload, carrying two small (15 kg) microgravity experiments for the French company Matra. The experimental boxes were handed back to Matra ten days after recovery. One of them involved the testing of food growth and algae in orbit. A Chinese microgravity experiment was also carried, involving the smelting and recrystallization of alloys and semi-conductors. It is not clear if the final FSW had any remote sensing role at all or if it was devoted entirely to microgravity experiments.

It is difficult to assess the quality of photographs returned to Earth by the imaging systems of the FSW. Although the Chinese have published photographs of China taken from space, the satellite concerned has never been identified and in some cases American pictures have been used. Years later, the Chinese claimed that the FSW series had returned good quality, broad-scale survey images which had made an important contribution to mapping, land use, forestry, water resources and problems of soil erosion.[10]

Fanhui Shi Weixing 0 series, 1975–87

Launch failure	5 Nov 1974	
Fanhui Shi Weixing 0-1	26 Nov 1975	First Chinese satellite recovered
Fanhui Shi Weixing 0-2	7 Dec 1976	Second test flight
Fanhui Shi Weixing 0-3	26 Jan 1978	Third test flight
Fanhui Shi Weixing 0-4	9 Sep 1982	First operational 5-day mission
Fanhui Shi Weixing 0-5	19 Aug 1983	
Fanhui Shi Weixing 0-6	12 Sep 1984	Land survey
Fanhui Shi Weixing 0-7	21 Oct 1985	Survey of Chinese land mass
Fanhui Shi Weixing 0-8	6 Oct 1986	Splashed down in lake
Fanhui Shi Weixing 0-9	5 Aug 1987	First materials processing mission

SHI JIAN 2

Shi Jian 1, in March 1971, was China's first scientific satellite and highly successful. It was eight years before China was again ready to launch scientific satellites. This time, the Chinese attempted to launch three satellites in one go. This was by no means unusual, for the Russians had pioneered 3-in-1 launches in 1964 and had even launched 8-in-1 missions (coincidentally, the first taking place the day after Dong Fang Hong was put in orbit in 1970). However, this is not to minimize the achievement, for the deployment of three scientific packages in this manner can often be accident-prone (as more advanced space nations have sometimes been reminded to their cost).

Project leader was Shi Jinmiao, lead design director Qian Ji and design engineers

Wang Zhenyin and Zhu Yilin in the Beijing Institute of Spacecraft Systems Engineering. The original Shi Jian 2 project dated to April 1972 when it was defined as a single space physics satellite to cover eight fields of work. In the course of refinement, three more areas were added. Design took place in the course of May 1973 to September 1974. An extensive instrumentation package was prepared.

Instrumentation for the Shi Jian 2 project

Magnetometer
Semi-conductor proton directional probe
Scintillation counter
Long-wave infrared radiometer
Short-wave infrared radiometer
Earth atmosphere ultraviolet radiometer
Solar ultraviolet radiometer
Solar X-ray probe
Thermoelectric ionization barometer

The orbit was planned for 250 to 3,000 km, inclination 70° with an operational lifetime of six months. Shi Jian 2 was a 257 kg, 8-sided 1.23 m diameter prism, 1.1 m high, with four small solar panels. Shi Jian 2 would send back telemetry both in real time and by tape recorder able to hold 520,000 bits of data at a time. This information would be dumped to the ground station during daily passes over China. It was the first Chinese satellite to store information for later retransmission. The Shi Jian 2 had a single ultra-short wave transmission system, much simpler and lighter than those carried on previous spacecraft. Shi Jian 2 would be the first Chinese satellite to use solar panels (as distinct from solar cells attached to the main body of the spacecraft). As a result, it was able to generate substantially more power than the body-mounted cells of Shi Jian 1. Each panel was 1.14 m long and 0.56 m wide, making for a total span of 2.55 m^2. The four solar panels contained 5,188 small solar cells, generating 140w which charged nickel cadmium batteries.

It was the first Chinese satellite to have a full solar orientation system. It had a device to point it toward the sun and thereby obtain maximum solar power to support the electric demands of the scientific instruments. A hydrazine-fuelled thruster system kept the satellite's panels pointed toward the Sun, rotating the spacecraft at 15 to 20 revolutions per minute. Shi Jian 2 made extensive use of the louvre system of thermal control so successful on Shi Jian 1. The satellite represented a substantial advance in satellite design.

NEW CARGO

This spacecraft was to have been the only cargo for this mission. It was almost certainly intended for launch on the Long March. However, if the Feng Bao launcher were used, experts calculated that there would be sufficient lift capacity to lift two other satellites at the same time. This was quite a complicated exercise, especially since Shi Jian 2 was half complete and work on the other two prospective satellites had not yet even begun. Specifically, it involved the construction of new support structures in the nose cone, small separation rockets and a new range of centre-of-gravity and vibration tests. The change of plan was decided in 1977. The change also meant an alternation to the orbit, to 59.5°, 240–2,000 km.

It was essential that the additional two satellites did not unbalance the nose cone, lead to collisions on deployment or that their radio frequencies interfered with each other. It was a tight fit, for the gap between Shi Jian 2 and Shi Jian 2A was only 4cm: a clumsy separation would crush the solar panels of Shi Jian 2 and wreck the satellite. Because of the rush, backup models were not made of the new spacecraft. The Chinese literature, whilst detailed in its account of the Shi Jian 2, has provided much less information about Shi Jian 2A and 2B.

FIRST 3-IN-1 LAUNCH ATTEMPT

The assembly was brought together for the first time in September 1978. However, the first Chinese 3-in-1 launch came to grief and failed to reach orbit on 28th July 1979. The Feng Bao's second stage vernier engine, designed for the final low-powered thrust to orbit, failed. In September, Ren Xinmin arrived in Shanghai arrived to head up another post mortem. He demanded a complete review of the second stage. In the ensuing changes, the turbo-pump system was rebuilt and more equipment was lifted from the Long March 2 to prevent such a setback taking place a third time. The new vernier engine was test fired six times, on one occasion for 60 min non-stop. The opportunity was taken to modify the satellite payloads. There was a spare Shi Jian 2, but the other two satellites had to be rebuilt.

A fresh attempt was organized and the new rocket was brought to Jiuquan in August 1981. Shi Jian 2, 2A and 2B were put in orbit in darkness at 5.28 am on 20th September 1981, the rocket being lost to sight after 3 mins. Feng Bao entered orbit after 7 mins 20 sec, separation being achieved in the planned 3.5 sec. No fewer than 59 separate operations had to be carried out perfectly in sequence for the separation procedure to work – and it did. The scientific satellites entered similar orbits, 240 by 1,610 km, 59.5°, 103 mins. Even though the Shi Jian 2 was the back-up model from the failed 1979 launch, its battery capacity had lost only about 6% of its power during the two years in storage. Although they began life in similar orbits, the three satellites were entirely different. Shi Jian 2A was heavier, bell-shaped, with two cones and antenna. It is known that Shi Jian 2A explored the ionosphere by transmitting radio signals at 40.5 MHz and 162 MHz to Earth stations and these signals were picked up in Sweden as soon as the third day into the mission. Shi Jian 2B was a combined metal ball and balloon, linked by a thin wire and designed to measure decay rates due to atmospheric drag. The three satellites operated for 332, 382 and six days respectively.

There appear to have been significant scientific results from the Shi Jian satellites. Shi Jian 2 provided details of the configuration, distribution and boundaries of the Earth's radiation belts. By flying during the period of an 11-year peak of solar

90 **Expanding the space programme**

Shi Jian 2 in assembly. (Courtesy Sven Grahn.)

activity, it was able to measure radiation from our Sun at its most violent and make predictions of solar storms. This time, the Chinese gave scientific details of the mission, marking a change from the earlier revolutionary rhetoric.

The mission, incidentally, marked the end of the use of the Feng Bao rocket. Its poor performance, low reliability and the availability of the superior Long March 2C led to its withdrawal from service. The Shanghai workforce was transferred to the ambitious Long March 3, then in the crucial stages of design and reintegrated back into the Seventh Ministry. Shanghai had lost its political clout after Mao's death. In the histories, the failures of the Feng Bao were retrospectively blamed on the Gang of Four.

Shi Jian 2.

The final postscript to this mission was not written until 2004, when Italian space writer Paolo Ulivi uncovered advanced plans for a mission that was to have been based on the Shi Jian 2. Called Tianwen Weixing, this would have been China's first astronomical satellite. Work on the 500 kg spacecraft began in 1976 and it was slated to carry at least five astronomical instruments. Some were developed by the Purple Mountain Observatory. The purpose of the mission was to observe cosmic rays, x-rays, gamma ray bursts and high energy solar emissions. The spacecraft never flew and the project was cancelled in 1984.

SHI JIAN 4, 5

Three other satellite received the Shi Jian designation. Shi Jian 3, which did not fly, was a Landsat (Earth resources) project in the 1980s never getting beyond design stage.

Shi Jian 4 was flown on the first flight of the Long March 3A launcher many years later (18th February 1994). Shi Jian 4 was a 410 kg drum, 1.6 m in diameter, 2.18 m high, with 11,000 2cm by 2cm solar cells whose mission was to study the spatial and spectral distribution of the Earth's charged particle environment. Six scientific instruments comprised stationery and dynamic single event monitors, a charging meter, electrostatic analyser, 5-channel electron spectrometer and high energy proton and heavy ion detector. The Long March also carried into orbit an unspecified 1,600 kg payload called Kua Fu, probably a technology demonstrator (in Chinese mythology, Kua Fu chased the Sun). Shi Jian 4 entered an orbit of 209 – 36,118 km, 28.5°, period 10.7hr, one suitable for researching charged particles because of its transit through the Van Allen radiation belts four times a day. It was designed to last for six months, a target which it apparently managed to reach before succumbing to the intense radiation of the belts.

Five years later, Shi Jian 5 was launched in May 1999, riding piggyback with a meteorological satellite, Feng Yun 1-3. Weighing 398 kg, its purpose was to study the terrestrial magnetosphere and single upset events which damaged satellites in orbit. Experiments comprised a suite of cosmic ray detection instruments: semi-conductor proton and heavy ions detector, static electrical analyser, electrical potentiometer, static single event monitor and dynamic single event monitor. The project was developed jointly with Brazilian cooperation. Intentionally or not, it had a short lifetime, the end-of-mission announcement being made in mid-August 1999, only two months later. The Chinese stated that this satellite would become the basis for a range of small, lightweight, low-cost scientific satellites to be flown over the next number of years. This is an American approach in which a basic satellite design or bus is mass-produced, different suites of scientific instruments being attached to the bus according to the type of mission flown. The bus is called the CAST968 (CAST for China Academy of Space Technology, presumably '96' and '8' for the design date). Its first adaptation after Shi Jian 5 was for the Haiyang oceanographic satellite launched 2002.

92 Expanding the space programme

Shi Jian 5.

Shi Jian series

Shi Jian 1	3 Mar 1971	Cosmic ray and X-ray detectors, magnetometer
Shi Jian 2, 2A, 2B	20 Sep 1981	3-in-1 mission with 11 scientific instruments
(Shi Jian 3	Not flown	Cancelled Earth resources project)
Shi Jian 4	18 Feb 1994	Cosmic ray satellite
Shi Jian 5	10 May 1999	Magnetospheric satellite

ASSESSMENT AND CONCLUSIONS

The main achievement in the Chinese space programme in the period after the first two launches was the recoverable satellite. The Chinese went straight from launching a basic satellite to the difficult challenge of orbiting and recovering cabins weighing over one tonne. The FSW satellite involved advanced techniques in space technology, such as heat shields, computers, sophisticated tracking systems and automatic control. The rocket used to support the FSW programme, the Long March 2, was a considerable advance over the Long March 1, involving new manufacturing techniques, gimballed motors and an inertial guidance system.

In addition to the recoverable programme, the Chinese maintained their commitment to space science. The Shi Jian 2, whilst possibly less sophisticated than Soviet or western scientific satellites of the same period, represented a significant investment in science. Like its predecessor, it appears to have returned a substantial volume of useful information. Shi Jian 4 and 5, subsequent scientific satellites, also appear to have been successful.

By contrast, it is difficult to comment usefully on the Ji Shu Shiyan Weixing until the Chinese disclose more of their purpose and function. Even to grisled observers of the classified and obscurer parts of Soviet space programmes, the JSSW series poses considerable problems of interpretation. It is not clear what purpose is served by retaining the classification of this programme. As in the 1960s, this decade saw a further period of disruption to the programme because of the unfinished business of the cultural revolution. The guiding figure of Zhou Enlai may have been decisive in ensuring the momentum of the programme in difficult times. With the crushing of the Gang of Four in 1976, the four modernizations which followed and the restoration of political stability in China by Deng Xiao Ping, the prospects for more balanced programme development improved significantly.

REFERENCES

1. Zhang Yun (Ed): *The Chinese space industry today*. China Social Sciences Publishing Co, Beijing, 1986.
2. Zhang Yun (Ed): *The Chinese space industry today*. China Social Sciences Publishing Co, Beijing, 1986.
3. Zhang Yun (Ed): *The Chinese space industry today*. China Social Sciences Publishing Co, Beijing, 1986.
4. Phil Clark: The Feng Bao I launch vehicle programme. *Journal of the British Interplanetary Society*, vol 55, #7/8, 2002.
5. Zhang Yun (Ed): *The Chinese space industry today*. China Social Sciences Publishing Co, Beijing, 1986.
6. Zhang Yun (Ed): *The Chinese space industry today*. China Social Sciences Publishing Co, Beijing, 1986. Later, the history refers to other satellite projects being abandoned in mid-course. However, some of them were not urgently needed and it was said of them that the preliminary work was a wasted investment. It would be interesting to know more.
7. Sven Grahn: *The satellites launched by FB-1*. http://www.svengrahn.pp.se
8. The main source of information on China's recoverable programme is Philip S Clark: *China's recoverable satellite programme*. Molniya Space Consultancy, 1994.
9. Mark Wade: *Shuguang 1*, www.astronautix.com.
10. Paper reports developments in recoverable satellite technology. BBC reports, SWB series, 1st April 1998.

5

Communications and conspiracies

The next main event of the Chinese space programme in the 1980s was the development of communication satellites. No sooner had China launched its first satellite than it embarked on the challenge of the recoverable satellite. With the FSW recoverable satellite programme well under way, Chinese scientists moved on to a new, ambitious goal: putting satellites into geosynchronous orbit 36,000 km over the Earth. This programme involved the building of a new launcher, the Long March 3 and a new launch site, Xi Chang. Overcoming these difficulties, China then made available its rockets on the world launcher market, with considerable initial success. International politics, three launch mishaps and allegations of spying meant that this did not last.

COMMUNICATION SATELLITES: PROJECT 331

The visit of President Richard Nixon in 1972 began the process of international recognition of communist China after years of isolation and siege. Whatever about the political significance of his visit, the Chinese were amazed by the satellite television crews who had followed the president's every movement and beamed pictures back live to admiring American homes. This the Americans did through their now well-established network of satellites in geostationary orbit.

There were several reasons why China should develop satellite-based communications. Communications satellites offered the possibility of providing advanced telecommunications for a large country quite quickly. Quality telecommunication links are now considered an essential element in any modernizing country. Communications satellites offer both direct television transmission (saving the establishment of elaborate systems of relays) and telephone lines (saving the setting up of land lines) or a combination of the two. In the 1970s the Chinese leased a number of western satellite lines to test the potential of a space-based communications system. They needed no further convincing.

Satellites circle at an altitude of 36,000 km orbit over the Earth's equator every 24 hrs, thus appearing to hover over the same point all the time. The value of such an orbital position was first appreciated by science writer Arthur C Clarke who described its merits in *Radio world* as far back as 1945. The Americans pioneered the use of the 24 hr orbit in 1965 with Early Bird. Now the 24 hr orbit is crowded and elaborate arrangements exist both for the allocation of slots there and for ensuring dead satellites are taken out of that orbit and sent to less densely populated regions of the sky (graveyard orbits).

However, the 24 hr orbit presents its own problems. First, to reach an altitude of 36,000 km and enter a circular orbit there requires a powerful launcher and upper stage able to reach the final destination. Second, the 36,000 km orbit is over the equator, which means that the rocket must not only reach a great height but carry out a dogleg manoeuvre southwards. While it is possible to reach 24 hr orbit on a conventional three-stage rocket, placing a sizeable payload there requires more powerful fuels and or a restartable upper stage.

The idea of a communications satellite for China was discussed and approved by the Central Committee in 1965. Exploratory research began. In June 1970, the China Academy of Launcher Technology (CALT) allocated staff for preliminary design studies of launchers and rockets. A project conference convened in November 1970 did not have the desired effect in moving the project forward, probably because of the chaos of the cultural revolution. Little further progress appears to have been made by 1974.[1] Responding to an appeal from the engineers involved, Zhou Enlai intervened on 19 May 1974 and demanded a report by the State Planning Commission. Three further project conferences were convened: the first, in June 1974 by the China Academy for Satellite Technology (CAST); the second by the Seventh Ministry in September 1974; and the third by the Ministry of Machine Building at the end of the year.

MAO'S LAST DECISION

The report of the State Planning Commission (*Report concerning the question of development of this country's satellite communications*) was received by the Central Committee on 25 November 1974 but it seems that it was not given final approval till 17 February the following year, 1975. The project was discussed by the Central Military Commission on 31st March. Intermediate options of a less ambitious low-Earth orbit system were examined carefully. The next month, in April 1975, the Central Committee and then Mao Zedong personally gave the go-ahead to the project (it must have been his last decision on the space programme) and gave it a code-name (project 331). Sun Jiadong was appointed chief designer and Liu Chuanshi director general of the project. Mao's endorsement finally unblocked the log-jam. The new rocket was named the Long March 3. None of this was publicly announced for many years.

The 24 hr orbit was an immense challenge. The Chinese considered the development of a low Earth orbiting system first, like the American Telstar. They also considered the idiosyncratic but highly effective Soviet Molniya system of satellites which orbit the Earth every 12 hr, but with an apogee slowly transiting the northern hemisphere. However, they opted to go straight for a 24 hr system, despite the difficulties.

Because of the complexity of the task involved, the Central Committee and state council set up a project leadership team the next month, May, so as to ensure the proper coordination of all the many different industrial and scientific groups involved. In June the Chinese Academy for Space Technology (CAST) held a planning conference which reviewed the progress of global satellite telecommunications to date. Locations for satellites in 24 hr orbit were formally requested from the International Telecommunications Union in March 1977. Locations were sought at 87.5°E, 98°E, 103°E, 110°E and 125°E. The project was listed as a state national priority in September 1977, which meant that it could command resources at will and that administrative log-jams would be quickly cleared. As was the case with projects 1059 and 651, the government convened a special national conference of scientists, engineers and industry in October 1977 to plan the management and development of the project, the launcher and the satellite (the conference also took in the launching of the DF-5 and the submarine-launched missile).

When political turmoil receded, the Chinese invited a high-level American aerospace team to tour Chinese space facilities, discussing with them ways of improving space communications. In 1983, the Italians were invited to establish a 3 m dish ground station in Beijing to receive signals from their satellite, Sirio. For the Chinese, this provided a further learning experience in advance of their own forthcoming comsat project.

The first mock-ups were built by 1977. The first electrical engineering model followed in 1979: tests revealed over a hundred problem areas which had to be sorted out. Next was a structural model, which was subject to both static and dynamic tests. Temperature tests were run in 1979, the model being cooked in a new hot vacuum chamber called the KM 4 at the Beijing Environmental Engineering Institute. A mockup was built in 1982 and used to test integration into the upper stage. The final, integrated flight version was constructed in the spring of 1983. Chinese officials attempted, in 1980, through NASA, to open negotiations for the purchase of a satellite system from American satellite manufacturers, but it is not known what became of this.

THE ROCKET

The need for a powerful launcher prompted the Chinese to consider the use of a hydrogen-fuelled rocket. Hydrogen, while having considerable advantages in terms of thrust (50% more than conventional rockets) and environmental friendliness, is a difficult substance to handle. It must be cooled at a temperature of $-253°C$ and its oxidizer at $-183°C$. This in turn requires very strong metals, for conventional alloys will turn as brittle as glass under such temperatures. The fuels and oxidizer evaporate quickly on the pad and have to be continuously topped up right to the moment of ignition. Liquid hydrogen has a rate of seepage fifty times higher than water. This area of work is sometimes referred to as cryogenics technology.

The Americans experienced great difficulty with introducing a hydrogen-powered upper stage in the 1960s (the Centaur) and the Russians did not operationalize such technology until 1987 (Energiya). Not only that, but restarting any rocket stage for a second burn has always proved a persistent problem in rocketry, for the engine must be restarted in zero-gravity, without the normal forces which push propellants into the combustion chamber. The Russians had a long series of problems with their Molniya and Proton upper stages failing to restart, expensive Moon, Venus and Mars probes becoming stranded in low Earth orbit as a result.

During a feasibility study in 1974, the Chinese weighed the options of using a conventional launcher and a hydrogen-powered third stage. Whatever the challenges, the Chinese decided in August 1976 to go for the most ambitious system – a hydrogen-powered restartable upper stage. However, progress was held up due to the political confusion associated with the rule of the Gang of Four.

CONFIGURING THE LAUNCHER

The Chinese began their first work on liquid hydrogen in the Liquid Fuel Rocket Engine Research Institute in March 1965. The first combustion test was carried out in January 1971. The first pumps for a liquid hydrogen engine were run in March 1974. Veteran Long March 1 manager Ren Xinmin was appointed Long March 3 project supervisor and Xie Guangxuan the chief research designer. The new rocket was the biggest yet constructed in China – 43.25 m tall, 3.35 m diameter, 202 tonnes in weight, with a take-off thrust of 280 tonnes.

The demanding third stage design was assigned to the Chinese Academy of Launcher Technology (CALT) while the first and second stages were given to the Shanghai Institute of Machine & Electrical Design (SIMED). Preliminary designs were approved in March 1978. Responsibility for the control systems of the satellite in 24 hr orbit fell to China's leading radio electronics expert Chen Fangyun (1916–). From Huangyan, he graduated in physics in Qinghua university before working in a radio factory in Britain.

NEW ENGINE: INTRODUCING THE YF-73

The new upper stage, named the H-8, required a new engine, the YF-73. This new engine comprised four combustion chambers, had a thrust of 4.5 tonnes and could gimbal through 25°. By 1979, the new YF-73 had passed its stand tests and had proved capable of reignition (at least on Earth). By 1983, the YF-73 had run for 8 hr in a hundred tests. These tests were not trouble-free and the design team experienced problems of over-heating, dangerous hydrogen leaks and materials unable to stand the strain. This was hardly surprising, for the engine turbine turned at a rate of 37,000 revolutions a minute. A particular problem, one which the Americans had encountered in the Saturn V, was the pogo effect, in which the entire rocket was likely to shake up and down with vibration, a snag overcome by the

Engines undergo tests.

installation of pressure accumulators. In addition to the YF-73, the H-8 upper stage required a set of control thrusters using hydrazine propellant. One of the principal designers of the YF-73 was Wang Zhiren, one of China's few prominent women rocket scientists.

ARRIVING ON STATION: THE APOGEE MOTOR

The transfer to geosynchronous orbit required a complex set of manoeuvres. First, the upper stage is placed in a low Earth orbit, typically at around 215 km. On its first southbound pass over the equator, generally about an hour later and over 160°E longitude, the hydrogen-power upper stage is fired to raise the high point of the orbit, the apogee, to 36,000 km, the altitude of geosynchronous orbit. This burn requires the highest level of energy and achieves what is called the Geostationary Transfer Orbit (GTO). Then the low point must be raised and the inclination changed from 28° to 0°.

Several of the key operations would take place when the rocket was well outside the line of sight of Chinese ground control. Chinese computers lagged far behind western and Soviet ones in the late 1970s and China was also embargoed from receiving the latest computer technology. One senior engineer, Hao Yan and a systems specialist, Song Jian, managed somehow to bring the computers up to sufficient standard to function autonomously and carry out the manoeuvres out of tracking range.

A key element in the new rocket was the apogee kick motor – the small, solid-fuel rocket that would be used to adjust the satellite's elliptical geostationary transfer

Communications and conspiracies

An apogee motor.

orbit into a circular geostationary orbit. The apogee motor was the third great challenge for the solid fuel engine research academy, after the Long March 1 third stage and the FSW retrorocket. The apogee motor would require a high level of both thrust and precision as well as being squat in shape to fit into the payload shroud. A total of 45 test runs were carried out between 1978 and 1982, the operational version being delivered to the Long March 3 team the following year.

TROUBLE, LEAKS, EXPLOSION

The three stages of the Long March 3 underwent their final all-up tests in summer 1983. The first test in May revealed a serious problem of hydrogen leaks but round-the-clock redesign of the sealant of the oxygen pumps bearings led to a successful test on 26th July 1983.

However, it seems that the programme suffered a serious setback when the motor exploded on the test stand. In January 1978, the motor blew up. Buildings were damaged all around, windows being blown out, technicians having their hair singed, some suffering broken eardrums. This disaster was reported only briefly six years later. Much later still, it was admitted that there had been 'martyrs', in other words, fatalities.

Designers of the YF-73 liquid hydrogen fuelled engine
Liu Chuanru
Wang Zhiren
Zhu Senyan
Wang Heng

The liquid hydrogen engine. (Courtesy China Astronautics Publishing House.)

NEW LAUNCH SITE NEEDED

A related problem to that of the launcher was how to reach equatorial orbit from a launch site with a high latitude (Jiuquan was 41.1°N). Other countries solved this problem by setting up launch sites on the equator, as France did in its colony of French Guyana, or, more exotically and very recently by converting Norwegian oil rigs into a launching platform and towing it to Kiribati in the mid-Pacific (Energiya and Boeing's project Sea Launch). The Chinese landmass lies some distance north of the equator. By establishing a new site much closer to the equator in southern China, some of the burden of reaching equatorial orbit could be reduced. Accordingly a new site was found, at Xi Chang, at 28.25°N, coincidentally at a similar latitude to Cape Canaveral.

Xi Chang, though much closer to the equator, had its drawbacks. The launch site is in hilly country, 1,826 m above sea level, which must have imposed additional construction costs. Climatically, temperatures are more clement than Jiuquan, ranging from −10°C to +33°C. The site has excellent dry weather from October to May which contributes to an annual average of 320 days sunshine. However, the

Communications and conspiracies

Early building at Xi Chang. (Courtesy China Astronautics Publishing House.)

weather in the period May to Septembers may see downpours and thunderstorms. It is also far from deserted, being surrounded by villages. Virtually all the other launch sites in the world are either on the coast (and far out to sea) or located in inland desert, thereby reducing the risk of civilian casualties to a minimum. This is not the case with Xi Chang. The first tests of mockup launchers began at Xi Chang in March 1982.

THE COMMUNICATIONS SATELLITE

The satellite itself was a drum 3.1 m tall and 2.1 m in diameter, with an apogee motor underneath and two receiving and broadcasting antennae on top. The satellite was identified subsequently as the Dong Fang Hong 2 series ('2', presumably in deference to the first Earth satellite which was '1'). Launch mass was 916 kg, but by the time the apogee kick motor had fired, the weight of the satellite on station would be 420 kg. Its essential function was to receive transmissions from the ground with a high-gain antenna, amplify and, using two transponders, retransmit them on a spot beam focussed on China itself. Solar cells generating 315w were fitted on the outside of the drum.

The design involved new challenges in devising a satellite which could operate faultlessly for several years, which could maintain its same position in the sky unchanging and continue to focus its transmitter accurately. The most difficult part was the despin system: whereas the satellite itself was spun at 50 revolutions per minute (to maintain its stability and ensure that it was evenly exposed to solar rays), the antenna system had to point in a fixed direction. This involved the development of a reliable, lubricated mechanical de-spin system, one which had caused the Americans much difficulty several years earlier.

PREPARING FOR THE FIRST LAUNCH

Zhang Aiping inspected the Xi Chang launch site on 27th October 1983 to check that all would be ready to receive the first Long March 3 rocket and its precious cargo. The new Long March 3 and its payload made their way to the pad on new year's day, 1984. The design team had already reached Xi Chang, planning the launch campaign even as its train sped south-westward. By early 1984, these preparations were complete. The first attempt to launch the Long March 3 was set for 8pm on 26th January. Just after fuelling had got under way, the guidance platform broke down, meaning that the rocket could not be controlled during ascent. This meant that the platform would have to be physically removed. As a result, the launch engineers missed their launching window to 24 hr orbit and the rocket had to be drained. They had an early experience of the downside of using liquid hydrogen.

Three days later, early on 29th January, the powerful new Long March 3 rocket was once again stacked for its maiden voyage. The storable fuels were loaded 16 hr before take-off. Five hours before lift-off, the supercold liquid oxygen and hydrogen were pumped aboard. The amounts had to be topped up every now and again as wisps of evaporating oxidizer blew away. China then launched its first geosynchronous communications satellite, Shiyan Weixing ('experimental satellite') at 8.24am. It was the first launch of the Long March 3 and the first time a rocket had taken off from Xi Chang. All seemed to go perfectly at first, the upper stage, with Shiyan Weixing entering an orbit of 308–448 km, 31°. This was a parking orbit. It was intended that the third stage would fire a second time over Xi Chang over its first pass to shoot the satellite on its way to geosynchronous orbit. However, the engine failed, leaving the satellite stranded in low Earth orbit. Pressure in the launching chamber reached only 90% of the level necessary and after 3 sec collapsed.

This was a great disappointment. The YF-73 had been run for over 30,000 sec in tests. It must have been little consolation to the Chinese that this problem – reigniting engines in weightlessness – was one of the most frustrating ones in astronautics and had taxed the other space powers as well. The Chinese decided they would salvage what they could of the Shiyan Weixing mission. They separated the satellite from the third stage and conducted the final on-orbit manoeuvres that should have been carried out at 36,000 km. In the event, the apogee motor moved the satellite into an orbit of 400 km by 6,480 km and changed the orbital inclination to 36°. They also made the satellite carry out the range of station-keeping adjustment manoeuvres that would have been necessary had the comsat entered the intended orbit. The satellite's various communications systems were tested rigorously, being turned on and off and tried in different régimes. The satellite experienced a number of problems in its electrical supply and heating systems which were remedied in good time for the next model. Further minor manoeuvres were made in the orbit in

104 **Communications and conspiracies**

Long March 3 launch.

February, March, May and July, to simulate the station-keeping operations that would have been required had it reached its intended orbit.[2]

BEATING THE SUMMER THUNDER

A second Long March 3 was already at Xi Chang. Ground tests were made to try sort out the problems that marred the first launch. There was added urgency, since summer storms affected Xi Chang from May onward. The launch teams worked day and night, right through the spring festival. Fax lines between Xi Chang and Beijing ran hot as launch teams and design teams tried to diagnose the earlier faults, working against time. To test for problems, the third stage was hot fired again at the Liquid Fuel Rocket Engine Testing Station in Beijing, the tests being completed on 20 March. The modifications had to be performed on the new rocket, which was already stacked on the launch pad. Wearing breathing apparatus, a team of engineers led by Ni Zhongliang and Ciu Yajie had to crawl into the third stage from the gantry. There was room neither to sit nor to stand and they had to work at awkward angles in the plumbing system, always being careful not to damage any delicate machinery. The work took them three days.

The next mission was flown four months later, on 8th April 1984. A new communications ship, *Yuan Wan,* took up position, 1,000 km off the Chinese coast.

The satellite was called Shiyan Tongbu Tongxin Weixing (experimental geostationary communications satellite). Despite their best efforts, the first of the spring thunderstorms had already arrived in Xi Chang, a potential cause of disaster should one attempt to launch a hydrogen-fuelled rocket. In the afternoon, Xi Chang launch site clouded over and it began to rain. The launch controllers went into emergency conference. The forecast predicted that the clouds would clear for the intended launch time, 7pm. An elderly local village sage stepped forward and predicted clear weather for the intended launch time 'or I'll never take a drink in my life again!' he swore. Now assured of fine weather, the controllers ordered the countdown to continue.

At dusk the clouds duly rolled away. Twinkling stars could just be made out up above. Spotlights played on the ready rocket. Fuelling began. Fifty minutes before launch, crowds began to gather on the surrounding hills. By 7pm, the last fuels had been loaded and the trucks dispersed. The engines were armed. At one minute the hilltop spectators could see the attachment plugs fall from the side of the Long March. The electric cable arm retracted. At 19.20.02 a technician pressed the red firing button and the Long March 3 headed skyward. But had the engineering team done the job?

As the Long March 3 upper stage cruised southbound over the Earth's equator, the rocket reignited and hurtled the one-tonne satellite skyward. Twenty minutes later, tracking ships in the Pacific reported that this time, the burn to geosynchronous orbit appeared to be alright, at 437–35,499 km, 31.08°. Two days later, the apogee kick motor fired to raise the perigee to 35,521 and the apogee to 36,383 km, bringing the inclination to less than 1°. Perfect: the satellite's own motor trimmed the orbit to let it drift the final distance to its destination hovering over 125°E. But would it work?

These were anxious moments: as the satellite drifted into position, the thermal control in the battery system broke down and the current began to fluctuate alarmingly. In a risky repair, the ground controllers in effect told the spacecraft's computer that the satellite had completely broken down and that it must make a full recovery, rather like restarting a computer system by turning it off and beginning again. It worked. At 18:27:57 on 16th April, the satellite arrived on station at 125°E longitude. The spacecraft was spun, the despin system turned on and transmissions begun. The first relay tests were carried out the next day and an hour's television transmitted to the most distant regions of the country. The first pictures were clear, the pictures stable, the colours realistic and the sound well up to standard.

ZHANG AIPING'S PHONE CALL

Zhang Aiping was one of the first persons to test out the new comsat by making a much-publicized telephone call to a distant party committee in Xinjiang, Wang Enmiao. The satellite established China's first 200 satellite-based telephone lines, connecting Beijing with Urumqi, Lhasa, Hohhot (inner Mongolia), Chengdu (Sichuan) and Guangzhou. The voice quality was good, with almost no background

Zhang Aiping's telephone call to Weng Enmiao.

noise or interference. The satellite was formally handed over to its new telecommunications owners on 25 April for tests. The system was declared operational on 24th May.

To mark the occasion, on 30th April a solemn conference was convened, presided by Hu Yaobang, in the Great Hall of the People, in Beijing. Subsequent accounts suggest it was far from solemn, for the Chinese had much to celebrate. It put China in the top league of applied space engineering. Ten years of hard work had been vindicated, though the official history noted that several of the engineers had gone grey, suffered ill-health from overwork and some had even passed on prematurely.[3]

NEW LANGUAGES FROM SPACE

Small thrusters were used occasionally thereafter to readjust the orbit. All went well and the Shiyan Tongbu Tongxin Weixing, although designed for only a short working life, worked perfectly for more than four years when it was taken out of service. Its 15 radio and television channels transmitted programmes in Cantonese, Amoy, Hakka, Japanese, Spanish, Russian, Burmese and Tagalog, some of these languages not hitherto familiar on the international satellite ether. These satellites carried scientific instruments to measure changes in the intensity of electrons and protons in 24 hr orbit, solar radiation and static electricity on the spacecraft. Typical instruments were a semi-conductor and electron detector, semi-conductor proton detector, solar x-ray detector and potentiometer.

Shiyan Tongbu Tongxin Weixing continued to operate for four years. Every two months or so, its small engines would fire to correct its orbit – this is called station-keeping – used to prevent it from gradually drifting away from its assigned spot. On 20th June 1988, a small manoeuvre was carried out to take the satellite out of position. This is called the 'retirement manoeuvre' and is used to place the satellite concerned in what is termed a 'graveyard orbit'. The 24 hr satellite location is much in demand and as soon as satellites there have completed their mission, it is considered good behaviour to take them off station to make way for newer, fresher satellites.

Following the success of Shiyan Tongbu Tongxin Weixing, the Chinese proceeded to the launch of the first operational geosynchronous communications satellite, *Shiyong* Tongbu Tongxin Weixing ('*operational* geostationary communications satellite'). The main difference with its predecessor was an improved 0.7 m diameter dish to beam its transmissions to Earth – though the actual main part of the satellite was the backup from the previous mission. It was launched two years later on 1st February 1986, entering transfer orbit of 438–35,546 km and reaching geosynchronous orbit two days later with a final operating position at 103°E longitude on the 18th. Its motor was fired monthly to ensure it stayed precisely in this position. The satellite took over many of the functions of Shiyan Tongbu Tongxin Weixing. It operated for four years, transmitting to 30 television stations, until it drifted off station in October 1990. It was the last in the Dong Fang Hong 2 series.

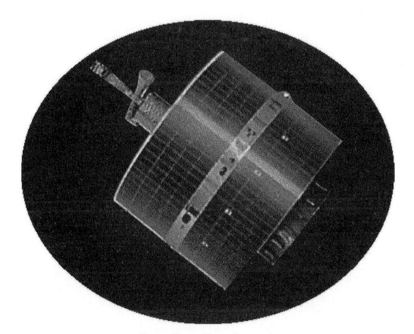

Dong Fang Hong 2 comsat.

DONG FANG HONG 2A: 3,000 PHONE CALLS AT A TIME

Following this, the Long March 3 was used for a series of commercial launches for other countries (these are described in the next sections). The next 24 hr domestic satellite saw the introduction of the Dong Fang Hong 2A series (which also may

108 Communications and conspiracies

have had the designation Zhongxing). The Dong Fang Hong 2A comsat was 3.68 m tall, weighed 441 kg on station and had a design life of four years, power being supplied by 20,000 solar cells. It had four transponders able to transmit five television channels and 3,000 telephone calls at a time. The first, Shiyan Tongbu Tongxin Weixing 2, was launched on 7th March 1988, took up position at 87.5°E on 23rd March and doubled its design life. It finally drifted off station nine years later in September 1987.

Equally successful were Shiyong Tongbu Tongxin 3 that December (110.5°E) and Shiyong Tongbu Tongxin Weixing 4 two years later (98°E). Its launch, on 4th February 1990, was observed by Chinese premier Li Peng. These early satellite were used to achieve complete television coverage for China with 30 channels and to permit telephone and fax services to be sent by satellite for the main governmental agencies and development bodies. 30,000 receiving dishes were built, education programming going out more than 30 hours a day, reaching over 30 m people. The Dong Fang Hong series also carried small scientific payloads – solar X-ray detectors, potentiometers, proton and electron detectors. Shiyong Tongbu Tongxin Weixing 3 operated until August 1999, over ten years, more than twice its design lifetime and was the last of the 2A series to cease functioning; 4 left for its graveyard orbit in summer 1998 after successfully transmitting for almost ten years.

A satellite dish in a rural area. (Courtesy Asiasat.)

The run of successes came to an end on 28th December 1991 when a Long March 3 launched what should have been Shiyong Tongbu Tongxin Weixing 5. Rather like the first attempt to send a satellite to geosynchronous orbit in 1984, the third stage failed, this time after burning for 58 sec. Apparently, the helium pressurizing gas in the third stage sprang a leak and pressure in the combustion chamber fell to zero 135 sec into the burn. Shiyong Tongbu Tongxin Weixing 5 entered orbit of 218 to 2,450 km. Once again, the Chinese separated the payload in an attempt to salvage something from the mission. They eventually fired the kick motor to raise the orbit to 2,003 by 34,057 km and continued to manoeuvre the spacecraft until 1994. It is not known what was accomplished, nor when the mission was finally abandoned, but presumably there was some return from the scientific instruments on board.

In the event, the following year the Chinese space authorities bought an American comsat, Spacenet 1, which was already in orbit and nearing the end of its useful life. Motor firings were used to move the Spacenet from its location at 240°E to what may have been the intended destination of Shiyong Tongbu Tongxin Weixing 5 at 115°E. They then renamed it Zhongxing 5 (Zhongxing means 'the star of China'), though Zhongxing 1-4 were never formally retrospectively identified (they were almost certainly Shiyong Tongbu Tongxin Weixing 1 to 4). Zhongxing 5 operated at 115.5° until December 1999 when it was retired.

DONG FANG HONG 3 SERIES: 8,000 PHONE CALLS AT A TIME

The third generation of Dong Fang Hong communications satellites was already in the pipeline. The purpose of the new generation was to increase twelvefold the capacity of the previous series and guarantee a working life of eight years. It was also China's intention to use a design called the DJS platform, which could be adapted for other satellites. The Dong Fang Hong 3 series, unlike its predecessors, had a minor (20%) western design contribution. In July 1987, agreement had been reached between the Chinese company responsible, the Great Wall Industries Corporation and Messerschmitt Bolkow Blohm (MBB) of Germany for the development of the Dong Fang Hong 3. MBB was responsible for part of the design and the solar array and the antennae.

There was an important design change. The drum shape of Dong Fang Hong 2 and 2A gave way to a box-shaped spacecraft with two solar wings. Dong Fang Hong 3 had double the weight of its predecessors – 2,200 kg at launch and 1,145 kg on station. It was 5.71 m tall and had a 2 m diameter communications dish with six spot beams. The satellite had 24 transponders to transmit six colour TV channels and take 8,000 telephone calls at a time, able to cover 90% of China. Its working life was eight years. The solar wings had a span of 18.1 m and were able to generate 2,000w. The assembly engineering team was led by Li Chunqi, a 48-year old engineer who passed up a university place for a chance to work on satellites. The welding of the Dong Fang Hong fuel tank involved the use of considerable electrical power supplies, to the extent that electricity was turned off in the small town in north-west China where it was being built in order to guarantee electrical power for the welders. The citizens were duly warned of the power cuts. Expecting a less than enthusiastic response, especially in mid-winter, the engineers feared the worst from irate citizens the following morning. Instead, to their surprise, callers to the factory gates were only interested to know if the welding job had been accomplished (which it had).

110 Communications and conspiracies

A satellite TV receiver.

However, all did not go well on the first mission. On 29th November 1994, the new Long March 3A rocket left the new satellite, Dong Fang Hong 3-1, in its transfer orbit of 181 by 36,026 km. The Chinese used its propellant over time to raise the perigee to 35,181 km by 29th December. By the time it reached that altitude, all the propellant had been used up and the satellite had to be abandoned. The problem is understood to have been in the satellite's propulsion system, not the launcher. The Chinese, then engaged in renaming their communications satellites, called it Zhongxing 6.

After the failure of Zhongxing 6, the Chinese bought a Hughes 276 spacecraft in the United States. They launched it on one of their own Long March 3 rockets in August 1996, but once again the transfer manoeuvre to geosynchronous orbit went wrong and it became stranded between 200 km and 17,230 km. Apparently, the pressurizing gas failed. This caused the thrust to stall a mere 48 sec before the satellite would have reached final orbit. The orbit was later raised to 21,667 to 46,507 km, so they got some use out of it, but Zhongxing 7 was then abandoned. It was the third such failure in a row in five years. The Chinese Telecommunications and Broadcasting Satellite Corporation, which was to have operated Zhongxing 7, was eventually paid $25.9 m by an insurance company for the loss.

Eventually, a Dong Fang Hong 3 satellite reached orbit successfully – put up on a Long March 3A on 8th May 1997. It was built by the Chinese Academy of Space Technology in Beijing with assistance from Germany's Daimler Benz Aerospace.

Dong Fang Hong 3 comsat.

The satellite, designed to replace Zhongxing 7, was alternately called Zhongxing 8, Zhongxing 6 (the term favoured by most Chinese) or Zhongxing 6B (by western experts) to replace the earlier 6 which had failed (it was also called it the Dong Fang Hong 3-2). On-orbit testing of its systems was successfully completed at the end of August 1997. In summer 2000, it celebrated three years in orbit by working as reliably as ever.[4] It was still operating at its location over 125° E in late 2001.

Eventually, it too will be phased out in favour of the 5.1 tonne Dong Fang Hong 4 series. The Dong Fang Hong 4, using what is called the DJS-2 common platform, will have between 22 and 52 transponders, support high capacity data links, generate 6 kw to 10 kw of electrical power and operate for up to 15 years. The French Alcatel company is directly involved in this project and one of the aims of the series was to ensure that Chinese communications satellites matched the highest western standards. The series will be used to support direct broadcasting and mobile communications. First launch, Sinosat 2, was due on the Long March 3B in 2005.

A January 2000 launch of a domestic communications satellite appeared to be routine at first. The launching on the Long March 3A went smoothly enough, but further difficulties came up with the satellite's name. Some reports called it Zhongxing 22 and others Chinastar 22. However, the '22' came from the location or slot allocated for satellites in 24 hr orbit (98°E, in the event): it was not the 22nd satellite in the series. Eventually the 2.3 tonne comsat acquired the name Feng Huo 1, 'fire and smoke' in Chinese, named after an ancient system of communicating using beacons along the Great Wall. Whenever invaders threatened ancient China, beacons had been lit all along the wall – a much faster method of warning than horseback. Built by the China Space Technology Institute, Feng Huo appeared to be a test of mobile frequencies for the Chinese military. Elsewhere, though, it was announced as a successor to the Dong Fang Hong 3 series.[5] American intelligence experts went further and said Feng Huo was part of a new command-and-control network for the Chinese military, providing targeting capability for ballistic, cruise, ship and aircraft-borne missiles.[6]

The new Feng Huo programme may also be an attempt to make good on the poor performance of some of China's domestic satellites. Writing in *Jing Ribao* on 30th

112 Communications and conspiracies

Satellites to computers. (Courtesy Asiasat.)

April 1998, Hang Wen described the quality of China's domestic satellites as 'pitiful', the quality of electronics being especially weak and lagging behind international standards. There was no point in having great rockets if basic industry is poor: it was like trying to cook a meal without rice, he said. Two years earlier, Zhu Yilin had drawn attention to the series of problems which had arisen in transferring satellites to their final orbit.[7] China began to put more and more effort into improving the reliability of the final stage motor.

How reliable were China's satellites? A feature of Soviet communications satellites during the 1970s and 1980s was their remarkably short lifetimes (only 25 months for Ekran). Whenever they ran out, they were simply replaced by another, an approach to economics not untypical of the period, but not sustainable in post-communist Russia (Ekran quickly improved to 72 months and launch rates fell). The following table lists the design and actual lifetimes of Chinese communications satellites.

On-orbit lifetimes of Chinese communications satellites (years)

Date	Satellite	Design life	Actual
29 Jan 1984	Shiyan Weixing	3	3
8 Apr 1984	Shiyan Tongbu Tongxin Weixing	3	4.3
1 Feb 1986	Shiyong Tongbu Tongxin Weixing 1	3	4.75
7 Mar 1988	Shiyong Tongbu Tongxin Weixing 2	4	9.5
22 Dec 1988	Shiyong Tongbu Tongxin Weixing 3	4	11
4 Feb 1990	Shiyong Tongbu Tongxin Weixing 4	4	8.6
28 Dec 1991	Shiyong Tongbu Tongxin Weixing 5	(Failure)	

Source: Adapted from Clark (2001)

The 2 series exceeded its 3-year design lifetimes by a third, the 2A series by more than double. These performances were better than Soviet ones, although not as good as American satellites – but were impressive for a new country with such a limited technological base. When Feng Huo was launched in 2000, it was announced that it was intended to last at least eight years. China's latest domestic comsat is the Zhongxing 20 launched by Long March 3A from Xi Chang on 15th November 2003, using a supersynchronous orbit, heading for a position over 103°E. Little information was given out about the new comsat, which could in reality be the second in the Feng Huo series.

BENEFITS OF COMSATS

By 1996, China claimed that the beams from its satellites were reaching 83% of Chinese people. 1.37 m college students had now been trained in distance learning programmes beamed down by China's comsats. China's development of its communications networks gathered pace during the 1990s. Speed has been more important than political imperatives: in addition to its own satellites, China has pragmatically bought communications satellites from the west and often leased lines on western satellites, rather than wait to establish a completely indigenous service. For example, in 1991, the Spar communications group won a €42 m contract to provide ten new Earth terminals using state-of-the-art 13 m diameter dishes operating on Intelsat's Indian ocean network, providing the highest quality communications in China's new economic development zones.

By 2001, China routed a standard 70,000 domestic and 27,000 international calls through satellite. Long range automatic dialing through 500 Chinese cities was now

Satellites in rural areas. (Courtesy Asiasat.)

Communications and conspiracies

First domestic communications satellites

Shiyan Weixing or Dong Fang Hong 2-1	29 Jan 1984	Stranded
Shiyan Tongbu Tongxing Weixing Dong Fang Hong 2-2	8 Apr 1984	125°E – experimental
Shiyong Tongbu Tongxing Weixing 1 Dong Fang Hong 2-3	1 Feb 1986	103°E – first operational
Shiyong Tongbu Tongxing Weixing 2 Dong Fang Hong 2A-1	7 Mar 1988	87.5°E – first of 2A series
Shiyong Tongbu Tongxing Weixing 3 Dong Fang Hong 2A-2	22 Dec 1988	110.5°E
Shiyong Tongbu Tongxing Weixing 4 Dong Fang Hong 2A-3	4 Feb 1990	98.5°E
Shiyong Tongbu Tongxing Weixing 5 Dong Fang Hong 2A-4	28 Dec 1991	Stranded – 3rd stage failure
Zhongxing 6 Dong Fang Hong 3-1	29 Nov 1994	Incorrect orbit – abandoned
Zhongxing 7	18 Aug 1996	Bought from US, abandoned
Dong Fang Hong 3-2 or Zhongxing 8 or 6 or 6B	8 May 1997	125°E
Feng Huo 1	20 Jan 2000	98°E
Zhongxing 20	15 Nov 2003	103°E

done through satellite. There were no less 30 companies providing small terminal communications satellite services with 10,000 terminals. There were 189,000 satellite TV receiving dishes. Thanks to comsats, television coverage has risen from 30% of the country to 84.5%. The Bank of China transferred data and dealings with 350 branches by satellite, and up to ¥600 m were transferred by satellite every day. 3 m people routinely received polytechnic courses via satellite, and agricultural educational programmes went out to 20 m small farmers. Newspaper pages were transmitted by satellite. Ordinary television programmes were distributed by cable

A future comsat. (Courtesy *Aerospace China*.)

Restrictions, prices and quotas 115

to 15,000 small station users. 33 satellite transponders were used for the relaying of the broadcasts of China Central Television all over the country.

COMMERCIALIZATION OF THE CHINESE SPACE PROGRAMME

Following the successful launch of the Long March 3 in 1984, China began to offer its Long March series of launchers to the west. The decision to do so was strongly driven by the reduction in space spending since 1978, accompanied by an authorization for the space agencies to generate as much external income as they could from abroad. The formal announcement was made by astronautics minister Li Xue in October 1985. The Chinese intention was to interest western communications companies in using the Long March to get their comsats into geosynchronous orbit. The offer generated little media interest at the time. Western companies did take up more modest opportunities to fly dedicated payloads on recoverable satellites. In the course of 1987–9, China flew commercial microgravity experiments for France (FSW 0-9) and Germany (FSW 1-2). Further opportunities were publicly advertised in the western press. The Chinese issued a *Long March 3 user manual* to encourage familiarity with its product.[8] In 1985, in Zukiha, near Tokyo, Japan, the Chinese contributed their first stand to an international science and technology exhibition, showing off the new Long March 3 launcher. Despite the lack of an immediate response, China continued to press its offer.

Little happened to the launch offer until 1986, when in the space of a few months, America lost the *Challenger* space shuttle and then two of its other leading rocket launchers exploded spectacularly (a Titan and a Delta). To compound the problem, Europe's Ariane went down as well. Western companies in a hurry to launch satellites faced increasing delays and were forced to turn to Russia or China, whose capabilities they knew little about.

RESTRICTIONS, PRICES AND QUOTAS

The Chinese however may not have reckoned with the battery of trade and defence sanctions at the command of the world's economic superpower. Both the Chinese and the Russian commercial space programmes were subject to American military and commercial restrictions, ostensibly designed to prevent them copying western satellites *en route* to the launch pad. These restrictions were enforced through export licences and what was termed the COCOM agreement between the United States, the North Atlantic Treaty Organization and neutral western countries. The Americans required any American company to get permission to launch a satellite on a Chinese launcher. The same requirements were equally enforced on non-American companies contemplating the use of American communications satellites, or companies building communications satellites which might include American components, however small. The Chinese tried to meet American concerns by

116   **Communications and conspiracies**

guaranteeing an inspection-free transit of the satellite from the States to the top of the launcher.

Even when American-built equipment was cleared from the military and security point of view, the Russians and Chinese faced trade quotas as to how many satellites they could actually launch each year and the price to be offered. The prices proposed by the Chinese for Long March 3 launches to geosynchronous orbit were generally below Ariane (Europe) and the American launchers, but more than the Russian (Proton). Under a Sino-American agreement signed in January 1989, for the period 1988–94, the Chinese were permitted nine commercial satellite launches. China had to agree to United States demands to 'ensure that the market was open and fair'. This quota was extended to 11 satellites for 1995-2001, provided that China did not offer prices less than 15% below western rates.[9] Whilst lecturing the communist and former communist nations about the virtues of free enterprise, the United States explained the need for these quotas as being to permit 'disciplined Chinese participation in the market'. At the time, China was aiming for about 4% of the world market.

Despite the agreements, there were repeated exchanges in the international trading world that China's launch prices were undercutting Europe and the United States. Such altercations were and are by no means unusual in international commerce and affect many other fields of trade (for example, aircraft). The United States were equally critical of the pricing policy of the Ariane launcher, though the difference was that the Europeans were better able to stand up to American economic pressure. Pricing in the international launcher business is far from transparent in any case, since all the main launch companies have had their development costs underwritten by their governments and are effectively guaranteed the launches of many of the satellites manufactured in their respective regions, especially the military ones. To complicate things still further, several companies vying for the launcher business often offered the first flight for a customer at an 'introductory rate'. The company offering Ariane routinely provided a range of financial packages as well. These rates became the subject for much contention. The Chinese were very much aware that they undercut western prices. They argued that they had objectively lower labour costs, lower prices for raw materials and that government investment for the programme was now very low, forcing space bodies to generate their budgets abroad.

EXPORT LICENCE CRUX

So, for an American-made satellite to fly on a Chinese rocket, an export licence was required. Because most of the world's comsats are made in the United States, this necessarily involved foreign countries like Australia pleading to the United States for permission to fly American satellites on Chinese launchers. Originally, export licences were decided by the State Department, but reference was also made to the National Security Council and the Economic Policy Council. When deadlock ensued, the matter was referred upward to the White House. Secretary of Defence

Casper Weinberger (who visited Xi Chang) and his successor Frank Carlucci both took the view that the risks of technology transfer to China were minimal and that the whole matter boiled down to political and economic considerations. In the former case, political considerations revolved around the desirability of doing business with a political system which many Americans repudiated. On the economic front, considerations revolved around arguments for free trade (favoured by the satellite manufacturers) and protecting the domestic American launcher market (favoured by American launcher companies). Thus the question of export licences for the Chinese became, at least in part, a domestic lobbying battleground for the giants of American industry. Many of the military arguments were proxies for domestic industrial politics.

The outcome of this domestic American political battle was that President Reagan ruled in favour of granting export licences. These licences were renewed by his successor, President Bush, though temporarily withdrawn in 1989 (due to political pressures following the summer events in Tienanmen square) and again in 1991 (to persuade China to sign international agreements on the proliferation of missiles). Overall, the régime became more relaxed and supervision of Chinese missions passed from the security-conscious Department of State to the more liberal Department of Commerce.

The Chinese offered several variants of their Long March – the 1D, 2C, 2E, 3, 3A, 3B and 3C. They put forward a range of possibilities between leaving the satellite in low Earth orbit, low Earth transfer orbit for a subsequent 24 hr orbit (with the western company providing its own rocket to ensure transfer to geosynchronous orbit), right up to deployment in geosynchronous orbit. To complete the package, Chinese insurance companies also offered insurance to the foreign customers.

FIRST COMMERCIAL MISSION

The first commercial mission, Asiasat 1, in April 1990, was entirely successful. Asiasat 1 was owned by the Asiasat company, based in Hong Kong. Originally Asiasat comprised Citic Technology Corporation, the British Cable & Wireless company and local financial interests (Hutchison Telecommunications), formed in 1988. Later, the main shareholders were Citic and the European operator of the Astra satellite system.

Asiasat 1 came with a history: it had originally been launched into orbit as Westar 6 by the space shuttle in February 1984 but had been stranded in low Earth orbit when its motor failed to fire. Nine months later, American shuttle astronauts on mission 51A had retrieved the errant Westar and returned it to Earth for relaunch. The much-travelled but still unused Westar had then been bought by the Asiasat company and this was its second – and more successful – journey into space. They paid China €35 m.

The second commercial launch was likewise a success – a Pakistani satellite called Badr, riding the Long March 2E on its maiden voyage. Badr was a 52 kg

118 Communications and conspiracies

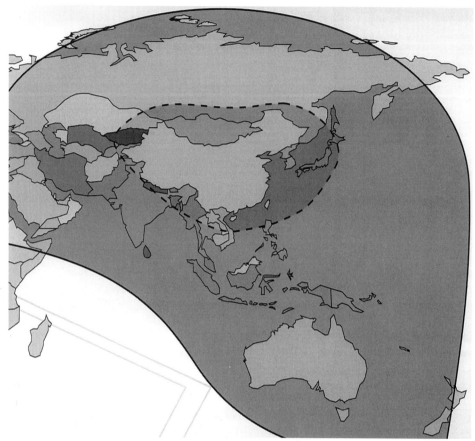

The Asiasat footprint. (Courtesy Asiasat.)

experimental satellite built by the Pakistan Space and Upper Atmosphere Commission and the first Pakistan satellite to enter orbit (China was paid €352,000 for the launch).

A Swedish satellite, Freja, rode into orbit in October 1992 accompanying the FSW 1-4 recoverable satellite mission. Freja, named after the Viking goddess of fertility, was a 259 kg spacecraft commissioned by the Swedish Board for Space Activities. A small American boost stage was used to place it in a high orbit from where its seven experiments could study the northern hemisphere's aurorae, electrical and magnetic fields, particles, plasma, electrons and magnetosphere. The spacecraft also carried an experimental communications payload called mailstar. Sweden paid China €5 m for the lift. It also provided one of the first opportunities for westerners to visit Jiuquan.

The launch of Freja. (Courtesy Sven Grahn.)

FIRE ON THE PAD!

The trouble started with Optus 1, whose launch was commissioned by an Australian communications company. The Australians paid €17.6 m each for two Optus launches, in effect a promotional rate. There was a near disaster on a pad abort on 22nd March 1992. The Long March counted down to liftoff and the main engines ignited. After 3 sec, the engines were turned off by the computers which detected a fault at a point a mere fraction before the scheduled liftoff. Three of the four restraints holding the rocket to the pad had already been released by this stage. However, fires had already broken out at the base of the rocket.

Reckless for their own safety, ground crews rushed forward to douse the flames and turn off the ignition systems. In their hurry, none donned oxygen masks and all inhaled the toxic gases swirling around the launch stand. The entire team had to be hospitalized later. All survived, though many had vomited blood from poisoning – a reminder of the dangerous fuels used. The rocket was not safe until 39 hr later when it was drained of all its propellants. In Russia, a similar hands-on attempt to deal with an errant rocket on the pad led to the world's worst launch holocaust in October 1960, an event that went down in the history books as the Nedelin disaster.

However, Optus 1 was eventually launched properly on 13th August, months later. A hundred Australians and other visitors attended the launch in Xi Chang and the event was broadcast live on Chinese central television. Cameras switched between the rocket lifting into the early morning sky and its hopeful customers eagerly watch it climb skywards. The purpose of the Optus Aussat system was to provide radio and television services for remote areas of Australia, air traffic control and educational and medical services by television.

DÉBRIS, RECRIMINATION

The earlier difficulty was nothing compared to what happened with Optus B-2 on 21 December 1992. Optus was launched with an American kick motor called the Star 63F to get it into geostationary orbit. 70 sec into the mission, a cloud of gas could be noted emerging from the shroud at the top of the launch vehicle. The rest of the launching proceeded normally, but there was widespread consternation when it transpired that all that had reached orbit was satellite wreckage. It seems that the satellite met with a fatal accident about a minute into the mission, but that the shroud had contained the explosion.

There was considerable recrimination afterwards between the Chinese, the Australians and the United States as to who was responsible. The Americans blamed the Chinese for a faulty shroud that failed under pressure; the Chinese blamed the Americans for failing to attach the satellite to its upper stage sufficiently to withstand vibration. Western press coverage laid the blame firmly on the Chinese. In the end, both the Chinese and the Americans agreed to paper over the cracks, eventually issuing a joint statement to the effect that neither the launcher nor the satellite was to blame!

The Chinese recovered somewhat with two successful launches in 1994 – Apstar 1, flown for a Hong Kong company in July and Optus B-3 in August, effectively replacing the satellite which had been destroyed two years earlier.

Long March 2E launch.

LONG MARCH CRASHES IN FLAMES

However, their new confidence did not last for long. On 26th January 1995, another Hong Kong satellite, Apstar 2, carrying another Star 63F kick motor, was lost. 51 sec into the mission, there was an catastrophic explosion and the launcher and satellite were lost. Television pictures showed the rocket crashing in an ugly billowing cloud of red, yellow and black toxic nitric smoke. Six villagers died and 23 were injured. The mission was an insurance loss of €188 m, the premium being over 18%.

Mysteriously, the explosion appeared to start at the top of the rocket, not the bottom part that was actually firing at the time. The Long March was grounded while the problems were sorted out and more recriminations flew back and forth. In a repeat performance of what had happened the previous year, the next two launches went smoothly – Asiasat 2 (Hong Kong, November) and EchoStar (United States, December). Asiasat used a small Chinese final stage, the EPKM kick motor, to reach final orbit. Asiasat 2 was a Lockheed Martin – built satellite stationed over 100.5°E with 33 transponders providing television, telephone and related services for small terminals in China, the rest of Asia, the middle east and Russia.

The causes of the two satellite losses – Optus B-2 and Apstar 2 – were never satisfactorily resolved by the launching company nor the customer. Neither the American nor the Chinese accident investigation reports were published. The most plausible explanation for the two failures has been published by Clark who put forward two explanations.[10] One is that the manner in which the American comsat should be placed inside the top of the Long March was miscalculated. It was unable to stand the high air pressure that rockets undergo about a minute into their flight. Between 45 sec, when a rocket goes supersonic, and about 90 sec, it experiences maximum dynamic pressure and the greatest structural strain (after that, the air is thinner and the pressure diminishes). A related theory is that vibration overstressed the shroud at this point in the ascent. Or second, that there was a problem with the American upper stage motor, called a STAR 63F, which on both occasions exploded about a minute into the mission. The technical aspects of fitting an American motor may not have been properly understood. Whichever case it was, the problem was something for both countries to fix, but 'blaming the Chinese' alone was probably the least helpful approach.

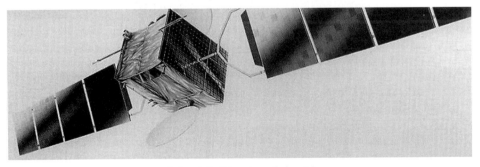

Asiasat 2. (Courtesy Asiasat.)

ST VALENTINE'S DAY MASSACRE

Then disaster intervened once more. 14th February 1996 saw the launch of a €70 m American Intelsat 708 advanced communications satellite. It was the first flight of the Long March 3B, a new version of the Long March 3 able to lift a record five tonnes to geosynchronous orbit. Although a new version, it relied heavily on well-tested rockets: the main stages were essentially those of the Long March 3A while the strap-on rockets had been verified on the Long March 2E. However, because the Chinese had received less than they had hoped in launch fees, they did not have the resources to make a test flight of the 3B before committing the new rocket to its first commercial mission. This decision proved calamitous.

Ground controllers were horrified as, a mere 2 sec after liftoff, the rocket began to tilt to one side, turned sideward and exploded in an enormous bang 2 sec later 1,500 m away, showering débris for miles around. There were two fatalities and a further 80 people were injured (later unconfirmed figures quoted as many as 56 fatalities). The crash was so shattering that no large pieces of débris were ever found. It was a very visible failure, screened instantly throughout the western world and provoking much comment about the temperamentality of Chinese rockets. Whatever about the Optus B-2 and Apstar 2 failures, this time no one could argue but that there was a fault in the launch vehicle.

Western investors and insurers predictably later called the episode 'the St Valentine's day massacre'. Two investigating committees were appointed and international experts invited to join. Four possible causes of the failure were indicated, including a broken wire supplying electrical power to the guidance system. By the end of the month, the China Great Wall Industry Corporation stated that the guidance platform had gone badly wrong, causing the accident. Another computer guidance problem was to cause the equally spectacular loss of Europe's brand new Ariane 5 on its maiden voyage less than four months later.

LOSS OF CONFIDENCE

With the crash of the Long March 3B, western investors lost confidence in the Chinese launcher system. At one stage, China had been heading for a 9% share of the international launch market. Now, satellites due for launch on Chinese rockets became uninsurable. Three American companies at once transferred their payloads to Atlas and Ariane launchers. So did an Argentinean company. Some customers went to the Russian Proton, which, after many years languishing in the commercial doldrums, was now enjoying a long order book. China made several approaches to India, offering the Long March at prices below Europe's Ariane, but without success.

China continued its efforts despite these severe setbacks. Some communications companies in the Asia Pacific region, especially those with direct links to China, had good political and territorial reasons to stay with the Chinese launchers, even if western companies bolted. Five months after the Intelsat disaster, in July 1996 the Long March 3 put Apstar 1A into its proper orbit. Then problems arose again. This

Long March 3B on the pad at Xi Chang.

time, only the following month, the Long March 3 was carrying an American-built communications satellite for the China Telecommunications Broadcast Satellite Corporation, the domestic communications supplier. During the burn to geostationary orbit, the third stage lost pressure and shut down 48 sec early, leaving the Chinasat 7 satellite stranded only half-way to its intended destination. The satellite was insured domestically for €304 m, which was paid up by the China Assets Insurance co that October.

Following this further failure, the Chinese instituted a rigorous programme for greater quality control and launch safety. By mid-May, engineers had managed to simulate the causes of the accident. While doing so, 44 further problem areas were uncovered. The programme was put under international quality standards (the ISO-9000 quality mark); there were additional quality checks; the guidance system was redesigned; and arrangements were made to evacuate areas near the pad as liftoff

Communications and conspiracies

Chinese commercial launches

Asiasat 1	Hong Kong	7 Apr 1990	First commercial launch
Badr	Pakistan	16 Jul 1990	Low Earth orbit
Optus B-1	Australia	13 Aug 1992	First western customer
Freja	Sweden	6 Oct 1992	Flown on FSW 1-4
Optus B-2	Australia	21 Dec 1992	Exploded
Apstar 1	Hong Kong	21 Jul 1994	
Optus B-3	Australia	13 Aug 1994	
Apstar 2	Hong Kong	25 Jan 1995	Exploded
Asiasat 2	Hong Kong	28 Nov 1995	
Echostar 1	USA	28 Dec 1995	
Intelsat 708	USA	14 Feb 1996	St Valentine's day massacre
Apstar 1A	Hong Kong	4 Jul 1996	
Chinasat 7	China	18 Aug 1996	Stranded
Agila 2	Philippines	20 Aug 1997	First LM-3B success
Apstar 2R	Hong Kong	16 Oct 1997	
Chinastar 1		30 May 1998	
Sinosat 1		18 Jul 1998	

neared. A quality control company was called in, the New Decade Institute. A 28-point quality control system was adopted, as was a 72-point regulation for management and production. The Chinese insisted that an international team of French, German and British experts approve the reforms. 256 specific changes were carried out. China sent a team to Britain and the United States to explain how the rocket had now been improved.

These measures obviously paid off, for on the early morning of 20th August 1997, the Long March 3B eventually made its début, lofting a comsat for the Philippines. The placing of the satellite in orbit raised a few eyebrows, for the apogee was 44,500 km, far above the 36,000 km norm. Was this a bad engine burn, another malfunction? In fact, this manoeuvre marked the introduction of what is called the supersynchronous orbit, a hitherto unadvertized feature of the Long March 3B (and the 3A as well). This is a technique of performing a very precise, carefully calculated, extra-thrust burn out to 44,000 km or so, one that produces a subsequent saving on the next manoeuvre, reducing the angle of inclination to the equator from 28° to zero. Several subsequent missions used this supersynchronous manoeuvre.

The following year, the Long March 3B put its troubled early history behind and effortlessly placed two more commercial satellites into orbit. More were in the pipeline, like Chinasat 8, a satellite built by Loral for a Chinese domestic communications company. Then all hell broke loose.

COX REPORT

Barely had the Long March 3B triumphantly returned to flight than China became embroiled in an acrimonious dispute with the United States, one heralding a long period of difficult relations.

On 18th June 1998, the House of Representatives voted 409 to 10 to set up a 9-strong special committee to investigate the transfer of space technology to China. Appointed as chairperson of the select committee was California Republican Christopher Cox. The congress gave him a budget of €3 m to facilitate his investigation, the largest since the Iran contra affair in the 1980s.[11] The investigation arose from rising concerns that China had taken advantage of its contacts with the American space industry to acquire information useful for the construction and targeting of ballistic missiles. Specifically, the Loral and Hughes companies were alleged to have insufficiently protected their satellites in transit to the launchpad. Hughes satellites used advanced technology arrays that could be used for electronic signals gathering. To add a partisan political element, President Bill Clinton had given export clearance to Loral's Chinasat 8 to fly on a Chinese rocket because, it was alleged, of a generous $1 m donation given to the Democratic party by Loral's president.

The setting up of the investigation prompted bitter but largely inconclusive debates in Washington. Strictly speaking, the debate revolved around whether China was engaged in spying, obtaining classified information, applying it to an aggressive military rocket programme and compromising security-slack American companies in the process. In practice, the debate was a proxy for a broader political debate about the Clinton administration's China policy and whether that should be one of hostility, isolation, containment, trade or engagement with China. To complicate matters further, American commercial launcher companies stood to gain from the revoking of satellite export licences to China. By contrast, the satellite manufacturers wished to deliver satellites on orbit to their customers at the lowest possible price. Here, Chinese prices were much lower. However, the lobby that the satellite manufacturers could organize was quite weak compared to that of the Republican hawks in the congress.

When the Cox committee began its work, Republicans specifically alleged that:

- The Department of Defence had failed to monitor three satellites in transit;
- When the Long March 3B exploded in 1996, the Chinese kept the Americans away from the crash site for five hours while they ransacked the American débris. Encrypted chips on the lost Intelsat 708 were never found;
- In the subsequent investigation, Loral officials sent their comments on the accident to the Chinese without US government approval – comments that inadvertently may have helped the Chinese to better design military rockets. Loral also held two meetings with the Chinese investigation team without the presence of US Defence Department officials;
- China had applied what it had learned from the Asiasat 1 launch in 1990 to build clean rooms that would be used to prepare military satellites;
- The lessons learned from the Long March 2E explosions about rocket fairings would be used to prepare the flight trajectories for ballistic missiles. Hughes had sent the Chinese its version of the accident without the approval of American national security agencies;
- In developing perigee kick motors for American communications satellites, China learned how to better target missile warheads;

126 Communications and conspiracies

- The techniques used to deploy the Iridium series of communications satellites (see later) were directly applicable to target nuclear warheads.

The White House, State Department and Democrats countered by saying that most of these claims were fanciful. The United States gained more from its contact with China and got a grandstand view of its space programme and its capabilities – something which just could not have been done if China were isolated, they said. The Cox committee was set up at the time of an upcoming visit by President Clinton to China and was intended to embarrass him, critics countered.

The temperature rose as the investigations got under way. American government agencies were accused of shoddy oversight of sensitive American satellites *en route* to launch in China. Loral was even accused by one congressman of engaging not in trade, 'but treachery'. Hughes responded with an elaborate description of how its satellites were kept under round-the-clock surveillance in China, protected by no less than 11 security guards from Pinkertons and explained how the Chinese could not have put as much as a finger on the precious cargo without their knowledge.

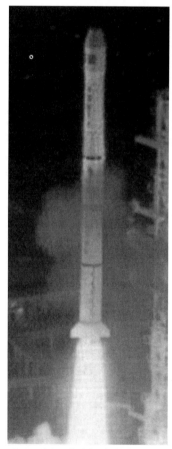

Long March 3A night launch.

In the frenzied atmosphere of the mid-summer, Congress acted promptly, without waiting for the outcome of the Cox report. The congress did not ban the export of satellites to China outright, but, reclassifying them as munitions of war, did transfer responsibility for their licensing from the Department of Commerce to the Department of State, to ensure that defence considerations were uppermost in licensing decisions, rather than trade. This in effect brought the arrangements back to where they had been in 1992. The congressional decisions nevertheless had the politically desired effect of slowing satellite trade with China almost to a standstill. The Department of State did not have sufficient officials to process export licences, so approvals slowed to waiting periods of 18 months or more, making flying satellites on Chinese launchers an unattractive proposition. Any export worth over $50 m to China also had to get congressional approval in any case. The Asia Pacific Mobile Telecommunications Company, which planned to fly two Hughes satellites on Long March, had to cancel. The Loral-built Chinasat 8, although authorized by President Clinton, was blocked by the State Department and remained on American soil.

In December 1988, the Department of Defence accused Hughes of revealing sensitive information to China, thus damaging to American national security when it passed on its analysis of the 1995 Long March 2E crash. The atmosphere of alarm intensified when a Los Alamos scientist, Wen Ho Lee, was charged on 59 counts of stealing nuclear secrets for China.

A classified version of the Cox report was eventually published six months later in January 1999.[12] An unclassified one, shorter by a third, followed that May. The Cox report painted a lurid picture of malevolent Chinese espionage going back to the day Tsien Hsue Shen fled the United States for communist China. According to Cox, the Chinese had used, over decades and in a systematic way, fair means and foul, neutral scientific conferences, licensing arrangements, dual-use military-civilian technologies and straightforward spying to ferret out information on nuclear technology, computers, rockets, submarines and atomic bombs for decades. Hughes and Loral were attacked for exceeding the terms of their export licences and carelessly giving away information that would enable China to improve the guidance systems of ballistic missiles.

China acted in a remarkably restrained way under the circumstances, merely releasing a statement that it developed its rocketry and upper stages through its own unaided efforts, without foreign assistance. It called the language of the Cox report unwarranted and inflammatory, mild comment under the circumstances.[13] The Clinton administration dismissed the report as alarmist and exaggerated. For all their alleged thieved secrets, Chinese military capabilities were still decades behind the United States, it said. China accused Republican hawks to trying to build a state of military tension. Observers of the Chinese space programme found the Cox report to be full of historical errors, which for them called into question the reliability of the analysis and conclusions reached.[14] The Cox report had a series of low-grade errors (e.g. wrong dates for space missions and political events; incorrect conversions between metric and imperial units) to more serious mistakes (e.g. that China would fly cosmonauts into space on Soyuz capsules bought in Russia). Some of the Cox

Hardest hit: the Long March 2E.

report's investigations of historical events were quite wide of the mark. Cox had Tsien Hsue Shen fleeing to the United States with classified documents (in fact, he left his papers behind) and his early familiarity with the design of the Titan rocket (five years before the tender for Titan was even issued). Five experts from Stanford University issued a 99-page rebuttal.

The Cox report had the effect – critics would say its desired effect – of halting the commercial exploitation of the Long March rocket in its tracks. Two final commercial launchings were made by China in 1998 – Chinastar 1 and Sinosat 1. Chinastar 1, also called Zhongwei 1, was built by Lockheed Martin for the China Orient Co and reached its desired orbit at 87.5°E. It was still operating several years later. Sinosat 1 was designed for use by the Civil Aviation Administration, People's Bank and Shanghai TV. It was launched in July 1998 and reached its destination of 110.5°E. Both had been given clearance before the changes instigated by the Cox report. For the time being, they would be the last.

The House committee concluded that Loral and Hughes had deliberately acted without the legally required license and violated U.S. export laws in helping China. Both companies denied committing intentional violations. Loral did acknowledged that it failed to get approval but insisted it was inadvertent and that it disclosed the mistake to the State Department.

In late 2000, relationships between the United States and China began to ease –

Ground comsat terminal.

though this was to prove temporary. On 21st November, following a Chinese commitment not to assist certain countries in the development of ballistic missiles, the State Department, announced that it proposed to lift restrictions on Chinese launchers and that the quota limits on Chinese launchings would be permitted to expire at the end of the year. However, as long as the export régime persisted, these changes were largely of academic value. Wen Ho Lee was eventually released, following a plea-bargain in which he admitted to one of the 59 charges: using the internet improperly to search for information.

Suggestions that the export régime had more to do with the domestic American battle between the launcher lobby and the satellite lobby were confirmed when the Centre for Responsive Politics published details of Republican campaign contributions, for they included American launcher companies Lockheed Martin and Boeing. However, the distinction between launcher and satellite manufacturer became less distinct when the launcher and aircraft company Boeing took over the main American manufacturer of satellites, the Hughes Corporation, a company dating back to the days of millionaire hermit flying boat designer Howard Hughes. In early 2001, divisions began to open up among the Republicans between free traders and national security devotees. In March 2001 the conservative think tank the Heritage Foundation issued a paper taking the side of the satellite manufacturers in the controversy. Information also came out that both Loral and Intelsat had submitted proposals for the exports of satellites to China. There were unconfirmed reports of other countries or companies hoping to go ahead with Chinese launches (e.g. Italy, Eutelsat).[15] Pressure was beginning to build for a change.

By summer 2001, the Congress was talking of rowing back on the legislation introduced three years earlier – with Republicans leading the move for change! Why the change of mind? Up to 25,000 people worked in the satellite business in California and there was a perception that satellite sales had now began to move to Europe where they would be less affected by American export clearance procedures.

Republicans said that they had never wanted to destroy the satellite industry and were now happy to trade with China if that did not undermine national security. The satellite building industry, not a powerful lobby in comparison to others, complained that it was losing business to Europe where export controls were lighter. In August 2001, the House of Representatives voted for responsibility for export controls to be returned to the Commerce department.

Despite this, in July 2001, four fellow Republicans countered by writing a letter to President Bush asking him not to relax the curbs on China, saying that China was still selling missile technology to rogue nations. At this stage, four satellites were reported to be awaiting export clearance to fly on the Long March. The precise nature of the commercial communications satellite launch queue has always been difficult to identify, as most communications companies naturally prefer to negotiate their deals privately. The Great Wall Industry Corporation stated that it had three firm contracts for the Long March 3B for the 2003–6 period and two possibilities. One was a French-built satellite (which may have legally eluded American export controls), but another was thought to be in its fifth year of waiting for an export licence.

However, it still wasn't over for the American companies involved in the earlier Long March episodes. Tidying up the legal consequences took years. In autumn 2001, Loral moved to a civil settlement with the State Department whereby it would pay a penalty of $14 m, admit it had made mistakes in passing information to the Chinese and give $6 m in commitments to improving its internal security.[16] Final settlement was reached in January 2002.

Chinese – American relations took another nosedive in 2002. When the world space congress took place in Houston, Texas in October 2002, the American government barred half the Chinese representatives from attending, either refusing visas or putting other bureaucratic obstacles in their way. Those who did attend were searched or closely followed by a dozen FBI agents hired to mind them. The head of the Chinese delegation, Luan Enjie, the director the China National Space Administration, was left stranded in Canada. At around this time, the Chinese had been taking soundings about the possibility of having some experiments fly on the International Space Station. NASA tried to respond, new administrator Sean O'Keefe reportedly having discussions with the State Department with a view to exploring possibilities. One leading congressman was having none of it and was quoted by the press as saying that he would not tolerate 'a bunch of nazis running around our space station'. When later that year O'Keefe was asked about cooperation with China, he told journalists that he was happy to cooperate, but, he added stiffly, was bound by the rules laid down by the State Department.

As if to rub in the point, in 2003 the State Department filed a $60 m fine against the Hughes Corporation for no less than 123 alleged improprieties many years earlier in the 1990s. Hughes was now part of the empire of aircraft maker Boeing, which stood to lose as well. Hughes was adamant that it had done nothing wrong and the company's refusal to plea-bargain had greatly angered the state department. One of the new charges was that a translator used by Hughes was the son of a top Chinese general. One of the ironies was that if the American company *had* been allowed to

Luan Enjie.

transfer sufficient data about its satellites, then it would have been properly integrated with its Long March launcher and at least one explosion might never have happened.

Relations between China and the United States continued poorly during the Bush presidency. There was eventually some evidence that they were beginning to hurt the United States. Several communications companies began to talk to European satellite manufacturers (principally Alcatel) with a view to having future satellites built 100% in Europe. If they could be built without a single American part, then they would not need to file for export clearance.

IRIDIUM

Missions to low Earth orbit were much less affected by the political drama. Here, China was successful in 1995 in negotiating a deal with the Motorola corporation. It booked a series of Long March 2C launches to low Earth orbit for 22 of its revolutionary new global communications system of Iridium satellites (Great Wall also took a 5% interest in the Iridium project). The idea behind Iridium was that people would use Iridium like a mobile phone, linking far-flung businessmen with their distant offices, messages being communicated from one satellite to another in relays encircling the world until they reached the appropriate downlink point. It would offer quality voice, global, mobile phone services, using satellites rather than masts.

The Long March 2C was one of China's most reliable launchers, having a success rate of 100%. It was adapted with a special dispenser (SD) for the Iridium system, able to launch Iridiums in pairs and was renamed the CZ-2C-SD. The Motorola corporation contracted China to launch 22 satellites in the 66-satellite Iridium constellation of comsats (the others being selected to fly the American Delta and the Russian Proton). The new fairing on top increased the height of the rocket to nearly 40 m. A test of the SD system was made on 1st September 1997 when the Long March CZ-2C-SD, flying out of Taiyuan launch site, discharged two mockup Iridium satellite at appropriate points in its orbit, clearing the way for the first operational mission.

The project then proceeded to operational missions. Over 1997–9, China launched its Long March 2C-SDs six times, putting into orbit 12 Iridiums from Taiyuan. Everything went completely smoothly – until the Iridium project collapsed in bankruptcy in 1999. People stuck to their ordinary mobiles, bad lines and all, and disliked the heavier Iridium handset. The Chinese had been paid by this stage. Great Wall must have counted itself lucky that its investment in the company was only 5%. In 2001, the US Department of Defense took over the Iridium lines for a military communications experiment, leaving the company free to attempt to attract paying customers again. Iridium's other legacy was of course visual. Although the Iridium satellites were not particularly large (between 650 and 670 kg), they had a big solar panel, which, as it turned in orbit, created a bright 3–4 sec flash frequently visible from the ground in evening skies to astronomers and casual skywatchers. They became known as 'Iridium flares'.

Iridium

Demonstration	1 Sep 1997	CZ-2C-SD	Taiyuan
Iridium 42, 44	8 Dec 1997	CZ-2C-SD	Taiyuan
Iridium 51, 61	25 Mar 1998	CZ-2C-SD	Taiyuan
Iridium 69, 71	2 May 1998	CZ-2C-SD	Taiyuan
Iridium 3, 76	19 Aug 1998	CZ-2C-SD	Taiyuan
Iridium 11A, 20A	20 Dec 1998	CZ-2C-SD	Taiyuan
Iridium 14A, 21A	11 Jun 1999	CZ-2C-SD	Taiyuan

In a sequel to Iridium, China signed a contract with the Korea Aerospace Institute on March 2001 to launch the KOMPSAT-2 satellite for Korea in 2004. The launcher will be a development of the CZ-2C-SD, the CZ-2C-CTS. KOMPSAT is a small Earth observation satellite with 1 m resolution for panchromatic images and 4 m resolution for mutli-spectral images and is to be put into a 685 km high sun synchronous 98.13° polar orbit.

China can still offer competitive prices for its launchers on the world market. In the face of American sanctions, its chances of breaking into the world launcher market, never good, became ever slimmer. By the mid-2000s, the world launch market was glutted by a wide range of American, Russian and European launchers, Russia being especially successful in marketing veteran Soviet rockets and converted missiles left over from the cold war. For the record, the following are the ballpark prices expected for Long March launches.

Prices of the Long March launcher fleet

Long March 1D	750 kg to LEO	€14 m
Long March 2D	3,300 kg to LEO	€35 m
Long March 2E	2,494 kg to GTO	€59 m
Long March 3	1,400 kg to GTO	€47 m
Long March 3A	2,500 kg to GTO	€59 m
Long March 3B	4,800 to GTO	€82 m
Long March 4	4,000 kg to LEO	€47 m

LEO = Low Earth Orbit
GTO = Geostationary Transfer Orbit

For comparison, costs for geostationary transfer orbit for the European Ariane 5 launches are around €140 m, depending on the requirement and payload. Russian Proton launch costs are in the €40 m to €82 m range and for the American Delta 3 around €100 m.

The Long March 3B disaster, followed by the breakdown in trading relationships between China, the United States and its allies, ironically led to a period of sustained technical success in the Chinese space programme. Following the St Valentine's day massacre, China had 34 straight orbital launch successes in a row in the following seven years. All the successful launchings were domestic missions which were of course unaffected by the international contretemps. However, the American sanctions meant that Chinese companies were no longer generating the foreign earnings that would otherwise have supplemented these endeavours.

ASSESSMENT AND CONCLUSIONS

The development of the Long March 3 series, the use of hydrogen fuels, the building of geosynchronous communications satellites were ambitious steps for the Chinese space programme at an early stage. The scale of effort involved was considerable. Careful development, rigorous testing and quality control lead to early success. By the early 1990s, Chinese launchers had become a part of the world launcher market and had established a niche in orbiting Asian, Australian and some American payloads.

These achievements were undone. First, there were two disasters in a row with the Long March 2E, in which externally fitted engines or shrouds appear to have led to the loss of the mission. Second, the explosion of the Long March 3B on its first mission was an own goal at the worst possible time. Third, the Chinese space industry became the victim of domestic American politics and later the on-going redefinition of America's foreign relationship with China, as a result of which the Long March commercial launcher fleet was effectively grounded. Although the global situation was not as extreme as the 1950s and 1960s, the Chinese space programme was very much forced back onto its own resources. This was a situation with which, as chapter 6 shows, it was well able to cope and demonstrate its prowess.

REFERENCES

1. Chen Hyi: *Into outer space*. China Pictorial Publishing Co, 1989, 156.
2. For a detailed analysis of the behaviour of the early Chinese comsats, see Phillip S Clark: China's Dong Fang Hong 2 and 2A communications satellite programme. *Journal of the British Interplanetary Society*, Vol 54, 2001.
3. Zhang Yun (Ed): *The Chinese space industry today*. China Social Sciences Publishing Co, Beijing, 1986.
4. Wei Long: Happy birthday *Dong Fang Hong 3* comsat. Spacedaily. http://spacedaily.com/news/china.
5. Phillip S Clark: *China's Shenzhou programmes reaches flight status*. Monograph, January 2001; New Chinese telecommunications satellite operational. BBC reports, 3 May 2000.
6. Bill Gertz: Chinese civilian satellite a spy tool. *Washington Times*, 1st August 2001
7. BBC reports, 20th May 1998; Peter de Selding: Chinese set ambitious space plans. *Space News*, 14-20 October 1996.
8. For example, How creative can you be in marketing your products? Now the sky's the limit! *Aviation Week & Space Technology*, 15th March 1993.
9. China faces commercial launch ban. *Spaceflight*, 32, April 1990; James Asker: China, US renew Long March pact. *Aviation Week & Space Technology*, 6 February 1985. The best summary of these issues may be found in Lawrence H Stern and Jack High: America takes a long march into space. *Spaceflight*, 32, April 1990.
10. Philip S Clark: *The Chinese space programme – an overview*. Molniya Space Consultancy, 1996, 124-7.
11. For a running account of these developments, see James Asker & Joseph Anselmo: China launch controversy expands to larger issues. *Aviation Week & Space Technology*, 22nd June 1998; US broadens probes of China tech transfer. *Aviation Week & Space Technology*, 29th June 1998; Joseph Anselmo: Hughes defends China security. *Aviation Week & Space Technology*, 6th July 1998; Congress nears decision on China satellite laws. *Aviation Week & Space Technology*, 13th July 1998.
12. China says satellite launch system developed without US firm's help. BBC reports, 19th April 2000.
13. A contemporary review of these events may be found in Johanna McGeary: The next cold war? *Time magazine*, 7th June 1999.
14. James Oberg: Mistakes can undermine conclusions. Special to ABC News, 4th June 1999. For a detailed and balanced treatment of the issue, see James Oberg: Year of the rocket. *IEEE Spectrum*, May 2001.
15. Christian Lardier: Levée de l'embargo sur les lanceurs chinois. *Air & Cosmos*, décembre 2000; Satellite exports getting review. *Los Angeles Times*, 4th June 2001; China to launch Alenia Aerospazio satellite. *Spacedaily*, 18th October 1999, http://www.spacedaily.com/spacecast/news.
16. Andy Pastor and David S. Cloud: Loral nears civil settlement with U.S. over sharing technology with China. *Wall Street Journal*, 31st August 2001.

6

Applying the space programme

Communications satellites were not the only area where China attempted to apply its know-how in space research to the domestic economy. During the 1980s and 1990s, China extended its domestic satellite space programme to weather forecasting, Earth observation and navigation satellites. The FSW series, developed in the 1970s, continued to provide direct benefits for the Chinese economy. Microsatellites became a new line of development and China amply demonstrated its ability to reach and even exceed high standards in evidence elsewhere in the world – all the more remarkable for having been achieved within an economy forced back largely onto its own resources.

METEOROLOGICAL SATELLITES (*FENG YUN*)

The development of weather satellites was a logical path for the Chinese. Accurate weather forecasting had always been important for a large country so dependant on agriculture in an area vulnerable to damaging storms and floods. China suffered heavily from storms, flooding and weather related natural disasters, with losses to the economy of up to ¥131 m a year – so anything that could be done to reduce that figure would be helpful.

The United States launched the first weather satellite in 1960 (Tiros) and the Russians followed with an operational system later (Meteor, 1969). The government approved the concept of a Chinese meteorological satellite in February 1970, but development was impeded by the cultural revolution. China set up its first station to receive internationally available meteorological data in 1970 in Beijing, stations being built subsequently in Urumqi and Guangzhou.

The Chinese first expressed a public interest in weather satellites when they revealed the existence of a programme for meteorological satellites called Qi Zing (though this name never reappeared). The first funds were not allocated to the programme until April 1978. The project had been supported by the Central Meteorological Bureau.

In the event, China's first weather satellite was named Feng Yun ('wind and

136 Applying the space programme

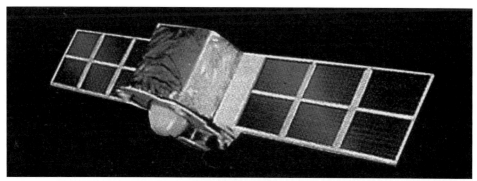

Feng Yun 1 series.

cloud'). Meng Zhizhong was appointed chief designer of Feng Yun. The satellite was built by the Shanghai Academy of Space Technology, SAST. The Feng Yun 1 satellite was hexagonal, 1.76 m tall, 1.4 m wide, weighed 757 kg and had two solar panels spanning 8.6 m. The satellite had a scanning radiometer designed to monitor clouds, water colour, crops, forests and pollution. Pictures were transmitted automatically in real time and by tape recorder. Small nitrogen-powered thrusters were used to ensure the satellite point the right way. The Chinese experienced much difficulty in devising a satisfactory system of gyros to orient the spacecraft and they eventually bought some American equipment. The satellite was designed to orbit the Earth 14 times a day.

Feng Yun 1 in preparation.

It was decided to fly Feng Yung into polar, sun-synchronous orbit. It is polar, because it crosses the planet on its south-to-north axis; and sun-synchronous, because it follows the same ground track each day and crosses the same point on the Earth's surface at the same time each day. Targets are illuminated by the same Sun angle. This makes it easier to compare weather data from one day to the next. A new launch centre was required for a satellite to enter polar, sun-synchronous orbit. Jiuquan was not suitable, for launching northward would take the departing rocket quickly over Russia. Xi Chang was too far south. Accordingly, a former missile site near the industrial coal town of Taiyuan, south-west of Beijing, was selected.

A new launcher was developed for the Feng Yun. The Long March 4 was designed by Sun Jingliang. Whilst the Long March 3 was too powerful for this mission, the Long March 2C had insufficient thrust to send the payload into polar orbit. The Long March 4 was based on the highly reliable Long March 2C, but with a new, more powerful upper stage. 41.9 m tall at liftoff, with a thrust of 300 tonnes, the Long March 4 had a small third stage using conventional fuels.

FIRST POLAR, SUN-SYNCHRONOUS, TAIYUAN, LONG MARCH 4 LAUNCH

The Long March 4 pad at Taiyuan. (Courtesy China Astronautics Publishing House.)

138 Applying the space programme

The first Chinese weather satellite some time to appear. Eventually, it was launched on 6th September 1988. The Feng Yun 1 launching achieved a number of milestones. It was the first use of the Long March 4 rocket, the first launch from the new launch site of Taiyuan and the first satellite to enter polar, sun-synchronous orbit. Feng Yun 1-1 entered an orbit of 99.12°, 881–904 km. It soon sent back pictures of cyclones, rainstorms, sea fogs and mountain snow. In addition to its meteorological role, Feng Yun-1-1 carried additional experiments. The main spacecraft carried equipment to detect cosmic rays, protons, alpha particles, and carbon, nitrogen, oxygen and ion particles in the Earth's radiation belts.

Feng Yun 1-1 was less than entirely successful. One of the radiometers did not work and the main spacecraft failed after 39 days. Apparently, condensation in the spacecraft had not been fully removed before it left Earth and this fouled up the sensitive radiometer. Until then, Chinese scientists had been very satisfied with its performance.

FENG YUN 1-2 CARRIES BALLOONS

Feng Yun 1-2 was launched two years later on 3rd September 1990. It featured a number of improvements and was heavier (881 kg). Feng Yun-1-2 carried two additional experiments – two balloons called Qi Qui Weixing 1 and 2 (the names Da Qi have also been used). Their purpose was to measure the density of the upper atmosphere between 400 km and 900 km. Measuring 2.5 m and 3 m in diameter, deployed in similar orbits, they decayed from orbit the following year, one in March, the other in July.

Feng Yun 1 panel test.

The weather satellite itself appeared to suffer radiation damage in February 1991 possibly from a solar flare, but after a 50-day struggle, ground control in Karshi performed a minor miracle by recovering the satellite fully. There was further radiation damage later in the year and the data eventually became unusable.

By way of a postscript, the upper stage of the launcher that had put Feng Yun 1-2 in orbit, the Long March 4A, exploded on 4th October 1990 when propellants leaked through the bulkhead into the oxidizer and ignited. This was an unwelcome development, for the world's space powers had begun to realize the threat which orbital débris of this kind caused to manned space station operations. The American space agency, NASA, had even formed an office in Houston, Texas, with the brief of trying to reduce space débris. In 1995, China joined the Inter Agency Space Débris Coordination Committee.

However, the Chinese always made it clear that the first two spacecraft were tests before the system became operational with Feng Yun 1-3. This satellite was successfully launched in May 1999. Its first images were returned the same day and by July good quality pictures were flowing in on all channels. A top priority for the mission was to ensure a working life measured in years, rather than months. Many precautions were taken against radiation damage and the satellite was first sent to the Lop Nor nuclear test site for checking against a recurrence of the problems that had plagued its predecessors.

In this the Chinese more than succeeded, for when Feng Yun 1-3 exceeded its design life two years later, it was still working, in a stable condition, with good power supply and still returning clear pictures, the longest of any Chinese metsat. Feng Yun 1-3 was crossing China every morning at 8.30am. The weather satellite carried a 10-channel multichannel visible and infrared scan radiometer – four in visible wavebands, three in near infrared, one in short-wave infrared and two in long-wave infrared, with a resolution of 1.1m. In its first two years, it collected in a range of information on floods, droughts, forest fires and ice, its cameras picking out the sandstorm that blew from Mongolia into China that April. In summer 2003, pictures from the satellite shaped the responses of the authorities to the worrying floods on the Huaihe river.[1]

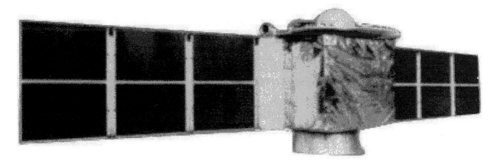

Feng Yun 1. (Courtesy *Aerospace China*.)

140 Applying the space programme

Because it had exceeded its design life and was continuing to work satisfactorily, its replacement satellite was delayed. Feng Yun 1-4 was eventually launched on the Long March 4 on 15th May 2002. Its purpose was to take over the role of Feng Yun 1-3. The new satellite, the last in the Feng Yun 1 series, carried a 10-channel scanning radiometer. On its second pass, it sent back its first pictures (of clouds in the event) down to the Urumqi station.

HAIYANG – NEW OCEAN SATELLITE

Feng Yun 1-4 was notable, for it also brought piggyback into orbit the first marine survey spacecraft, Haiyang 1. This was a small, 365 kg oceanographic satellite that used the same CAST968 bus as Shi Jian 5. The original orbit with Feng Yun 1 was not suitable for Haiyang, so during the last week of May a motor lowered Haiyang's altitude to an operational height of 792–795 km, 100.7mins. It carried a 10-band, 3-dimensional ocean colour scanner and a 4-band charge coupled device scanner.

Haiyang 1 was designed with a two year lifetime to observe chlorophyll concentrations, surface temperatures, silting, pollutants, sea ice, ocean currents and aerosols. The scanner covered the same path over the Earth's surface every three days in 1,164 km wide swaths and data were transmitted both in real time (22 mins at a time) and by tape recorder. The satellite was a success and relayed back high quality images, from the Strait of Qongzhou to Mexico Bay.

Haiyang was developed for the Science and Technology Department of the State Oceanic Administration and planned as the first of a series of regular launchings of

Haiyang 1.

observation satellites able to photograph the ocean in 3-dimension colour images. The Haiyang series will monitor the seas, tidal zones, offshore sandbanks and the marine environment, picking out pollutants and sand pouring into the sea. The Chinese announced that there would be three Haiyang series, 1, 2 and 3, a total of eight satellites in all by 2011.

The next meteorological programme is a new generation of at least two satellites, called the Feng Yun 3 series. These 2,500 kg observation satellites will carry a 15-channel scanning radiometer, infrared and micrometer radiometers, microwave imagers, microwave humidity detectors, medium resolution imaging spectrometer, altimeter and spectrometer, with a resolution of 250 m. Launch dates were set for 2005, 2006 and 2008. Images will be sent back 48 times daily.

GEOSYNCHRONOUS METEOROLOGICAL SATELLITE

For the Chinese, the next stage was to operate a weather satellite in geosynchronous orbit. Called the Feng Yun 2, this would complement the Feng Yun 1 series. The concept was that Feng Yun 2 would send back constantly scanning pictures of China and the western Pacific from its high vantage point 36,000 km out, while Feng Yun 1 would send back detailed weather maps from its lower, regular 100 min passes over China from an altitude of 900 km.

Geosynchronous meteorological satellites are expensive, requiring a big launcher, and high operating standards of the satellite concerned. However, the vantage point

Feng Yun 2.

of 36,000 km can provide quality weather coverage of large land masses round the clock. The United States operated its first geosynchronous metsat in 1974 (Synchronous Meteorological Satellite 1), Japan and Europe in 1977 (Himawari and Meteosat respectively) and Russia not until 1994 (Elektro).

Feng Yun 2 was a drum-shaped satellite 4.5m tall, with a diameter of 2.1 m and a weight of 1,380 kg. It carried a multi-channel scanning radiometer made by the Institute for Technical Physics in Shanghai, a cloud coverage information system and data collection translator. It was intended to provide cloud, temperature and wind maps from a vantage point of 105°E, the location planned for the series. It was developed by the Shanghai Aerospace Technology Research Institute of the China Aerospace Corporation and built in the Hauyin machinery plant.

After several years of operation, the two satellites of the Feng Yun 2 series will be replaced by a new generation, the Feng Yun 4, each with 5-channel instruments, two more than at present.

FUELLING DISASTER IN THE PROCESSING HALL

The Feng Yun 2 series got off to a disastrous start. When Feng Yun 2-1 was being loaded with propellant in the processing hall at Xi Chang launch site on 2nd April 1994, the satellite exploded, killing one technician and injuring 31 others. The satellite itself, valued at over €88m, was of course a write-off. It took over three years to redesign the propellant tank system so as to make sure this accident would never happen again.

Some American electronic components were installed in the replacement satellite which was eventually launched on the troublesome Long March 3 rocket from Xi Chang on at 9 pm Beijing time on 10th June 1997. 23 mins after launch, the hydrogen – powered third stage fired to send the 1.38 tonne metsat on its way to its permanent position at 105°E, with a scheduled lifetime of three years. It was intended to provide cloud maps, temperatures and wind movements, leading to much improved forecasting. It also carried a solar x-ray spectrometer and space particle detector. By September it had completed its full range of systems testing and was ready for handing over to the state meteorological administration.

Feng Yun 2-1 lasted until 10th April 1998, only six months of full operations, when problems arose. Ground controllers managed to regain control at the end of the year, but the resumption was limited to six images a day. Contact was off and on during the year, good images being returned from time to time. In March 2000, meteorological operations with the satellite appear to have ended and the satellite was moved to 86°E by the end of April. Station-keeping manoeuvres continued there so it must have continued to return some data.

The gap in operations did not last for long, for a replacement satellite was lofted into orbit by Long March 3 on 25th June 2000. 2-2 soon arrived at 2-1's old station. Its main instrument was a visible and infrared radiometer able to scan the globe every 25 mins in visible, infrared and water vapour. Data were also collected from automatic weather platforms at sea. 2-2 was placed carefully on station at 105°E. A month later, following on-orbit testing, the imaging systems were turned on by the National Satellite Monitoring Centre. 25 minutes later, after they had completed a full scan of the Earth, full disk images in colour, infrared and water vapour came flooding in to the centre, showing clouds swirling over south east Asia, a clear view of southern Australia and tropical storm Tembin menacing Japan. Resolution was as sharp as 5 km, which, from 36,000 km out, wasn't bad. Three images in each format were, from thereon, sent to the monitoring centre every 25mins. Feng Yun 2-2 was also designed to monitor solar radiation – a solar x-ray spectrometer and space particle detector. It is expected to work until 2004 when it will be replaced by 2-3 (or 2C) in 2004.

By the new century, the metsats had provided a high return on their investment. Advance weather warnings had led to the following benefits:

- Advance warnings on when typhoons would cross from sea onto land
- Flood diversion measures were avoided (involving the preventative flooding of agricultural land) when metsats showed that rainfall had already passed its peak
- Weather warnings led to ships returning to port in time, rice being harvested early and river banks being reinforced
- Satellite images enabled river managers to better control the flow of water through dams
- Instruments on Feng Yun permitted the prediction of Earthquakes

Chinese weather satellites

Feng Yun 1-1	6 Sep 1988	Failed after 39 days – poor data return
Feng Yun 1-2	3 Sep 1990	Carried Qi Qui Weixing balloons
Feng Yun 2	2 Apr 1994	Exploded during fuelling
Feng Yun 2-1	10 Jun 1997	Positioned 105°E.
Feng Yun 1-3	10 May 1999	
Feng Yun 2-2	25 June 2000	Arrived at 105°E
Feng Yun 1-4	15 May 2002	Haiyang launched piggyback

RECOVERABLE SATELLITES: THE NEW FSW 1 SERIES

Chapter 4 examined the introduction of the recoverable satellite programme, the FSW (Fanhui Shi Weixing). The FSW 0 series, first flown in 1975, had made nine missions by August 1987, all successful.

The FSW 1 series was introduced in September 1987, barely a month after the conclusion of the FSW 0 series. Compared to the FSW-0 series, the '1' series was heavier (2,100 kg), with a greater payload (180 kg) and able to orbit up to ten days (although eight days was the norm). A digital control system was introduced, new gyroscopes were added to help control attitude, new sensors were added, the satellite could be re-programmed when in orbit, a control computer was installed and the pressure inside the cabin could be regulated. According to an assessment of the

144 Applying the space programme

programme ten years later, the FSW 1 series 'was a photogrammatic satellite [which] could take photos with high geometrical accuracy, accurately locate ground targets and draw a map'. Later, the FSW was confirmed as a cartographical and mapping satellite.[2] A comparable Russian satellite was the Kometa.

The first mission, FSW 1-1, was devoted to microgravity experiments – seeing how algae would grow in orbit, processing gallium arsenide – and it is not known if the remote sensing package was even carried. FSW 1-2 was dual-purpose, carrying both the Chinese remote sensing package and a German protein crystal growth experimental package called Cosima. The German experiment, developed by Messerschmitt Bolkow Blohm and the German space agency, DFVLR, was intended to find new ways of producing the medical drug interferon from large and pure protein-based crystals. Germany paid China €440,000 for the mission. The Germans were handed back their package the day after landing. Subsequently, a joint company was formed between German Aerospace (DASA) and the China Aerospace Corporation, for the development and marketing of Earth observation satellites.

CHINA FLIES ANIMAL PASSENGERS INTO ORBIT

Guinea pigs and plants were carried on FSW 1-3 as part of a microgravity experiment. In doing so, China became the third nation to send animals into orbit and recover them. FSW 1-4 carried a Swedish satellite, Freja, piggyback into orbit while the main spacecraft carried Chinese and Japanese microgravity experiments (the latter being a 710° C microgravity furnace). The Chinese ones involved testing how rice, tomatoes, wheat and asparagus would grow in orbit (apparently, they grew much faster).

In 1988, the Chinese revealed that one of the missions in the FSW series was effectively lost when the second stage vented used propellant over the cabin. Although the mission was carried out perfectly, the photographic results were useless, due to the windows being covered in rocket fuel. The Chinese would not identify the mission in question, but it could have been FSW 1-1. Control of this mission was lost at one stage but recovered due to the efforts of the number one tracking station in Karshi in western China.

ROGUE SATELLITE ON THE LOOSE

The last mission in the FSW 1 series went wrong, the only one to have done so. FSW 1-5 failed to return to Earth when commanded to do so in October 1993. Launched on 8th October 1993, FSW 1-5 should have returned to Earth on 16th October after the normal 8-day profile. However, the satellite failed to rotate downward for the return to Earth. Instead, it fired in the direction of travel, firing the FSW into a much higher orbit. Whereas it had been circling the globe at 196 by 251 km, the burn had the effect of making the upper point of the orbit swing far out, to 3,023 km. However, the low point, or perigee, was slightly lower, 181 km, meaning that the satellite would, sooner or later, burn up. Except that, being designed for reentry, there was a real possibility that it might survive the fireball reentry and crash somewhere on Earth. FSW's orbital inclination, 56°, meant that anywhere on Earth between 56°N and 56°S was at risk – in other words, the most inhabited zones of our planet.

These developments were especially unfortunate, for FSW 1-5 was flying a number of unusual cargoes. In addition to scientific equipment, the cabin had 1,000 stamps, 3,000 first day covers, credit cards, photos, 194 calling cards and 235 ornaments and gold-studded medallions of Mao Zedong – apparently destined for sale to Japanese collectors where such items reached premium prices.[3]

The media, as ever, warmed to the apocalyptic prospect of a rogue satellite plunging to Earth, as it had earlier with Skylab, Cosmos 954, 1402 and Salyut 7. Although tracked carefully by United States Space Command and other sophisticated radars, the fact that it could fall on most inhabited zones on Earth added to the excitement. Xian control centre kept the FSW under observation, for it had built up its own training by tracking Skylab and Cosmos 1402 much earlier. Experts predicted the size of the crater it could make if it survived reentry and exploded on impact. Eventually, after several days of bouncing on the upper layers of the atmosphere, FSW 1-5 came down on that part of the Earth where it was most likely to end up, the sea, crashing into the southern Atlantic on 12th March 1996.

FSW 2 SERIES: MANOEUVRABLE, HEAVIER, 18-DAY PROFILE

The FSW 2 series was introduced in August 1992, even before the FSW 1 series had come to an end. The principal innovation of the FSW 2 series was the ability to manoeuvre in orbit, but there were other improvements. Compared to the FSW 1 series, the '2' model had a greater weight (3,100 kg), 53% heavier payload (350 kg), 20% greater cabin volume and could stay in orbit up to 18 days. The length of the spacecraft was increased by a third to 4.6m, with a much larger service module, part of which was pressurized. FSW 2 carried a more sophisticated attitude control system and an advanced computer. Overall, it was 13 times more capable than the first, FSW 0 series.[4] The much-increased size of the FSW 2 meant that it required a larger launch vehicle, the Long March 2D, an improved version of the 2C. This now replaced the old 2C.

FSW 2-1 (August 1992) was a dual-purpose mission, with both remote sensing and microgravity experiments (in this case, cadmium, mercury, tellurium and protein crystal growth). The cameras could take 2,000 m of film and had an imaging capability of 10 m. FSW 2-1 carried semi-conductors and ten protein growth experiments in 48 cells. Crystals were grown in a multipurpose finishing stove, a furnace able to provide heating of 813°C. FSW 2-1 used its manoeuvring system to change orbit three times during the mission. The first attempt to reenter, on the 12th day of the mission, failed, but after going through the procedures again carefully, ground controllers were successful on the 16th day.

FSW 2-2 flew in July 1994. Like its predecessor, it manoeuvred in orbit. FSW-2-2

146 **Applying the space programme**

FSW readied for the mission.

carried an even more exotic cargo of rice, water melon, sesame seeds and more animals. The most significant change was the reentry manoeuvre. The equipment module remained attached until retrofire. Clearly, this had something to do with improved reentry procedures following the loss of FSW 1-5.

FSW 2-3 carried Japanese microgravity experiments for the Japanese Marubeni Corporation with Waseda University. They involved the development of indium and gallium monocrystals. China also carried its own microgravity materials processing experiments, as well as a biology package of insect eggs, algae, plant seeds and small animals for the Shanghai Institute for Technical Physics. This was the 16th successful recovery out of 17 attempts. For the first time, the information collected by the satellite in orbit was stored on compact disk. The Chinese materials processing experiments concerned the production of monocrystalline silicon, photoconductive fibre with impurities of 10^{-7} and medicines to prevent haemophilia. Perhaps the most intriguing feature about FSW 2-3 was that its equipment module, left behind in orbit when the descent craft came down, itself manoeuvred in orbit. FSW returned to Earth on 31st October 1996, leaving, as normal, the equipment module to decay

An experiment with silkworms.

naturally. Instead, on 11th November, it was fired from 167 by 293 km into a higher orbit of 212 by 299 km. This was not announced by the Chinese at the time, nor has it been explained.[5] This was the last mission for seven years.

Rumours circulated in the late 1990s of a new version to come, the FSW 3 series. China then announced its intention of flying an orbital mission to test improved varieties of grass, shrubs and trees. Wang Yusheng of CAST explained how the exposure of grasses to radiation could be used to develop different types of grasses – those that could spread quickly, or grow more slowly or be capable of resisting harsh climatic conditions. Later, China announced plans to fly a silkworm experiment on board a forthcoming recoverable satellite.[6] The silkworm experiment was contrived by Jingshan High School, Beijing, to follow the entire lifetime cycle of the silkworm from egg to adult in the course of a mission. The original aim was to compare the results with a similar experiment carried out on the last, lost flight of the American *Columbia* space shuttle.

This mission seems to have been put on hold, although the FSW series resumed in November 2003 but in a different form. This time it was initially given the Jian Bing 4 designator, earlier Jian Bing flights being associated with unspecified photography missions. Most western accounts labelled the mission FSW 3-1. It was launched by the Long March 2D from the pad adjacent to the one where, two weeks earlier, Yang Liwei had made history as China's first astronaut. The cabin returned to Earth with its payload on 21st November after 18 days circling the Earth.

RESULTS FROM THE FSW SERIES

A progress report was issued on the early outcomes of the FSW materials processing and biology missions to date.[7] By this stage, six such missions had been flown within the series. Tests on alloys, tellurium and gallium arsenide had yielded positive results, crystals having high purity. Rice seeds brought back to Earth and crossed with Earthly grains produced high yield rates, some giving 53% more protein. Space-grown yeast offered higher and faster fermentation rates, opening up new prospects for a space beer industry. Algae flourished in orbit.

148 Applying the space programme

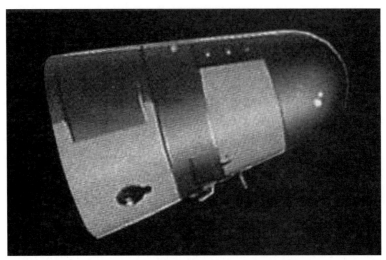

FSW in orbit.

Altogether, 300 varieties of seeds and 51 kinds of plants were carried in seven different biology packages. Once back from space, with its zero gravity, seeds from the plants grown on board – rice, carrot, wheat, green pepper, tomato, cucumber, maize and soya bean – were planted out by the Institute of Genetics to observe the effects, further note being taken of succeeding generations over the following years. The results varied. Some strains of rice improved from their space experience, while others did not. Some grains grew faster and were fatter, heavier and sturdier. Wheat experiments produced new strains that had short stems and grew fast. Space-exposed rice were set on a field of 667ha to test their yields. One strain of green pepper, called the Weixing 87-2, demonstrated an increased yield of 108%, 38% less vulnerability to disease and an improved vitamin C content of 25%, bearing fruit long after terrestrial peppers had lost their leaves. Bumper 400g green peppers were bred, twice as much as normal ground size. A fifth generation space tomato had a yield 85% higher than its terrestrial rivals and doubled its resistance to disease. Space-grown cucumbers demonstrated a surprising ability to withstand greenhouse mildew and wilt. Female cucumber flowers were observed to flourish in the space environment. Asparagus seeds flown in space also thrived on Earth.

The seeds experiments were funded by what was called 'project 863', a project that was to subsequently reappear in other parts of the space programme. Project 863, really a programme rather than a project, was authorized in March 1986, hence the '3' and '86' designators. This was a horizontal scientific research and applications programme, designed to stimulate Chinese technology across a broad range of fronts. The programme followed in the wake of the four great modernizations of Deng Xiao Ping, one of them being science. Project 863 was proposed by Wang Daheng, Wang Gangchang, Yang Yiachi and Chen Fangyun as a means of bringing China up to the level of the European Union, Japan and the United States in research and development. Project 863 won quick endorsement from Deng Xiao

Caterpillars – before and after.

Ping and was coordinated by Zhu Lilan's Ministry of Science & Technology. It was a horizontal programme with seven categories (biotechnology, information, aerospace, laser technology, automation, energy and new materials), 15 themes and 230 sub-categories. Between 1986 and 2001, €780 m was invested in the programme in 5,200 individual projects. The idea was to fund small cutting edge projects (e.g. digital mapping, internet libraries in Chinese) that could be applied across wide areas of the economy. By the time an exhibition was held in Beijing to mark the first 15 years of the programme, China had indeed closed the gap with its industrial rivals in some key areas and 2,000 patents had been filed. 863 funded a series of studies, exploratory projects and missions in the space programme, the seeds project being one of the first.

The project 863 – funded seeds project involved three types of seeds on FSW cabins. They were replanted on their return to Earth. Exposure to weightlessness created a genetic variation which meant that the replanted seeds doubled their weight and grew twice as heavy, taller fruit with higher resistance to disease, with a higher proportion of vitamin C and a longer shelf life. In 1998, following these experiments, a Space Vegetable Foundation was established in Anning by the Academy of Sciences, where it further developed and sold to the open market 'space fruit'. By 2002, space vegetable gardens had been established in Hebei, Gansu and Sichuan and were growing 12 varieties of wheat, rice, tomato, peppers and cucumber. The space-developed cucumbers were especially successful, growing 20% longer than the purely Earthbound variety and having a strong disease resistance (as well as tasting better, according to the experts).

150 Applying the space programme

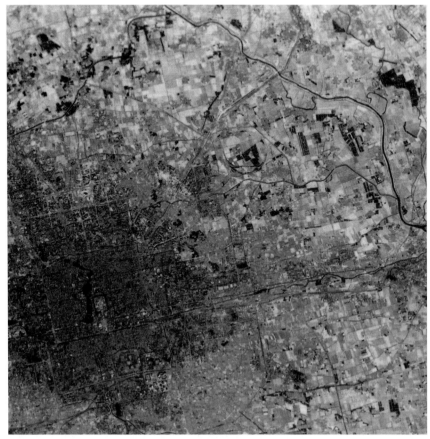

FSW imaging from orbit.

Following the return of FSW 2-3 in November 1996, the Xinhua news agency reported on some of the results of the second series. Among the Earth resources results from the FSW series was a recalculation of the total number of islands off the Chinese coast (5,000, instead of 3,300). The country's farmland had been recalculated at 125.3 m ha rather than 104.6 m ha. The Chinese claimed a resolving ability of FSW cameras of 19 m which would be at least as good as comparable American Landsat data at the same time.

The following year, further results were announced. The FSW satellites had compiled detailed Earth resource maps of Beijing and its eastern environs, Tianjin and Tangshan. Oil deposits had been discovered in Tarim, chromium and iron deposits in inner Mongolia and coal elsewhere. The FSW satellites had discovered remnants of the Yuan dynasty's ancient city of Yingchang. They even uncovered buildings erected by the first Yuan emperor, Kublai Khan, for his daughter, Princess Luguo Dachang, in 1270. The Chinese claimed that the imaging systems had done much to help complete the mapping of China. A new map of China was commissioned in 1949, but only 64% of it had been finished by 1982. However, 600

A Feng Yun 1 image. (Courtesy *Aerospace China*.)

FSW pictures were able to finish the job in a matter of months. FSW satellites had been used to prepare geological survey maps; identify the optimum routes for railway lines; and track the patterns of silting in the Huang (yellow), Luan and Hai rivers. Images tracked the path of the Great Wall across northern China and found the old walls of the Chengde summer palace. They tracked water and air pollution, observed soil erosion and identified geological fault lines. The FSW satellites located goldfields in Mongolia; and oil and natural gas in the Yellow river delta and offshore.

Data from the FSW and Feng Yun series, combined with information from the American Landsat and the French SPOT satellites, provided a worrying picture of desertification in Qinghai in the north-west. Dynamic changes were taking place, according to the satellite data: dunes had advanced, grassland was damaged, water resources had been misused. Elsewhere, soil erosion had been noted. Positively, the rate of afforestation had been assessed and was seen to be growing. The use of windbreak forests in northern China had already regenerated the ecology of the area. Earth resources satellites carefully tracked the evolution, speed and impact of the Yellow river: as a result, timely warnings about floods were given during the inundations in 1991, minimizing damage. Satellite tracking of the 1987 forest fires in Xinanlang enabled firefighters to save up to 10% of the forests from further damage.

152 **Applying the space programme**

To mark World Space Week in 2000, China recorded some of the accomplishments of the FSW series:

- Mapping of the sand deposited to sea by the Yellow river (Huang He)
- Finding of seven mineral deposits for the Capital Iron & Steel Co
- Four new oilfields in Xinjiang
- Completion of a general territorial survey
- 80 material science experiments
- Improved tomato yields of 20% with 40% reduction in disease

Some data from the FSW missions were still being analyzed many years after their missions. Early in 2001, Chinese scientists released the results of experiments from gallium arsenide superconductors. The experiments, developed by the Chinese Academy of Sciences and the Hebei Semiconductor Research Institute, found that

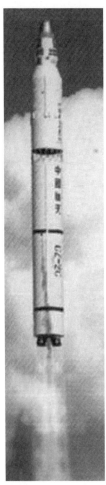

Long March 2 carried FSW series.

electronic devices made from crystals in space outperformed those developed on Earth. Space-manufactured crystals were more sensitive, carried more current, were less prone to voltage noise or suffer leakage.[8]

At one stage, when it was flying French and then German payloads, the series appeared to hold out the prospect of being a serious foreign exchange earner for China. The United States and Europe both lacked dedicated spacecraft able to fly microgravity experiments. The only other country with such a capability was Russia (the Foton series).

As military reconnaissance satellites, possibly its original role, the FSW series had definite limits. FSW film was sent back in recoverable cabins and military analysts had to wait until the film was developed before they could analyse points of interest. In 1977, the United States introduced downlink imaging – satellites could transmit photographs to the ground directly from orbit. Analysts would have their intelligence information immediately and could even command the spy satellites to adjust course to investigate new and more interesting targets later in the mission (with its fifth generation of photoreconnaissance satellites, the Soviet Union had a digital imaging capacity from 1982). However, it seems that China did not develop this series for military photoreconnaissance in the long term, adapting the series instead for civilian Earth resources studies, microgravity experiments and for commerce – with evident success as witnessed by the outcomes cited.

FSW 1 and FSW 2 series

Fanhui Shi Weixing 1-1	9 Sep 1987	Gallium arsenide, algae
Fanhui Shi Weixing 1-2	5 Aug 1988	3 German experiments
Fanhui Shi Weixing 1-3	5 Oct 1990	Guinea pigs on board
Fanhui Shi Weixing 1-4	9 Oct 1992	Semi-conductor, protein
Fanhui Shi Weixing 1-5	8 Oct 1993	Recovery failed
Fanhui Shi Weixing 2-1	9 Aug 1992	16 day mission
Fanhui Shi Wcixing 2-2	3 Jul 1994	13 days, new return procedure
Fanhui Shi Weixing 2-3	20 Oct 1996	15 days, Japanese cargo
Fanhu Shi Weixing 3-1 (Jian Bing 4)	3 Nov 2003	18 day mission

Programme summary

Fanhui Shi Weixing 0	10 attempts, 1 launch failure
Fanhui Shi Weixing 1	5 attempts, 1 recovery failure
Fanhui Shi Weixing 2	3 attempts, no failures

CBERS AND ZI YUAN EARTH RESOURCES PROGRAMMES

In the late 1980s, reports began to filter out of China of a joint applications project between China and Brazil. Its roots actually went back further, to the abandoned Shi

Jian 3 concept for an Earth resources satellite. Brazil had an active space programme (its first satellite in 1993) and indeed had put some efforts into trying to develop its own launcher (though with little success). The project was developed on a 70/30 China/Brazil basis.

The project was called CBERS (China Brazil Earth Resources Satellite) and involved the building of two satellites, one in China, the second in Brazil. CBERS was 1,450 kg in weight, box-shaped, with one large 6 m solar panel able to generate 1,100 w. It was designed to cross the equator at 10.30 am every day, so as to set a standard point of reference for sun angles on the targets observed. One hydrazine and two other thrusters were used for manoeuvring and pointing control. Cost of the CBERS programme was estimated at €330 m.

CBERS was designed to enter an orbit of 774 km, 78.5°, similar to that of the Feng Yun 1 and provide detailed images of the Earth in five channels using linear charge couple devices (CCD) with a resolution of 20 m, able to tilt up to 32° to either side for oblique shots. CBERS was also built to carry a multispectral infrared scanner (resolution 160–180 m) and a wide field imager (resolution: 258m). A data collection system could pick up and retransmit information from unmanned inland stations and buoys. Chief designer was Chen Yiyuan.

CBERS 1 was eventually launched in October 1999, riding a Long March 4B from Taiyuan launch centre. 22 mins 40 sec after launch, CBERS 1 was ejected from the now burned out third stage and, 25 sec later, a Brazilian subsatellite, SAC-1, or Scientific Applications Satellite 1. The successful deployment of both was soon confirmed by a ground tracking station at Nanning. The launch was attended by the Minister for Science & Technology in Brazil, Ronaldo Sardenberg. By way of an

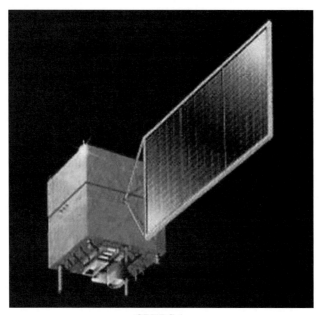

CBERS 1.

endnote, the fuels left in the Long March 4B upper stage combusted accidentally on 11th March 2000, blowing the stage apart, scattering 300 fragments in orbit and spewing out a dustcloud of tiny fuel particles. Ironically, the Chinese had become increasingly aware of the problem of orbital débris and the Long March 4 was the first Chinese rocket fitted with a system of venting residual propellants, to prevent this very situation from arising, obviously without much success.[9]

Seven hours later CBERS was over Brazil and picked up by the main tracking station at Sao Jos dos Campos. In China, CBERS was tracked by stations in Beijing, Guanzhou and Urumqi as well as the main mission control centre in Xian which was responsible for on-orbit checkout. The initial orbit of CBERS 1 was 728 to 745 km, inclination 98.55° (polar orbit) and period 99.6mins. Six days later, the satellite began to use its motor to gradually raise its orbit, in order to reach a perfect sun synchronous operating altitude. Over a month, in the course of 13 manoeuvres, it pushed its altitude to a circular orbit at 774 km, period 100.32 min, an altitude it reached on 9th November, one which enabled it to circle the Earth 14 times a day and revisit the same track over the ground every 26 days. CBERS carried out small but regular manoeuvres to raise its orbit: whenever it fell to 100.315mins, the motor would fire briefly to lift it back up to 100.322mins. CBERS 1 operated for double its two-year planned lifetime, sending back over 8,000 images of China, covering 99% of the country.[10] CBERS 1 took several months in calibration tests before its Earth resources equipment was declared operational in February 2000.[11] CBERS images covered a swath about 120 km wide, with a resolution of 20 m, comparable to the French SPOT 4. The work of this versatile satellite was initially successful. A mission report four months later recorded that data were being collected for agriculture, forestry, water quality, urban planning and environmental protection. Among the commercial users of CBERS data were cellulose manufacturers (for information on eucalyptus trees), and government agencies trying to prevent slash-and-burn farming practices in the Brazilian jungle. However, the wide field instrument was reported to have failed after 177 days because of a short circuit.

The small 60 kg Brazilian subsatellite, with four experiments weighing a total of 28 kg had began to circle the Earth at 761 km. Called Plamex, Magnex, Photoex and Orcas, these were intended to study cosmic rays, the magnetic field and atmospheric phenomena. The minisatellite cost €5.4 m and the project was overseen by the Brazilian Science Academy. It was completely developed and built within Brazil itself without reliance on foreign equipment. SAC 1 was tracked by the Brazilian tracking station of Natal Rio Grande do Norte. However, the solar panels on the subsatellite quickly failed and it was lost.

In October 2001, the Space Mechanical and Electrical Research Institute in Beijing held a seminar to mark the first two years of the mission. CBERS 1 had sent back more than 200,000 images, of which 3,000 had been customized. They had picked up anything from coal mine fires in Ningxia to landslides along the Yigong river in Bomi, Tibet.

CBERS 1 retired in August 2003 and was soon replaced by the 1,550 kg CBERS 2, built in Brazil and put into orbit by China on 21st October 2003. CBERS 2 carried the Chuang xin ('creation') microsatellite. CBERS 2 was technically similar to

156 Applying the space programme

CBERS 1, although it carried a number of technical improvements. CBERS 2 carried a high resolution camera (20 m), a wide field camera (resolution 260 m) and a multispectral camera (80 m resoluion). Downlink data volume was doubled and the revisit time shortened from 26 days to 13 days. CBERS 2 was expected to concentrate on observing deforestation, land changes, natural disasters, pollution and underground resources.

Both sides moved quickly to confirm plans to launch future missions. In September 2000, Chinese foreign trade minister Tang Jiaxuan met in the Amazonian jungle city of Manaus with Brazilian government ministers to reach agreement for CBERS 3 and 4 for 2003–5, this time splitting the €250 m costs 50/50. CBERS 3 will have improved resolution of 5m.

To much surprise, in September 2000 China launched its own version of CBERS called called Zi Yuan (the Chinese word means 'resources'), an entirely domestic satellite. Another Long March 4B put Zi Yuan into polar orbit on 1st September 2000. Then the plot thickened. Zi Yuan's original orbit was a full 40% lower than CBERS 1, at 468–493 km. By 10th September, Zi Yuan had used its engine three times to raise its perigee to 484 km and between then and the end of October fired a further three times to maintain an almost circular orbit of 488–496 km. Zi Yuan continued to trim its orbit every nine to 47 days. Whenever the orbit dipped to 94.41mins, a small burst from the engine would send it back to 94.45mins. Its orbit was never quite circular, there being about 10 km between apogee and perigee. This pattern continued into 2002.[12]

The Republic of China on the island of Taiwan alleged that Zi Yuan was flying the same cameras as CBERS 1 but at a much lower altitude in order to gather images for the military. The Chinese denied this and insisted that Zi Yuan was gathering civilian Earth resources information from a slightly different orbit. Evaluation of the nature of the Republic of China's claims was inconclusive. We know that the ground resolution of CBERS 1 was 20 m, so that for Zi Yuan, at a lower altitude, its resolution was likely to be about 12m, which was much poorer than what could be obtained from the FSW satellites and well below the standards necessary for quality photoreconnaissance.[13] Rumours of a military association persisted and Zi Yuan was connected by the *Washington Times* newspaper to a reported Chinese plan for an electro-optical reconnaissance programme called Jian Bing 3, pictures of 5 m resolution being sent back digitally (Jian Bing 1 was a code associated with military photography under the FSW programme).[14] The *Washington Times* suggested that, whatever about spying on the Republic of China, Zi Yuan was eyeing American forces in Japan and the rest of the Pacific. The fact that the Chinese published so little information about the mission compared to CBERS 1 suggests that these rumours may be well founded.[15] The orbital paths of CBERS 1 and Zi Yuan were quite different from one another, something the Chinese never satisfactorily explained.

Just to confuse the designations in the traditional way, China then renamed CBERS 1 Zi Yuan 1 and the new satellite Zi Yuan 2. Zi Yuan 2 was joined by a companion in a similar orbit on 27th October 2002. The second set of Zi Yuans then became known as Zi Yuan 2A and 2B respectively. By 12th November 2002, Zi Yuan 2B had manoeuvred into an orbit of 475 km to 504 km while Zi Yuan 2A continued in a similar orbit of 488 km by 492 km. Each covered the same ground path every five days, so between them they could cover any ground location every two and half days. According to Chinese space officials, the two craft were operating in tandem. Zi Yuan 2B had an improved imaging system.

Whatever about the speculation of the intelligence community, both Zi Yuans had an immediate impact on those living under the ground path of their ascending Long March 4B rockets. Zi Yuan 2A startled farmer Li working on a hill slope in Xiangzhen cultural village in the Wuxi district near Chonging in Sichuan. A loud bang was followed by the strange sight of a large white object rolling down a nearby hill. The object left a crater and he found, at the bottom of the hill, a cylindrical object 12 m long, 2.5 m in diameter, weighing about a tonne. It turned out to be part of the upper stage of the Long March 4B.[16] When Zi Yuan 2B was launched a year later, no less than 19 pieces of the Long March 4B rained down on Yanghe in Dangen county in Shaanxi. A nine-year old boy was injured and had to be treated in hospital.

Earth resources satellites

CBERS 1	14 Oct 1999	Long March 4B	Taiyuan
Zi Yuan 2A	1 Sep 2000	Long March 4B	Taiyuan
Haiyang 1	15 May 2002	Long March 4B	Taiyuan (piggyback)
Zi Yuan 2B	27 Oct 2002	Long March 4B	Taiyuan
CBERS-2	21 Oct 2003	Long March 4B	Taiyuan (with Chuangxin)

NAVIGATION SATELLITES: STARS OF THE HEAVENS

There had been rumours, from the late 1980s, that China would build a navigation satellite system. It even acquired a name, Twinstar, Twinsat also being used. Nothing much came of the story and it was assumed that it was a paper project. In the meantime, the United States operated the Global Positioning System for its military (with a degraded but enormously popular version for civilian use); Russia had a system called GLONASS in 19,000 km orbits over the northern hemisphere; and Europe planned its own system, Galileo, scheduled for 2008. We now know that the Chinese system was proposed by Chen Fangyun and approved in 1983. To save development costs, the Dong Fang Hong 3 communications satellite design was used.

With no forewarning, China's first navigation system appeared on 31st October 2000. Following a midnight launch, a Long March 3A placed the satellite at 140°E at 36,000 km with complete precision 10 days after leaving Xi Chang. It was given the name Beidou, the Chinese word for the plough constellation, or the big dipper. Two months later, on 21st December, Beidou 2 followed. The last satellite to be launched that year, it reached its final destination at 80°E three days before the end of the old year and the millennium. With satellites at 80°E and 140°E, the Chinese system had followed a third way different from the United States and Russia, providing a very long baseline triangular navigation system. Users triangulate their positions using, as

158 Applying the space programme

Beidou ready for launch. (Courtesy *Aerospace China*.)

reference points, the two Beidous and the signals between Beidou and Beijing. China said that Beidou would provide accurate navigational fixes for ships, road and rail transport and presumably also for aircraft. The system was intended to be fitted to Beijing's 7,000-strong bus fleet and a 22-floor control system for this purpose was completed in 1999.[17] China also booked an orbital location at 110.5°E to store an on-orbit spare Beidou, ready to replace one of the originals should it break down.[18] A third Beidou duly arrived there on 25th May 2003, Beidou 3. Like its predecessors, Beidou 3 reached the point following a 200-41,991 km super-synchronous orbit. Then, to confuse things, China renamed the series, calling the new satellite alternately Beidou 1-3 or Beidou 1C.

The Chinese had in fact been preparing the project for some time, evidently quite secretly: most of their own experts seemed not to know about the project. Preliminary tests had been done in orbit in 1989 using the Dong Fang Hong 2 and 2A satellites.[19] Separate reports suggested that China had developed sophisticated atomic clocks far in advance of western ones. Under the guidance of Gua Guantan of the Chinese Science & Technology University, China had built a quantum computational centre in 1999, learned how to use lasers to cool atoms and conquered the problem of the atomic fountain.[20] These clocks, much more advanced than the old western caesium clocks, had an accuracy of 1 sec in 30 m years. True or not, the completion of a navigation satellite satellite system in less than two months was impressive by any standards.

Preparing Beidou. (Courtesy *Aerospace China*.)

Beidou control. (Courtesy *Aerospace China*.)

Chinese navigation satellites

Beidou 1	31 October 2000	Long March 3A	Xi Chang
Beidou 2	21 December 2000	Long March 3A	Xi Chang
Beidou 3	25 May 2003	Long March 3A	Xi Chang

In autumn 2003, China reached agreement with the European Space Agency and the European Union for participation in the European Galileo programme. The precise implications of this were unclear, though they suggested some degree of interoperability between Galileo and Beidou. China paid €200 m for its stake in Galileo and appeared hopeful of gaining some of the launch contracts for the 30-spacecraft system.

Another type of satellite which must enter the conjectural class is a data relay system. The Americans introduced what they called the Tracking and Data Relay Satellite System (TDRSS) in the 1980s to support shuttle operations. Hitherto, the shuttle had communicated with the ground as its crew flew over tracking stations around the globe – an inefficient system which required continuous retuning to each new ground station. With TDRSS, the shuttle sent its signals outwards and upwards – to the nearest of three TDRSS communications satellites in 24hr orbit, which then relayed signals back to mission control. Russia had a similar system, Luch, for communicating with Mir. It was an expensive system, but one which produced comprehensive round-the-clock communications between mission control and its astronauts. There have been reports that China has been considering a TDRSS system. However, the investment by China in a fleet of ocean-going communications ships to support manned missions suggests that this option is still some distance away. As it is, the tracking ships are able to relay data directly back to Beijing mission control through the existing satellite network.

MICROSATELLITES: SMALLER AND SMALLER

Even as the space powers built ever bigger and more powerful launchers, the 1990s saw, paradoxically, the introduction worldwide of ever-smaller satellites. This development was made possible by electronic microcircuits and ever more sophisticated computers. These computers not only permitted satellites to be much smaller but much smarter – able to do more and more without the intervention of ground control. These new versatile satellites could be launched on much smaller, less expensive rockets (e.g. the American Pegasus), or as piggyback payloads on existing rockets, thus cutting costs even further. Technically, this new generation could be divided into small satellites (less than 500 kg), microsatellites (less than 100 kg), nanosatellites (less than 10 kg) and even picosatellites (less than 1 kg!).

China was not backward in joining the microsatellite revolution. Qinghua University was the centre of microelectronics in China and hosted the National Aerospace High Technology Space Robotic Engineering Research Centre. The centre obtained project 863 funding for the development of microsatellites and set up the Quinghua Satellite Technology Company in 1998 as a joint enterprise of China Space Machinery and Electrical Equipment Group, Qinghua University Enterprise and Qinghua Tongfang Company. The engineers there turned to the world leaders in microsatellite technology, the University of Surrey in England (Surrey Satellite Technology Limited, SSTL).

China's first small satellite entered 700 km high orbit on 20th June 2000, lofted by a Russian Cosmos 3M rocket from Plesetsk. The 75 kg microsatellite, called Qinghua 1, was 1.2 m high, with a volume of only 0.07 m^3. No sooner was Qinghua in orbit than it sent back its first photographs of the China Sea. Qinghua carried a

Microsatellites: smaller and smaller 161

The Tsinghua team (Courtesy SSTL.)

camera system able to image the Earth in three spectral bands with 39 m resolution so as to monitor vegetation, floods, wild fires, desertification and red tides. Within a month it had sent back over 100 images, which the university made available free to anyone requesting them. Its design life was 10 years and, according to senior engineer at Qinghua, Xu Xin, was a serious attempt to close the gap with Indian and western imaging systems. In orbit, it also took part in rendezvous manoeuvres with another Surrey satellite, the 6.45 kg nanosatellite SNAP-1 (Surrey Nanosatellite Applications Platform).

The University of Qinghua hoped that this would pave the way for a constellation of microsatellites, also to be developed with SSTL, for a fleet of 70 kg disaster-warning satellites. These would fly on a Russian Dnepr rocket into a 772 km polar orbit, the others run by Nigeria, Algeria and Thailand. In 1996, Zhou Sumin of the China State Seismological Bureau outlined how a fleet of small satellites, could, by storing and then forwarding signals picked up from 800 ground seismological stations the world over, provide advance warnings of potential Earthquakes.

China's next microsatellite was the 100 kg Chuang xin, meaning 'creation' or 'innovation'. This was a store-and-forward communications satellite built by the Shanghai Academy for Space Technology for the Academy of Sciences and Shanghai Telecom. It was launched in October 2003 piggyback on the CBERS 2 Brazilian-Chinese Earth resources satellite. This mission was quite a success and a report was given on its first 120 days of operation in February 2004. Tracked by terminals in Shanghai, Beijing, Xinjiang and Hainan, it showed off its digital communications capacities and was unaffected by two strong solar flares and 29 single particle incidents.

162   Applying the space programme

Tsinghua rendezvous. (Courtesy SSTL.)

Tsinghua in orbit. (Courtesy SSTL.)

At this stage, another microsatellite project was also under way: Tansuo ('Exploration'), a high – resolution imaging satellite at the University of Technology in Harbin. Tansuo was a 150 kg microsatellite with a 10 m stereo resolution camera and was a development by Harbin with the European company Astrium. At a symposium held in Shanghai, devoted to small satellites, there were speakers from

An environmental microsatellite. (Courtesy *Aerospace China*.)

Harbin University of Technology (Zhang Yinchun) and Qinghua University (Zhang Xiaomin) who outlined the various small satellite projects in China, from microsatellites down to picosatellites. Qinghua University also had a 10 kg nanosatellite under design for a piggyback launch. Called THNS-1, it will be for Earth observation work. In 2003, China announced the establishment of a National Research Centre for Small Satellites and Related Applications and first ground for the 8,000 m² site was broken on 20th April 2003.[21]

Quinghua Space & Satellite Technology Company is building a 10 kg microsatellite dedicated to the China Children and Teenagers Fund. The idea is to involve Chinese children in tracking the satellite from 300 sites across the country. The idea is that the microsatellite, to be named *Olympiad* in honour of the upcoming Olympic games, will orbit piggyback from a Long March 4B from Taiyuan at 600–700 km, 100 mins, accompanied by a shiny 5 m diameter balloon. The children will learn to pick up signals relaying details of altitude, velocity, latitude, longitude and time and make comparisons between the radio signals and the visual observations of the balloons. The microsatellite will also carry a digital camera, store-and-forward recorder and global positioning system receiver.

Microsatellites are a focus of Chinese international cooperation with other countries in the Asia-Pacific region. In 1992, China sponsored the first Asia-Pacific space symposium. This became an annual event for the region in which the development of small satellites was the main focus and led to the Asia-Pacific Multilateral Cooperation in Space Technology and Applications Programme which involves Iran, Pakistan, Thailand, Mongolia, South Korea and Bangladesh. In 1998 the participating countries signed the *Memorandum of understanding on cooperation in small multimission satellite and related activities* for the exchange of information on microsatellite development.

ASSESSMENT AND CONCLUSIONS

Following the development of communications satellites in the 1980s, China broadened its space programme into an impressive range of space applications: meteorology, Earth observations, materials processing, Earth resources and navigation. By the new century, China had a full fleet of operational weather, resources and navigation satellites girdling the Earth.

The original space applications programme was the FSW series. Designed before China had even orbited an Earth satellite, the FSW satellite was a demanding project for a space programme at such an early stage of development. The series was ridiculed in the west, where it was misrepresented as low-tech, using a tourist camera with regular film and even equipped with a heat shield made out of wooden oak planks.[22] In fact, the development of recoverable spacecraft was a challenging engineering task, one which few other minor space powers have undertaken. The FSW series has shown a high level of reliability, only one failing out of 18 missions. It has brought in a rich haul of results. The new Zi Yuan Earth resources satellites built on the success of FSW and meant that China's Earth observing capability was at least on a par with the other leading space powers.

The success of the FSW series in returning Earth resources data probably encouraged the Chinese in the development of meteorological satellites. Although the Feng Yun series has experienced operational problems, the quality of data and imaging returned has been very promising. With the Feng Yun 2 series of geosynchronous weather satellites, the Chinese again chose an ambitious option, but one guaranteed to present a maximum economic return from investment. The Beidou navigation system, introduced without forewarning in autumn 2000, demonstrated the technical capacity, efficiency and reliability of the Chinese space programme and warned of the dangers of underestimating Chinese potential. Finally, China's capacity to be on the cutting edge has been underlined by its commitment to microsatellites. These satellites promise innovation, application as well as commercial opportunities and they will likely be an important line of development in the future.

REFERENCES

1 Wei Long: Chinese weather sat exceeds design life. *Spacedaily*, 20th May 2001. http://www.spacedaily.com/news; Sun Qing: FY-1 satellite monitoring flood in Huaihe river. *Aerospace China,* vol 4, #3, autumn 2003.
2 Paper reports developments in recoverable satellite technology. BBC report, SWB series, 1 April 1998. China Space News: *World Space Week*, 4th–10th October , commemorative publication, Hong Kong, 2000.

3. EP Grondine: The Chinese manned space programme – behind closed doors. *Encyclopaedia Astronautica*, http://www.friends-partners.org/-mwade/articles.
4. Paper reports developments in recoverable satellite technology. BBC report, SWB series, 1 April 1998.
5. Phillip S Clark: *Chinese space activity 1996-2000*. Molniya Space Consultancy, 2001.
6. Satellite d'observation chinois. *Air & Cosmos*, 8 septembre 2000;. China plans mutant grass in space. BBC reports, 27th September 2000. For details of the silkworm experiment, see Liu Je: Recoverable satellite to piggyback silkworm experiment. *Aerospace China*, vol 4, #3, autumn 2003.
7. Zhu Yilin: Space microgravity scientific experiments in China. *Spaceflight*, 35, October 1993; Zhu Yilin: *Some results of Chinese space microgravity life science experiments*. Monograph.
8. Mark Schrope: Whispering wafers: crystals grown in space make quieter microprocessors. *New Scientist*, 24th February 2001.
9. China strives to improve space vehicle design to reduce débris. BBC reports, SWB series, 1 April 1998.
10. See China–Brazil EO bird alive and well in orbit. *Spacedaily*, 3rd February 2000, http://www.spacedaily.com/news. For a progress report on the programme, see Liu Je: CBERS 2 to be launched on LM-4B. *Aerospace China*, vol 4, #3, autumn 2003.
11. Phil Clark: Further analysis of Zi Yuan manoeuvres. Correspondence, *Journal of the British Interplanetary Society*, vol 55, #7/8, 2002
12. Phil Clark: Orbital manoeuvres of China's Zi Yuan satellites. *Journal of the British Interplanetary Society*, vol 55, #7/8, 2002. Subsequent figures on Zi Yuan orbital manoeuvres are also taken from Clark.
13. Phillip S Clark: *Chinese space activity, 1996–2000*. Molniya Space Consultancy, 2001.
14. Bill Gertz: Chinese civilian satellite a spy tool. *Washington Times*, 1st August 2001.
15. Phil Clark: Chinese satellites – status report at the end of 2001. *Journal of the British Interplanetary Society*, vol 55, #7/8, 2002
16. Wei Long: Latest Chinese launch leaves débris in local village. *Dragon space – Space Daily*, 25th September 2000, http://www.spacedaily.com/news/.
17. Satellite-based system to guide buses in Beijing. BBC reports, 29th September 1999. For the origins and development of the series, see He Ying: China putting third Beidou navigation satellite into orbit. *Aerospace China*, vol 4, #2, summer 2003.
18. Phil Clark: Chinese satellites – status report at the end of 2001. *Journal of the British Interplanetary Society*, vol 55, #7/8, 2002.
19. Phillip S Clark: *Chinese space activity, 1996–2000*. Molniya Space Consultancy, 2001.
20. Anders Hansson: *Quantum communications – the Chinese contribution*. Paper presented to the British Interplanetary Society, London, 1st November 2000.
21. See: Les nouveautés du centre spatial de Surrey. *Air & Cosmos*, 20 octobre 2000; China Space News, 7th June 2001, *Go Taikonauts!* website; Phillip S Clark: *Chinese space activity, 1996–2000*. Molniya Space Consultancy, 2001; Sun Qing: National research centre of small satellites and related applications to be established. Aerospace China, vol 4, #2, summer 2003.

22 Roisin Ingle: Satellite now set to land on Limerick. *Sunday Tribune,* 11 February 1996. The claim that the heat shield was made of wood appeared in 'In orbit' in *Spaceflight News* (October 1988) pp. 28–9 and in other reputable journals, but this does not appear to be borne out in the technical descriptions of the design of the spacecraft as published by Clark, where the heat shield is described as made of molybdenum alloys and other composite materials. None of the Chinese literature suggests the use of wood in the FSW series. The 'wooden heat shields' story may have its origin in poor translations.

7

Behind the scenes

This chapter goes behind the scenes of the public face of the Chinese space programme. It explains the architecture and the infrastructure – how the programme is organized, who is in charge, the lines of command. The chapter starts with the role of the chief designers of the Chinese space programme, how they were recruited, how they related to the political leadership of China and how they survived the many political upheavals of the early years. The main organs in the Chinese space programme are described, outlining the respective roles of the Chinese National Space Administration, the China Aerospace Corporation and its leading academies. There is a description of the main ground testing infrastructure, whose aim is to produce space hardware to the highest possible standard. Other important parts of the programme are reviewed, such as the Institute for Aviation and Space Medicine and the Academy of Sciences. China has an extensive system for tracking spacecraft, starting with the main mission control centre in Xian and including four communications ships at sea. The benefits of the space programme to the Chinese economy are reviewed. The chapter concludes by taking a brief look at China's international links and contacts.

DESIGNERS AND THE ENGINEERS: THE PROBLEM OF THE INTELLECTUAL AND THE REVOLUTION

A space programme is a reflection of the quality of its leadership. It is doubtful if the American space programme could have conquered the moon without the genius of rocketeer Von Braun or Apollo designer Maxime Faget. Similarly, the Soviet space programme attracted the genius of Sergei Korolev, Valentin Glushko, Vladimir Chelomei and others.

In China, the scientific community faced particular political problems. Here, Nie Rongzhen was charged by the Central Committee in the mid-1950s with the recruitment of the men and women who would build the Chinese missile and space programme. His first efforts were to persuade the many Chinese people who had gone to study abroad in the 1930s, 1940s and early 1950s to return home. Aided by

Zhou Enlai as foreign minister and then as prime minister, he appears to have been spectacularly successful, for the biographies of China's space designers look like a graduation list from Massachusetts Institute of Technology and California Institute of Technology, with smaller numbers from Britain.

Beside Tsien and many other rocketeers already mentioned, he brought home materials specialist Yao Tongbin, aerodynamics expert Zhuang Fenggan, automatic control designer Yang Jiachi and electronics expert Huang Chang. Zhou Enlai went to some lengths to ensure that the Chinese communist party welcome them home and not treat them with suspicion because they had been born in what was termed 'the old society' or were not communists. His liberal approach was to come under severe strain during the anti-rightist struggle of 1957, the campaign against experts in 1958 and during the cultural revolution, 1966–76. Zhou Enlai, Nie Rongzhen and Zhang Aiping were put to considerable efforts to defend the scientists from those who saw all intellectuals as likely class enemies. Nie Rongzhen felt obliged to explain that 'real socialism' involved scientific achievement more than political correctness; Zhou Enlai weighed in with the axiom that 'if we serve scientists, the scientists will serve socialism'.

During the great leap forward, many scientists (though not, apparently Tsien) made clear their contempt for the principles underlying the campaign against the four pests, such as sparrows. During the cultural revolution, there was strong anti-intellectual sentiment. Long tracts were written on the class status of the intellectual in building communism: indeed, it was one of many themes that obsessed the red guards. The intellectuals had to make practical compromises with the revolutionaries: these mainly took the form of extra ideological and political classes for the intellectuals, so that, in a pithy expression of a slogan of the period, they could be both 'expert' and 'red' at the same time. During rectification, ideology class times were cut back to no more than 1/6th of working time (one wonders what proportion it was before). In 1978, Deng Xiao Ping effectively declared this debate to be closed by offering a theoretical and strategic 'clarification' of the problem: he calmly announced that China's intellectuals were now an integral part of the working class and that was the end of the matter. Job titles were restored and the pre-1966 system of ranks and promotion resumed.

In addition to recruitment from abroad, Nie Rongzhen drafted university graduates into the Fifth Academy. Using the powers which are available in a command economy, graduates were requisitioned from the military engineering academy – a state order was issued, called simply *Notice concerning the transfer of university graduates to the Fifth Academy* and quotas were set. In 1960, a thousand senior military officers were drafted into the space programme. Some were grisled veterans from Mao's army. Most knew nothing of science but they studied furiously and learned to get on with the intellectuals (i.e. the scientists), some even becoming friends.[1] Nie Rongzhen's drafting of graduates, with promotion by merit and fast-track promotion of talented workers, was politically contentious. The policy, officially termed, 'the cream of the crop', was challenged as inegalitarian and suspended for the cultural revolution. In a further move, he sent students to study abroad, in the Soviet Union and eastern Europe (though this did not last) and commandeered the services of those who had studied there earlier.

FIRST GROUP OF CHIEF DESIGNERS

The first group of chief designers was appointed when the Fifth Academy was set up in 1956 (though for some reason not formally established as a system until much later, in May 1962). Their leaders were:

The first group of chief designers, and their responsibilities, 1958

President	Tsien Hsue Shen
General programmes	Ren Xinmin
Aerodynamics	Zhuang Fenggan
Structures	Tu Shoue
Engines	Liang Shoupan
Propellants	Li Naiji
Control – systems	Liang Sili
Controls – components	Zhu Juingren
Radio systems	Feng Shizhang
Computers	Zhu Zheng
Technical physics	Wu Deyu

Like their counterparts in the United States and the Soviet Union, most of these men were well in their forties, some in their fifties, when they led the Chinese space programme in its formative years. They have since retired, though in keeping with the Chinese tradition of longevity, many have lived and still live to old age and retain a lively interest in their former occupation. They had a powerful sense of mission, enormous dedication and will to overcome difficulty. For those who had left – or been forced to leave – the United States, the change in their living conditions and remuneration must have been dramatic.

From the 1970s, a new generation of leadership emerged. Many were the young graduates transferred into the space industry in the 1960s. They came principally from Harbin Military Engineering College, the National Defence Science and Technology University, Beijing University, Qinghua University, Beijing Aerospace College, Beijing Industrial College, Harbin Industrial College and Northwest China Industrial College. The first of them began to move into leadership positions during the period of rectification – people like Song Jian (promoted to chairman of the state science and technology commission) and Li Baijong (promoted to president of CALT). A third generation of very young engineers was drafted in to serve the manned programme in the late 1990s.

DESIGN BUREAUX AND ORGANIZATION

The quality of management and organization of a space programme is one of the keys to its successes. One of the least understood achievements of the Apollo space programme which put American astronauts on the Moon was the way in which clear objectives and lines of command ensured speedy, effective decision-making. By contrast, among the reasons why the Russians lost the race to the Moon were confused objectives, unclear lines of command and rivalry between design offices that Soviet politicians and bureaucrats were unable to tame. Likewise, the Chinese space programme spent most of its earlier period in some organizational turmoil. The 1960s and 1970s saw rivalry between the institutes based in Shanghai and Beijing.

From its foundation in 1956, the Chinese space programme was coordinated by the Fifth Academy; from 1964, this became the Seventh Ministry. In neither case was the true nature of the respective body revealed outside the bureaucracy itself. Not until May 1982, when the Space Ministry was formally established with Zhung Jun as its first minister, did the Chinese space programme have an identifiable public identity.

Who's in charge? Ruling bodies of the Chinese space programme

8 October 1956	Fifth Academy
23 November 1964	Seventh Ministry
9 April 1982	Ministry of the Space Industry
10 June 1993	Chinese National Space Administration

In the United States, rockets and satellites are built by the large aerospace corporations on contract to NASA. In Russia, space programmes were developed by design bureaus (OKBs) that were awarded responsibility for particular missions and areas of development. The broad lines of the Chinese organization were laid down from the late 1950s and followed a pattern broadly in line with the Soviet system.

CHINESE NATIONAL SPACE ADMINISTRATION (CNSA)

The Chinese space effort now comes under the Chinese National Space Administration, established 10th June 1993, the direct equivalent of NASA in the United States or the Russian Space Agency (RKA) and the main policy-making body. The CNSA is responsible to the prime minister, government and party. The CNSA is an executive agency which replaced the space functions of the old Ministry of the Space Industry. CNSA director is Luan Enjie. CNSA's operations are tied to and must conform to the work of the Space Leading Group.

The Space Leading Group of the State Council, set up in 1991, has, as its primary purpose the coordination of relationships with foreign governments, especially in China's efforts to attract foreign contracts, though it has a broader role in policy-making. Its members include the prime minister; the chairman of the State Commission for Science, Technology and the National Defence (sometimes called COSTIND); the vice-chairman of the State Committee for Science and Technology; the minister of the aerospace industry; the vice-minister of foreign affairs and the vice-chairman of the state committee for central planning.

The work of the CNSA is also required to conform to the work of COSTIND, which has a policy-making and directional role in the organization of China's scientific, research and industrial progress (COSTIND is broadly similar to the Military-Industrial Commission of the former Soviet Union). The Commission is directly responsible for the three main launch centres.

CHINA AEROSPACE CORPORATION

The Chinese space industry is brought together by the China Aerospace Corporation (CASC) (sometimes, it has also been called the China Aerospace Science & Industry Corporation). Founded in July 1989, it has 103,000 employees and up to 140,000 people working in sub-contracted industries. General manager is 40-year-old Zhang Qinwei, appointed 2001. The China Aerospace Corporation has overall authority for the main industrial groups concerned with spaceflight. These comprise a sometimes bewildering mixture of organizations variously termed academies, commercial companies, promotional and export agencies (e.g. China Great Wall Industry Corporation), research institutes, bodies and technology development organizations (some university based). CASC is divided into eight academies:

1 Chinese Academy of Launch Technology (CALT), in Nan Yuan, Beijing;
2 Chinese Academy of Mechanical and Electrical Engineering (CCF), in Beijing;
3 Chinese Electromechanic Academy (CHETA), in Haiying;
4 Chinese Academy for Solid Rocket Motors (ARMT), in Xian;
5 Chinese Academy for Satellite Technology (CAST), in Beijing;
6 Shanghai Academy for Spaceflight Technology (SAST), in Shanghai;
7 Chinese Academy for Space Electronics (CASET) in Beijing;
10 Chinese Academy of Aerospace Navigation Technology.

Some of these are of great importance. The most eminent are CALT, CAST and SAST.

Academy 1: Chinese Academy of Launcher Technology
The Chinese Academy of Launcher Technology (CALT), originally called the Beijing Wan Yuan Industry Corporation (estd 1957), is located in Nan Yuan, 50 km south of Beijing and was once, like equivalent Soviet facilities, secret and closed to visitors. CALT employs 27,000 people: of these, 100 are professors, 2,000 are senior engineers, 7,000 engineers and 10,000 technicians in 13 institutes and six factories, mainly in Beijing and Shanghai. President of CALT in 2003 was 39-year-old rocket expert Wu Yansheng.

CALT has overall responsibility for the Long March 1, 2C, 2E, 2F and 3; the Dong Feng 5; and the hydrogen-powered upper stages of Long March rockets. In practice, the Long March 2, 3 and 4 rockets are assembled in Shanghai in a horizontal position on a factory floor. Four complete rocket assemblies may be handled there at any one time. The stages are then transported by rail to the appropriate launch site. The Chinese Academy for Launcher Technology and the

172 Behind the scenes

Vibration test tower.

Shanghai plant between them have the capacity to manufacture up to 16 rockets a year, but demand has never reached such a level of possible activity.

CALT has its own railway terminus, linked to the national railway grid, connecting its plants in Shanghai and Beijing to the launch sites of Jiuquan, Xi Chang and Taiyuan. Adjoining the institute are residential blocks for the many scientists, engineers and technicians who work there. The research units test out new materials, parts and components. CALT is responsible for a 6,000 m² static test hall, a 50 m tall vibration test tower, engine test stands and a moored test stand. These are vital facilities for a space industry and are described in more detail later.

Society of Astronautics
The Chinese Academy of Launcher Technology includes an amateur Chinese Society of Astronautics (CSA) which attempts to bring together engineers, scientists, amateurs and enthusiasts for space flight. It is the body affiliated to the International Astronautical Federation, though in the best traditions of science and politics there is a rival Chinese Society of Aeronautics and Astronautics.

Academy 5: China Academy for Space Technology (CAST)
Formed on 20th February 1968 by Zhou Enlai, with Tsien Hsue Shen as its first

head, the China Academy for Space Technology, Beijing consolidated the range of bodies then engaged in the most advanced aspects of space research. It is the primary body which designs and manufactures scientific and applications satellites. It has 10,000 workers in 10 institutes and three factories. CAST has lead responsibility for the FSW programme and sounding rockets. President is Xu Fuxiang, with Zhu Yilin the secretary general of its science and technology commission.

One of its principal bodies is the Institute of Control Engineering in Zhongguancun, Beijing, set up in 1956, which has played a key role in satellite design and construction. Director in the 1970s was Tu Shancheng, though it has since been led by Zhou Gangrui, an expert in the control of satellites and the development of sensors. He and his colleagues had to work by night during the cultural revolution.

China's spacecraft are brought together at the Satellite Assembly Plant in Beijing, which until 1967 had been a precision instrument factory. This had originally been the Science Instruments Plant of the Chinese Academy of Sciences but in 1967 Zhou En Lai turned it into a military plant in order to protect it from the cultural revolution. In 1994, it was formally renamed the Beijing Satellite Manufacturing Plant, located in Haidan district, Beijing, the bright characters of its banner contrasting with its otherwise unattractive exterior. Satellites from Shi Jian to Zi Yuan have been built there and the backup models are on display in the plant's museum.

A substantial expansion of CAST facilities was reported from 1994-8, with the construction of spacecraft integration hangars, test facilities and laboratories. The

Vacuum testing. (Courtesy China Astronautics Publishing House.)

new facilities were on a 100ha site in Tangjialing, north west Beijing.[2] CAST director is Gao Shenbin.

Academy 6: Shanghai Academy of Space Technology (SAST)
The Shanghai Academy of Space Technology (SAST), sometimes loosely referred to as 'the Shanghai bureau', was set up on 1st August 1961 by Mao Zedong as part of an attempt to develop technological engineering in the industrial centre of Shanghai. Located in Minhang, Shanghai, SAST had overall responsibility for the Feng Bao and now the Long March 2D and 4. SAST makes the guidance and attitude control systems of the Long March 3 and builds the actual rocket in the Xinxin machinery factory, a converted tobacco hall. SAST has 16 institutes and 13 factories, employing up to 30,000 people, including 6,000 engineers. President is Zhang Wenzhong and the current director Yuan Jie. For the Shenzhou, SAST is responsible for the propulsion module (designer: Liu Zhongying), the electrical systems, command and communications and the main engine (there is a full-scale mock-up of Shenzhou on display). Rockets are not the only product of the academy, which has ten commercial companies and has branched out into defence equipment, cars, office equipment, machinery, electrical products and even property management.

Shanghai Institute for Satellite Engineering (Hauyin)
Also there, the Shanghai Institute for Satellite Engineering builds the Feng Yun metsats and previously built the JSSW series of satellites. The institute may have the best technical facilities in Shanghai, with three vacuum chambers and a centrifuge. This institute was part of CAST until 1993 when it became independent. Director is Prof Meng Zhizhong. Imaging systems are made by the State Meteorology Administration and the Shanghai Institute for Technical Physics (est. 1958).

The other academies: ARMT, CASET and CHETA
The third academy of CASC, the Chinese Electromechanic Technology Academy (CHETA) was set up in 1961 and now employs 15,000 people in ten institutes and two factories building cruise missiles.

The fourth academy of CASC, ARMT (estd. 1962) makes solid rocket motors in Xian city. The president is Ye Dingyou. ARMT employs 1,200 people making solid rocket motors for the EPKM kick stage, retrorockets for the FSW recoverable cabins and other small rockets. The fourth academy has 4,000 staff and 120 research fellows. By the turn of the century it had made more than 70 rocket motors, many of them small in size but central to the space programme. Achievements included the retrorocket for the FSW recoverable capsule, the apogee engine for the geostationary comsats and the escape tower for the Shenzhou. In 2000, a new company was set up in Beijing to develop solid-fuel rockets and eventually sell them on the international market. Called the Space Solid Fuel Rocket Carrier Co, it was set up originally as an affiliate of the China Space Machinery and Electronics Group. The fourth academy, and within it the 41st institute, led up the development of China's first solid fuel rocket carrier in the early years of the new century.

The seventh academy of CASC, the Academy of Space Electronic Technology

(CASET), is in Beijing. The president is Tao Jiaqu. CASET has 10,000 people in nine institutes, two factories and five technical centres.

A 10th Academy was formed in autumn 2001, the Chinese Academy of Aerospace Navigation Technology. The idea behind the academy was to bring together a dispersed and uncoordinated range of small companies and institutes involved in the design and production of inertial instruments, optoelectronic products, electrical and electronic components, precision instruments, computer hardware and software in navigation and guidance systems. The new academy had an initial complement of 5,200 people, a valuation of ¥1.1bn, the authority to aware degrees in navigation, guidance and control and a post-graduate training programme. The first general manager was Xu Qjiang.

Finally, one should mention Beijing University of Aeronautics and Astronautics. The university, which has 23,000 students, is one of the main research centres supporting the Chinese space effort.[3] About a thousand research projects are under way and several members of the Chinese astronaut corps studied there. The university made theoretical studies of the docking of two Shenzhou spacecraft. Practically, the university has an altitude chamber, which has been used to test Chinese suits for spacewalks.

Rocket engines
Rocket engines are made by what is called the '067 base' of the China Space Science and Technology Group, set up in 1970. 95% of China's liquid-fuelled rocket engines have been made there (the YF series) and the base is responsible for the entire process from design through to manufacture, testing and flying. Broadly speaking, its role would correspond to that of the Gas Dynamics Laboratory or Energomash in the Russian space programme. The director is Lei Fanpei. Like SAST, the company has diversified in recent years in this case into such areas as fire-fighting, environmental protections, electronics, machinery and electronics. A subsidiary, the Shaanxi Space Dynamics High Technology Co, was set up to apply rocket engine technology across a broad range of economic sectors.

Great Wall Industry Corporation
The Great Wall Industry Corporation is the promotional agency at home and abroad for the Chinese Aerospace Corporation. Its offices in China may be found in Beijing (Haidan), Chongqing (Sichuan), Guangzhou, Shanghai and Jinan; and at one stage abroad in California, Washington DC and Munich, Germany. The Great Wall is a multi-product promotional agency, its current portfolio including, as well as space rockets, bicycles, beer, safes, home-made ice-cream machines and electric fans. Its last president was Zhang Tong, but, on a visit to the European space base in French Guyana in March 1997, while posing for a photograph near the Devil's Island former penal colony, was tragically swept away by a freak wave and drowned. His successor was Zhang Xinxia.

New company for small satellites
The Qinghua Satellite Technology Company was set up at the end of the 1990s with the specific task of developing micro-satellites and space imaging. Sounding like a typical western university – commercial company, it is a joint enterprise of China Space Machinery and Electrical Equipment Group, Qinghua University Enterprise Group and Qinghua Tongfang co. Located in Zhongguancun Science & Technology Park, Qinghua Satellite Technology Company quickly found a western partner to work with – the University – based Surrey Satellite Technologies Limited (SSTL), which operates on a broadly similar basis. Its aim is to develop China's autonomous microsatellite research capability in a short period of time and build high-performance, low-cost space applications satellites, especially in such areas as weather observations, disaster prevention, environmental monitoring and cartography.

THE TESTING INFRASTRUCTURE

Testing and quality control have been an essential element of the Chinese space programme and considerable resources have been devoted to building up the machinery of testing and verification. Other parts of the space industry are scattered over China in a range of facilities, institutes, bodies and companies, large and small. All work closely with the China Aerospace Corporation (CASC) and its relevant academies.

Static test hall
CALT's static test hall was, when built in 1963, the largest building in China. It took eight months to construct and involved the driving of 1,300 piles – some as long as 10 m – and two pourings of more than 5,000 m^3 of seamless concrete. Entire rockets can be tested there at a time. For the development of the communications satellite, a large vertical dynamic equilibrium machine was developed. Construction of the machine began in 1976 and it was operational five years later.

Vibration tests
Vibration tests are essential if the strains put on a rocket during the ascent are to be anticipated. No amount of theory prepared the rocketeers of the 1950s and 1960s for the stresses imposed on rockets as they were shaken by the vibration of their engines and the strains of passing through the dense layers of the atmosphere. CALT's vibration test tower is unprepossessing on the outside, looking like a shabby yellow-and-orange brick grain mill, but able to test all the likely stresses an ascending rocket is likely to experience. Built in 1963, the vibration tower is 50 m high with 13 floors and 11 working levels. Entire rockets are hoisted into place on the stand, gripped by bearing rails on the floors and then shaken to exhaustion by 20-tonne hydraulic vibration platforms.

Testing a satellite.

Engine tests

An essential element of any rocket development programme is a comprehensive facility for engine testing. The main site for testing rocket engines is the Beijing Rocket Engine Testing station, now called the Beijing Institute of Test Technology, established 1958. It is part of CALT, located 35 km from Beijing (the Russian equivalent is in Sergeev Posad) and has sometimes been called the Fengzhou Test Centre. The director is Zheng Jiwen. The first rocket test stands were completed in 1964 under the direction of Wang Zhiren. She designed them to run up to four engines at a time and simulate high altitude tests. The institute aims to provide a comprehensive range of tests for engines, be they at ground level or simulated for altitude.[4]

The largest engine test stand, completed in 1969, is 59 m high and has a cooling system which draws on a tank holding 3,000 tonnes of water which cools the engines with 35,370 nozzles, dousing the rocket with 7.9 tonnes of water a second. The engine tests are fed by eight tanks which hold propellants and oxidizer.

In the early 1980s, the station employed 750 people, including 250 engineers and technicians and had a ground area of 5 km^2. It is guarded by a strong military presence. The first stand was built for initial rocket tests in the 1950s and stands 2 and 3 in anticipation of the Dong Feng rocket developments in the 1960s. Stand 1

Rocket engine test stand.

now handles attitude thrusters. Stand 2 was subsequently modified to test the liquid hydrogen third stage of the Long March. Stand 4, built in 1963, is designed for all-up testing of a complete launcher. Stand 5 is available for horizontal tests of rocket engines and hydrazine thrusters. The centre has facilities available to supply the full range of fuels required for long engine tests.

The moored test stand is 30 m tall and has a 33 m flame trench. This is designed for all-up, full-power tests of entire rocket systems, not just individual engines. For a rocket in development, this is the last phase of static testing. Rockets under test are held between two 33 m high concrete pillars set in reinforced concrete 23 m deep.

Other specialized facilities
Wind tunnels are a speciality of the Aerodynamics Research Institute, Beijing. The first wind tunnels were built in China in the defence ministry in 1959. They played an important role in determining the air flow and pressure on climbing rockets and the successful execution of staging.

To test against leakages in satellite, the Beijing Satellite General Assembly Plant developed a highly sensitive leakage detector using krypton-85, able to pick up a leakage of 50 microns (half the width of a human hair). For the Long March 3 third stage, the Lanzhou Physics Institute developed a helium mass spectrum leakage detector.

One of the strangest test facilities is that used for testing out satellite radio systems. In order to eliminate the possibility of interference, the main requirement of

Thermal vacuum testing.

the test hall is that it have no metallic components. Accordingly, it is made entirely of glued red pine wood.

Thermal vacuum tests are essential if a satellite is to withstand the intense cold of the space night and the great heat of the space day, complicated by vacuum (10^{-13} torr at geostationary altitude). A series of vacuum chambers were built in Beijing and Shanghai in the 1960s, able to simulate up to 10^{-7} torr. The largest vacuum chamber for testing spacecraft before flight, the KM 4 may be found in the Environmental Simulation Engineering Test Station in Beijing. Spacecraft are lowered by crane into the 7 m diameter, 12 m high chamber where they may be alternately frozen, heated, shaken and exposed to a vacuum. Supercooled helium is the chief agent for freezing the chamber while pumps are used to suck the air out. There is another vacuum chamber in the Huayin Machinery Plant in Shanghai.

The Beijing Institute of Spacecraft Environment Engineering has five vacuum chambers.[5] Their dimensions are as follows:

	Diameter	Length or height
KM 1	2 m	3.2 m long
KM 3	3.6 m	7.3 m long
KM 4	7 m	12 m high
KM 5	5 m	10 m high
KM 6	12 m	22 m high

The KM 6 was completed in 1998, with the severest levels of vacuum, temperature and solar light under the supervision of Shenzhou deputy commander Yuan Jiajun. The first Shenzhou was tested there in September 1998.

180 Behind the scenes

Large vacuum test facility. (Courtesy China Astronautics Publishing House.)

New testing facilities were opened by Harbin Polytechnical University in 2000 after four years of construction. The Environmental and Engineering Space Laboratory, directed by Professor Yang Dezhuang, initially had 32 researchers and was designed to simulate the space environment, with its vacuum and electronic radiation, offering conditions for the testing of elements, materials and full-scale spacecraft. Harbin Industrial University was made the national leader of the Project 863 industrial robots project for the development of industrial robots, many of which could be applied to work on space stations and in automatic lunar exploration.[6]

INSTITUTE FOR AVIATION & SPACE MEDICINE

The Institute for Aviation & Space Medicine (many variations of this name appear) in Beijing is the main centre for the preparation for manned spaceflight and life-support systems. It was established as far back as 1968. The person mainly responsible for the development of aviation and space medicine in China is Cai Qiao (1897–). From Jieyang in Gungdong, he studied psychology and then medicine in

California, Chicago and subsequently in London and Frankfurt. He received senior appointments in China soon after 1949. He directed research into aviation and space medicine from the very beginning, designing test centres and carrying out research into survival at altitude. He has been published in six major books and over a hundred papers, his main text being the *ABC of aviation medicine*.

In 1984, the institute developed a Gemini-class space suit. In the early 1980s, five men spent a month in a high altitude chamber at a pressure of 7psi. In spring 1997, there were reports that an underwater tank had been built in Shanghai to train astronauts in simulated weightlessness – though underwater tanks are normally associated with space walks. The president is Min Guirong.

There are two large centrifuges in China. The most advanced is the large centrifuge operated by the Institute for Aviation and Space Medicine which has an arm of 12 m, is computer-controlled and can reach a maximum acceleration of 25 g. It is used to test the effects of high gravities on spacecraft, during ascent or descent, but can also be used for training astronauts. In addition, the Shanghai Research Institute of Satellite Engineering has a 15 m long centrifuge, the biggest in Asia, which can achieve 17 g.

CHINESE ACADEMY OF SCIENCES (CAS)

As in the Soviet, now Russian, space programme, the Academy of Sciences makes an important contribution by the provision of advice, personnel and facilities. A space committee was set up in the academy in 1963. The Academy of Sciences has five departments – mathematics, physics and chemistry; technology; biology; Earth sciences; and philosophical and social sciences. The first two are of the greater importance for the space programme, the former having institutes of mathematics (Beijing), physics (Beijing), dynamics (Beijing), chemistry (Beijing), applied chemistry (Changchun) and observatories in Beijing and Nanking. A Space Science and Technology Centre was formed in the late 1980s with departments of space science, space technology, space programmes, remote sensing and ground stations. One of the institutes of the Chinese Academy of Sciences is the Xian Institute of Radio Technology (XIRT), established in 1956 and transferred to CAST in 1968. XIRT has played a lead role in developing the electronic capabilities of Chinese satellites. The department of technology includes institutes of electronics (Beijing), cybernetics (Beijing), metallurgy (Shanghai), new metals (Shenyang) and optics (Changchun). Two Academy of Sciences institutes in Beijing are of particular importance – the Centre for Space Research and Applied Sciences; and the Laboratory for the Detection of Microwaves, Technology and Information. In 1994, the academy broke the first ground of the new Beijing Centre for Payloads and Applications. These are the other main institutes in the programme.

182 **Behind the scenes**

Key institutes and bodies in the Chinese space programme (and their specialities)

Shanghai Electronic Equipment Factory	Spacecraft electronics
Xinyue Mechanical Electronics Plant, Shanghai	Gimballing systems Precision instruments
Scientific Instrument Factory and Institute of Technical Physics, Shanghai	Sensors
Shansi Electronic Equipment Factory	Spacecraft electronics
Tung Fang Scientific Instrument Plant, Beijing	Testing devices
Lanzhou Space Research Institute	Low temperature, vacuum resting
Institute of Space Physics, Xian	Computer software
Institute of Radio Technology, Xian (XIRT)	Radio communications
Institute of Space Technology, Xian	Charge coupled devices
North West Institute for Electronics, Xian	Antennas, ground tracking
Shaanxi Liquid Rocket Engine co	Engines
54th, South West, Institute of Electronics Technology, Shijiazhuang	Tracking systems
Institute of Power Sources, Tianjin	Solar panels

Finally, a number of institutes have been set up to support the applications work of the Chinese space programme. These were:

- National Remote Sensing Centre
- National Satellite Meteorology Centre
- China Resources Satellite Application Centre and the
- Satellite Oceanic Application Centre.

MISSION CONTROL: XIAN

Proper control centres are essential elements in national space programmes. The Americans, for example, use facilities in Houston, Texas for manned missions and the Jet Propulsion Laboratory in Pasadena, California for deep space missions. China first began space tracking soon after the Soviet Union launched Sputnik 1 in 1957. Telescopes and radio receivers were installed in Beijing, Nanjing, Guangzhou, Wuhan and Shaanxi. An artificial satellite motion theory laboratory was founded by Shang Yushe of the Purple Mountain observatory (Nanjing) in 1958 where he collected in data from observations and predicted future satellite passes. In early 1965, Nie Rongzhen and Zhou Enlai conferred on establishing a ground tracking network in anticipation of China's own first satellite. In 1967, the existing network of satellite observation stations made a series of radio observations of foreign satellites to refine their observational techniques. Since then, the tracking network was established as an organizational command under the State Commission for Science, Technology and the National Defence. It has 20,000 workers, including 5,000 engineers.

Later in 1967, construction began of the network that would track the first Chinese

Mission control: Xian

Mission control, Xian. (Courtesy China Astronautics Publishing House.)

Earth satellite, the Dong Fang Hong. It was decided that there would be seven ground stations at strategic points on the Chinese land mass, the main control centre being in Weinan, Shaanxi (though in the event, Jiuquan fulfilled this function for the first launch). Each station would be equipped with tracking radars; tachymeters to receive the signals from the satellite; data processing systems; and control systems to send commands to the satellite. They relied largely on optical systems. Since then, more sophisticated and accurate laser trackers have been introduced.

The most westerly station was Karshi in the high western desert and this became designated the 'number 1 tracking station' because it was the first station Chinese satellites would overfly on their west to east paths across the sky. Karshi was built in the course of 1968–70 to be ready for the first Earth satellite. When Dong Fang Hong duly appeared, Karshi gave predictions for when the satellite would overfly the 244 largest cities in the world.

China's tracking system

Mission control:	Xian
Tracking sites:	Weinan, Shaanxi province, near Xian
	Min Xi, Fujian
	Xiamen, Fujian
	Changchun, Jilin
	Karshi, Xinjiang
	Nanning, Guanxi
	Yilan, Nanning
	Guiyang, Nanning
Additional observation sites:	Jiaodong, Lhasa, Mount Wuzhi (Hainan)

184 Behind the scenes

A tracking station.

For the later series of communications satellites, China built more ground stations: in Nanjing (1975), Shijiazhuang (1976), Kunming (Sichuan) and Urumqi (1982), Beijing (1983, the central station) and Llasa, Tibet (1984). Optical observations of satellites continue to be made from Nanjing.

China's main mission control is located in Weinan, 90 km north east of Xian in Shaanxi province. Xian, at 34.3°N, 108.9°E, is fed information from fixed tracking sites and up to four tracking ships at sea. Xian control was set up on 23rd June 1967, though it was then called the Satellite Survey Department. Construction appears to have begun straight away and the first computers were installed in 1968. Xian actually consists of two centres – a satellite telemetry and control centre and an adjacent ground station.

The control centre is located in 5,000 m high mountains whereas the ground station is located 10 km distant on a high plateau. In between are antenna farms and masts. In the autumn, as the communications dish swivels to track satellites in the sky, Chinese farmers may be seen gathering in wheat by hand on the terraces they share with the station. Mission controllers live nearby in modern apartment blocks. The mission control room comprises television screens, consoles, plotters and high-speed computers. The centre also includes a section devoted to following, calculating and predicting the orbital paths of all Chinese satellites in orbit at a given time. A core of 120 engineers runs the centre.

The ground station has a further hundred engineers and technicians. In the 1980s, Zhou Enlai slogans were hung on the wall, reminding engineers to go about their tasks carefully and seriously. The ground station has a key role in maintaining contact with geosynchronous satellites, with which it holds 30 mins communication sessions each day. Young scientists dominate the crew of Xian control – the average age is under 33 years. The director is Zhang Fengxian.[7]

A Landsat Earth resources station has been operating in Miyun (100 km north-east of Beijing) since 1986 with a processing centre in Beijing itself. It also receives data from the European Earth Resources Satellite (ERS) and the Japanese Earth Resources Satellite (JERS). Ground stations in Guangzhou and Urumqi receive data from the Feng Yun metsats and the American NOAA metsats. Later, China has picked up data from the French SPOT system (Satellite Pour l'Observation de la Terre).

In addition to controlling Chinese satellites in orbit, there is an important task in identifying and following satellites in orbit, be they Chinese or belonging to other countries but overflying China. Responsibility for tracking in China lies with the China Satellite Launch and Tracking General Control, set up 1966–7. This comprises the Beijing Aerospace Command and Control Centre, the Satellite Maritime Tracking and Control (comships), the Institute for Special Equipment in Beijing, and the Institute for Tracking, Command and Control in Luoyang, down the Yellow river from Xian. The system monitors satellites in low Earth orbit and in 24hr orbit. About 20,000 people are involved in the whole tracking system, including 5,000 engineers.

In 2000, COSTIND was given ¥30 m (€3 m) to begin systematic study of space débris in advance of the first manned spaceflight. By this time, about 26,000 objects had been sent into orbit, of which 9,131 were still in orbit, under surveillance by Chinese tracking and considered large enough to pose a danger. By the new century, space débris had become a serious problem and NASA had routinely tweaked the orbit of the space shuttle to avoid small pieces of potentially dangerous débris. In August 2003, the city of Shanghai hosted the second Chinese national space débris workshop. There, Li Benzhen told the workshop that an automatic collision avoidance system had been designed for the Shenzhou manned spaceship to move it out of harm's way in time.

COMSHIPS

One of the main weaknesses in China's tracking system was a lack of overseas bases. One of the first achievements of the United States space programme was the construction, with the assistance of many friendly nations, of a worldwide network of ground-based tracking stations to assist in the manned space programme and deep space missions. Lacking strong allies overseas, the Soviet Union and China had to rely on communications ships, or comships. These were especially important when satellites were flying over southern latitudes, away from the northern Russian and Chinese land masses. Comships were first used to track Chinese satellites from the sea in the 1970s, two oceanographic ships being used, the *Xiangyanghong 5* and the *Xiangyanghong 11*.

The current tracking ships are called the *Yuan Wang 1,2*,3 and 4 (the words mean 'looking far into the distance', or 'long view' for short in Chinese). The first two *Yuan Wang* were ordered in September 1977 and delivered in 1978, being prepared

186 Behind the scenes

A *Yuan Wang* tracking ship. (Courtesy China Astronautics Publishing House.)

for the tests of the Dong Feng 5 missile in May 1980. They are ocean-going in size, each being equipped with two 20 m wide communication dishes. They carry out oceanic survey work when not participating in space tracking tasks. They were brought into dry dock for extensive refitting in 1987. Their range is 21,000 km and they can stay out to sea for 100 days at a time. Both ships were completely renovated in their home port of Shanghai in 1998–9 during preparations for the first flight of Shenzhou.

Yuan Wang 3 was commissioned in March 1995. This is a big ship of 20,000 tonnes displacement and 190 m long. Described as a scientific city at sea, it looks like a cruising liner. It has nine decks. The *Yuang Wang* top deck is equipped with antennae, arrays and satellite dishes with a helideck where weather balloons are launched. Below deck are computer and control rooms, much like mission control on land. A fourth ship, originally the *Xiangyanghong 10*, joined the fleet as *Yuan Wang 4* in July 1999, just in time for the first flight of the Shenzhou.

Comships have drawbacks. They are expensive. For the ocean-going tracking fleet, conditions in the southern hemisphere's seas are quite poor during April – October, which has the effect of limiting missions like the Shenzhou tests to the southern summer when they are kinder. For these reasons, China began to consider overseas ground stations. In 1997, China opened a very small tracking station on South Tarawa Atoll in the Pacific with the permission of the government of Kiribati. The dish was so small it barely peeped over the boundary wall, never mind reaching the height of the surrounding palm trees. The arrangement ended in tears in 2003 when Kiribati, bucking a global trend, decided to recognize Taipei, Taiwan as the government of China rather than Beijing with whom it already had diplomatic relations. Apparently, the Taiwan government had offered Kiribati more development aid. China retaliated with the decision to dismantle its station and began to do so within a month of Yang Liwei returning to Earth. In January 2001, China signed an agreement with Sweden for mutual access to its tracking network. Here, either side could purchase access to the other country's network whenever need arose. The Swedish Space Corporation operates receiving dishes in Sweden and in neighbouring Norway in Tromso.[8]

In September 2000, China began construction of its first overseas land satellite station, in Swakopmund, Namibia. For observers, this might appear to be a strange location. However, the retrofire manoeuvre for China's manned spaceship, the Shenzhou, takes place soon after it passes over the coast of south-west Africa. On the FSW series, the Chinese had always done the retrofire over China, making a crude, nose-down dive. This had the merits of ensuring a descent into Chinese territory, with little danger of straying into neighbouring Russia, but made reentry a violent event with high G forces. For Shenzhou, a more gentle descent path was required, but one that would involve a retrofire manoeuvre far away from Chinese territory.

During the flights of Shenzhou 1 and 2, China positioned a *Yuan Wang* tracking ship off south west Africa to prepare for and monitor these crucial manoeuvres. A nearby land station, requiring fewer personnel and not being affected by rough seas, offered a cheaper and more secure alternative. Accordingly, China and Namibia signed an agreement for a tracking station to be built beside the Swakopmund salt works, on the road to Henties Bay. A company called Windhoek Consulting Engineers got the $N12 m contract (€1.5 m) to build the 12,675 m^2 complex. The ground station was completed in July 2001 and Zhang Jiefu was appointed station director. The station comprised satellite dishes, control rooms, administration building and support facilities. The two dishes – one of 5 m, the other 9 m – were set 16 m above ground. The station had five permanent staff, expanding to 20 when missions were under way. Not only would the station play a key role in the retrofire manoeuvre, but it would receive television pictures from the astronauts in their cabin, much as the Soviet Union had done during its early manned flights. In June 2003, another overseas land station was constructed, this time in Pakistan, to follow the descent to Earth of the Shenzhou. A couple of months later, the Italian Space Agency offered the use of its Malindi, Kenya site, providing an additional point of coverage during the reentry.

THE SPACE PROGRAMME AND THE CHINESE ECONOMY

There is no doubt but that the development of a space programme forced the pace in a wide range of Chinese industrial technologies and products. Many western commentaries have questioned whether, for all its poverty, the Chinese economy can afford a space programme. The evidence is that without a space programme, economic development would prove more difficult.

The space programme drove the development of computers, transistors, modern electronics, precision engineering, chemicals, welding, pumps, high-strength materials, to name just a few items. Evidence of the application of the space programme to the economy may be found in a variety of areas, such as important aspects of material innovation, whether they came from the experience of handling low-temperature fuel or building high temperature resistant heat shields: 220 new materials alone were used in the development of the Long March 3 third stage. In metallurgy, rocket development led to significant advances in high-temperature soldering, chemical milling of fuel and storage tanks and the machining of titanium alloys. The welding skills used in rocket casings led to major advances in new forms of welding like electron beam, plasma, laser and diffusion welding. Glass fibre reinforced plastics, used for the first time on the FSW recoverable cabin, had been an infant industry in China. The techniques used on developing the FSW were later applied in bottles and cylinders, tractors, locomotives, boats and household appliances.

The gains were not just in innovation but in standards. The requirements of quality control in rockets and satellites, in which parts were required to work flawlessly for years under extreme conditions, drove up standards in a range of industries. For example, the tough quality requirements for rubber seals in rocket engines improved the standards of rubber seals used in more humble industrial and domestic situations. By 1986, the Chinese were able to identify 1,800 industrial items of space spin-off, where techniques developed for the space industry had been used in the rest of the economy. By the same time, almost five thousand awards had been made for space products and processes for standards, new products and inventions. These were some of the highlights:

- techniques developed for the automatic control of spacecraft at the Beijing Institute for Control Engineering were used to develop control systems for industrial film production;
- advance control technology was adapted directly to the production line system for the Beijing Packaging Company Cardboard Box Factory and for the Capital Machinery Factory's bread production line;
- telemetry control systems developed for the space industry were adapted for the navigational control system at Beijing airport and for automatic control of electric systems and reservoirs;
- satellite telemetry monitoring systems were adapted for telemetric monitoring of heart patients in hospitals and for an x-ray medical television system;
- temperature control systems devised for communication satellites have been used in cotton mills, swimming pools and for the manufacture of ceramics, with considerable energy savings;
- vacuum test chambers, used to prepare satellites for the rigours of the space environment, have been used for the nuclear industry and in advanced optics;
- cryogenic technology, mastered for the Long March 3 upper stage, has been used in the development of refrigerators and night-vision optical equipment;
- television transmission systems, used on communication satellites, have been adapted for miniature television monitoring systems for the police, harbour, traffic control and airport authorities;
- the use of titanium as a material for rocket engines has been adapted to gas turbine engines. Plasma spray coating developed for satellites is used as a permanent spray on electricity lines which previously had to be repainted every three years;

- Precision machinery developed for rockets was used by the Chinese carpet industry to make new, automatic carpet-weaving machines with intricate designs;
- inertial guidance systems for rockets have been modified for installation on oil drills to measure their precise location under the ground;
- infrared satellite sensors have been fitted to trains to warn of bearings overheating and compromising safety.

The value of a space programme in driving the economy has long been a matter of debate – generally inconclusive debate – between economists. Intuitively, many feel that the development of cutting-edge technologies as a result of space investment must spin off to the rest of the economy. Critics take the view that building space cabins is an inefficient means of developing household appliances. Some take a broader view. State-led investment in areas such as the space programme will propel national development, they say. If China were to be led by world market forces alone, China would be purely a supplier of goods and materials to the European and American trading blocks. During Deng Xiao Ping's period of market-led development, Chinese technology actually fell back in some areas. Now, with state-led investment in the space programme, China has the opportunity to lead in distinct areas of regional and world technology.[9]

INTERNATIONAL CONTACT

From 1956 to 1977, with the exception of the brief period of the Sino-Soviet accord, China developed its space programme relying almost entirely on its indigenous resources. During the period of rectification and reconstruction, Deng Xiao Ping led a policy of openness and cooperation. International cooperation was subsequently promoted at a commercial level by the Great Wall Industry Corporation; at a scientific level by the Chinese Society for Astronautics; at a ministerial level by the Ministry for the Space Industry.

A series of exchange visits and meetings kick-started the process in 1977–9, such activities taking place with Japan; the United States; and collectively and individually with the members of the European Space Agency. China's first international agreement was with France. The protocol agreed between the two countries covered cooperation in the areas of communications satellites, the surveying of natural resources, launchers and balloons. The Chinese were invited to watch the launch of an Ariane rocket. An agreement with Italy shortly afterwards involved the use by the Chinese of an Italian communications satellite called Sirio which was moved from its normal position in geosynchronous orbit (15°W) to 65°E, to test out ground stations in anticipation of China's first comsat. A memorandum was then signed with the European Space Agency in 1986.

In the course of time, links were built up with over 40 countries. The standard procedure was for the first contacts to lead to bilateral visits, the exchange of minutes of meetings, followed by a protocol for cooperation initialled by the two

Cooperation with Italy. (Courtesy *Aerospace China*.)

governments. This led to structured contacts thereafter functioning at varying levels of intensity (permanent working groups, as with Russia, being possibly the most intensive). China has now entered cooperative arrangements with a number of countries. These are detailed in the following tables. China has entered into bilateral collaborative programmes with Brazil (e.g. the Ziyuan project) and Germany (future solar observatory) and the European Space Agency (Double Star). In the first years of the new century, the cooperation with Russia was probably the most intense. This work was done through a bilateral commission

Cooperation with Brazil. (Courtesy *Aerospace China*.)

which met alternately in Moscow and Beijing. The commission covered 21 cooperation areas with eight priority themes.

The Chinese signed an exchange agreement with the United States in 1978 and an Understanding on Cooperation in Space Technology in 1979. A Joint Commission on Scientific and Technological Cooperation was meeting by 1980 and working groups were set up. Later, two small Chinese student chemical and materials experiments flew on board the space shuttle (mission STS-42) in January 1992. A Chinese alpha magnetic spectrometer flew on the space shuttle *Discovery* mission to the Mir space station in June 1998. Two Chinese universities – Southeastern in Nanjing and Jiaotong in Shanghai – signed plans to jointly fly another alpha magnetic spectrometer, this time to the International Space Station, through the Massachusetts Institute of Technology.

The level of Chinese international collaboration was evident in September 2001 when 300 delegates attended, from 20 countries throughout the region, the 6th Asia-Pacific Conference on Space Cooperation. The Chinese delegates arrived direct from a day-long meeting of French and Chinese specialists from the French space agency, CNES, Alcatel, SPOT image and other leading French companies.

International cooperation agreements

Joint working groups	United States, Italy, Russia
Bilateral cooperation agreements	Germany, Italy, United States, France, Germany, Britain, Brazil, Russia, Ukraine, Europe, India, Pakistan, Argentina
Multilateral agreements	European Space Agency
Commercial agreements	France, Japan, Germany, Russia
Joint programmes	Brazil, Germany, Europe
Other contracts or agreements	Netherlands, South Korea

In 1992, China joined the COSPAS/SARSAT international satellite-based sea and land distress and rescue system. This uses American and Russian satellites to relay distress calls from ships foundering at sea (most famously to rescue stranded yachtsmen). In 1995, a €17.5 m agreement was made for France to supply China with an Earth terminal for COSPAS/SARSAT.

A cooperation accord was signed with Ukraine 1997 – Ukraine being the home of the biggest rocket factory in the world (NPO Yuzhnoye, Dnepropetrovsk), producer of the Zenit rocket and a range of satellites (e.g. Okean). Part of the deal saw the Harbin Polytechnic Institute buy a thermal simulator from the Ukrainians for €1.17m. The agreement went ahead despite an incident just beforehand in which Chinese space experts were accused of spying in the Pivdenmash plant in Dnepropetrovsk.

Xi Chang launch centre has even received tourist visits. From 1985, China began to exhibit its commercial and scientific achievements in space at international exhibitions and the big aerospace shows, like the biennial Paris air show.

Participation in the International Telecommunications Union, a specialized United Nations agency, was particularly important for China, since the body

192 Behind the scenes

regulated international television and radio frequencies, including those of geostationary comsats. China did not obtain its seat there until 1972, at the expense of the Republic of China (Taiwan). It asked for and received three orbital positions for its geostationary comsats in 1977 and more subsequently.

China joined COPUOS, the United Nations committee on the exploration and peaceful uses of outer space in 1980, specifically attending the meetings of its Science, Technology and Law subcommittee. At the first meeting, China announced an allocation of €59,000 toward a conference on space science and technology in the Asian-Pacific region which it would host (it was duly held in 1985). China has signed the main international outer space treaties of the United Nations – those for the exploration and use of outer space, the return of stranded astronauts, responsibility for damage caused by space objects and the registration of objects launched into space.[10]

Finally, the Beijing Marine Communications & Navigation company is the Chinese member of the International Maritime Association (INMARSAT). It sells and leases INMARSAT lines and equipment to foreign and domestic customers in China.

Chinese membership of international, space-related organizations

International Astronautical Federation
International Telecommunications Union
United Nations Committee on the Peaceful Uses of Outer Space
International Maritime Satellite Association, INMARSAT
International Organization for Standardization

Perhaps the moment which marked the end of China's coming of age in the international space community was the 47th International Astronautical Federation (IAF) Congress, held in Beijing in October 1996. Attended by two thousand domestic and over a thousand foreign delegates, the IAF congress was opened by Chinese President Jiang Zemin and hosted by prime minister Li Peng. The Chinese presented their space industry on show, brought westerners around Chinese space facilities, unveiled plans for their own space future and appealed for greater international cooperation between China and its international partners.

ASSESSMENT AND CONCLUSIONS

China now has a well-developed space infrastructure and architecture. The current architecture emerged from the Soviet-style academies and post-box institutes of the 1950s and the political upheavals of the 1960s into a hierarchy that involves an executive space agency, CNSA, working with the Space Leading Group and COSTIND. Beneath these is the main industry umbrella body, the Chinese Aerospace Corporation (CASC), which brings together the leading academies in the areas of launchers (e.g. CALT), satellites (e.g. CAST), solid fuel technology (e.g. ARMT), electronics (CASET) and navigation (10th academy). They are flanked by the wide range of facilities, institutes and bodies used for rocket engine testing, space medicine, vibration testing and static engineering. Since the time of Sputnik 1, China has built up a national network for the optical tracking of satellites and a national control system based in Xian. This has been supplemented by communications ships at sea and, recently, by the development of land-based tracking stations abroad, for example, in Swakopmund. There is evidence that China sees its space industry as a driving force in industrial development, citing many diverse ways in which different types of space technology have been put to economic purpose. Finally, China is, despite its degree of diplomatic isolation, an active participant in bilateral and international space cooperation programmes.

There is evidence of an on-going programme of improvement, updating and development in the Chinese space infrastructure. There are reports of new capital investment, for example in tracking facilities, solid fuel rockets, new facilities for CAST, payloads and applications. New investment can be seen in cutting-edge space technologies such as robotics and minisatellites, as evidenced by developments in Qinghua University and Harbin. All this is apart from the extensive top-of-the-range new facilities built to support the manned space programme which will be described later (chapter 9, 10). The benefits from this on-going investment are likely to be seen over the next ten years or more. Developments in China must make for a dramatic contrast with Russia, where fresh investment in facilities, research and development has, due to its difficult economic situation, been virtually absent since the end of the Soviet period.

REFERENCES

1 Zhang Yun (Ed): *The Chinese space industry to*day. China Social Sciences Publishing Co, Beijing, 1986.
2 Agency details report on visit to Beijing satellite manufacturer. BBC reports, 23rd September 1998); New space centre in Beijing. *Go Taikonauts!* site, 8th November 1999, http://www.geocities.com/CapeCanaveral/Launchpad/1921/news
3 Beijing university key to space development. *Aviation Week & Space Technology*, 12th November 2001.
4 Beijing Institute plays important role in rocket, satellite testing. BBC reports, 6th May 1998.
5 Théo Pirard: Chine: Simulators et lanceurs. *Air & Cosmos*, 22 juin 2001.
6 China sets up first space engineering laboratory. BBC reports, 8th March 2000.
7 Young scientists dominate space instrumentation. BBC reports, 12th May 1999.
8 Chinese and Swedish satellite operators sign long term access agreement. *Spacedaily*, 24th January 2001, http://www.spacedaily.com/news.
9 Susan Lawrence & Sadanand Dhume: Houston, we have company. China leads Asia toward the final frontier – but is prestige worth the cost? *Far Eastern Economic Review*, 3rd August 2000.
10 For a recent description of Chinese international cooperation and a view on the American blockade, see Lan Xinzhen: Expanding international space cooperation. *Beijing Review*, 11th July 2002.

8

Launch centres, rockets and their engines

One of the most important parts of the infrastructure of any space programme is its launch sites. China has three main launch sites – Jiuquan, Xi Chang and Taiyuan – with, in addition, a military base, a sounding rocket site and the prospect of a new spaceport on the large island of Hainan. Here the development and role of each of the space centres is reviewed. Then there is an account of the main rockets of the Chinese space programme, their technical capacities and prospects for future development. There is an examination of the crucial rocket engines that make the achievements of these launchers possible.

LAUNCH SITES

Perhaps the most important and expensive infrastructural element of any space programme is its launch site facilities. China has three launch sites. The first, Jiuquan, was built in northern China for China's first satellite, Dong Fang Hong and is the base for the manned space programme. The second, Xi Chang, was built in Sichuan in south-western China for launches to equatorial orbit. The third, Taiyuan, was built for launches to polar orbit. There is a minor launch site for sounding rockets, Haikou, on the island of Hainan. For the sake of completeness, one should mention the military-only launch site in Harbin, Manchuria, used for the Dong Feng 4 and 5, and presumably for its later replacements, the DF-31 and DF-41 (location: 45.8°N, 126.7°E. This site has also been named Jingyu). There have also been occasional discussions about the construction of equatorial Long March launch sites abroad, though these have never come close to project design. These discussions centred on a collaborative project with Brazil for an equatorial launch site and with Indonesia on Biak island. A new, domestic southern site is more likely to happen, on Hainan island.

Of the three existing launch sites, Jiuquan and Xi Chang are best known. Both have been seen by western visitors attending the launches of scientific missions (Jiuquan) or commercial communications satellites (Xi Chang). The Sichuan site is even on some tourist itineraries, while the Great Wall Industry Corporation provides considerable technical detail on Jiuquan in the *Long March 2C launcher manual*. Although some aspects of the Chinese space programme are still secret, details of these launch sites relatively plentiful.[1]

Launch sites: orbital and sounding rockets

Centre	Location	1st flight	Launchers
Jiuquan	40.57.4°N, 100.17.4°E	April 1970	Long March 1,1D, 2, 2C, 2D, 2F, 2EA, 5, Feng Bao
Xi Chang	28.2°N, 102.02°E	Jan 1984	Long March 2E, 3 series, LM-2C-CTS
Taiyuan	38.8°N, 111.5°E	Sep 1988	Long March 4 series, 2C-SD
Haikou	19.31°N, 109.5°E	Dec 1988	Sounding rockets

Although like Russia, the Chinese launch centres rely on rail, in practice rocket stages are often brought the final stages to the pad by road. Although like the Americans, rocket stages are assembled vertically in doors, in China this is done out of doors (with the exception of the new Long March 2F). Of the three main launch sites, the busiest is Xi Chang. The following table lists the total number of launches from each:

Launches by centre, 1970–2003

Jiuquan	29
Xi Chang	31
Taiyuan	15
Total	75

JIUQUAN

Jiuquan launch site is located in the Gobi desert in north-west China – the environment is similar to Russia's Tyuratam – Baikonour cosmodrome in Kazakhstan. To make two important distinctions, the town where the engineers and specialists live is sometimes called Dong Feng ('East Wind') and there is an East Wind Park in the downtown area; while the launch site is sometimes called the East Wind Centre (*Shuang Cheng-tzu*). The actual launch centre is some distance to the north-east of the real town of Jiuquan, which can now be found on the most recent tourist maps. For the sake of convenience, the launch centre will be given the generic designation of Jiuquan.

Storms whip up the desert sands from time to time, dunes often creeping toward the railway line serving the site. Being desert, the average rainfall is very little – only 44mm annually. When it comes to temperatures, this is a place of extremes.

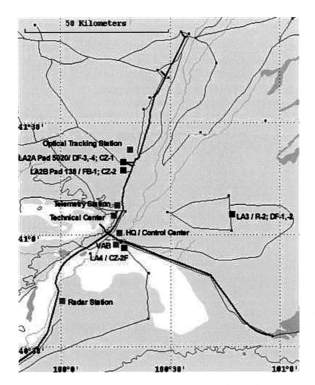

Jiuquan: an overview. (Courtesy Mark Wade.)

Temperatures range from $-34°$ C in December to $+42.8°$ in July. Averages are kinder: from $-11°$ C in January to $+26.5°$ in July. The winter nights are bitterly cold, down to $-30°$, but the skies brilliantly clear. The precise location of Jiuquan is given as 40.57.4°N, 100.17.4°E although the cosmodrome covers an area of 5,000 km². There were two big building phases: 1958–62 and 1992–9. Launch site director is Zhang Jianqi with Lu Sicheng its political commissar.

Jiuquan is an oasis, the word meaning, literally, 'liquor springs'. Jiuquan is on the river Ruoshui which is very seasonal and runs north-south. The thin soil is a light dusty brown shale. Two thousand years ago, Jiuquan was a battleground where the Han dynasty erected fortifications to keep out the invaders from the west. Subsequently it became part of the north-western silk road – a long trading route which, thousands and thousands of kilometres further west, passed by another cosmodrome in the Kazakhstan steppe. There are few bushes there, only brown camelthorn and a few wild animals, mainly yellow goats and wild deer. Later, some elms and red willows were planted. Not far from the launch site, Mongolian herdsmen may be seen minding their sheep. Camels wander past from time to time. In the very early days, conditions in Jiuquan must have been very primitive. Now, the centre has a fully equipped railway station, hospital and library. The town's main street, Chang an street, has lamps sculpted in the shape of rockets. A total of 20,000 people live and work there now. Water is supplied by a reservoir of 10 km² just outside the town. There is a graveyard

198   Launch centres, rockets and their engines

Rail transport, Jiuquan. (Courtesy Sven Grahn.)

to the north east and five hundred people now lie buried there, including one of the great military directors of the Chinese space programme, Marshal Nie Rongzhen.

The cosmodrome is spread out from the railway station toward launch sites and facilities to the north, connected by railways and a bumpy road. Although railway plays an important role in the centre, the actual launch sites are served by surface transport as well as by railway. The site is bordered by a rim of desert mountains to the west. It is a very spread-out site, with large distances between facilities, like the Russian cosmodromes at Baikonour and Kapustin Yar, and quite unlike much the much more concentrated facilities in Xi Chang or Russia's Plesetsk. The river runs between the town and the launch pads. During the summer the river runs dry and Swedish space engineers used to cycle to picnic there while awaiting the launch of their Freja satellite.

Construction of Jiuquan began in April 1958 for the launch of the R-2 Soviet rocket that eventually took place in November 1960. A branch line was constructed northward from the Gansu-Xinjian railway and the Gansu-Xinjian highway. Although official histories suggest that most of old Jiuquan was built in the mid-1960s in advance of the first Earth satellite, the American intelligence reports of 1962 described what could only be called a substantial rocket base much earlier on. Corona satellite photographs from the period showed a large housing area, home to an estimated 9,700 people, with barracks sufficient for a further 2,500 rocket troops. The airfield was a fully equipped site able to handle any modern plane of the period.

Ordinary western experts became familiar with Jiuquan in 1976 when the first Landsat Earth resources satellite photographs became available. American aero-

The Swedish team with Freja. (Courtesy Sven Grahn.)

space experts visited the site in 1979. Western scientists followed to accompany scientific payloads in the 1980s and 1990s, but foreign visitors have been small in number.

There are two launch zones now: the original north site with three pads and south site, the south site being built for the manned programme at the end of the 20th century. When north site became known to American intelligence in 1962, it comprised three pads. Running south to north on the west of the river, these were labelled pads A, B and C. A, the most developed, was a double pad, with two concrete sites each measuring 60 m by 60 m, with a concrete bunker to command launch operations. The Chinese later called this area pad 2 and pads B and C to the north were, it seems, used primarily for military ballistic test firings.

Both area 2 pads are served by the same 55 m tall, 1,400 tonne moveable service gantry running on 17 m wide rails between the two. The two concrete pads are only 416 m apart. One pad is called #5020 (designed for the Long March 1) and the other #138 (designed for the Feng Bao, DF-5 and Long March 2 series). The first pad was used for the Dong Feng, Dong Fang Hong 1 and Shi Jian 1 only (though it may be used for Long March 1D). The second was used for most subsequent missions (Long March 2A, 2C and 2D).

A crane on the tower can lift weights of 15 tonnes to 44.5m. Each pad has a flame trench. #5020 has a 37 m umbilical tower which provides fuel, gas and electricity right up to the final moments of the countdown. Pad #138 has a 40 m umbilical tower with 11 floors, a rotary launching stand with four supporting arms for holding the rocket and a 19 m flame trench. The gantry has a clean room for final payload installation. #138 can handle hot tethered test firings, in which the rocket fires all its engines but does not take off. A domed launch control blockhouse is 200 m distant,

Convoy on way to the pad.

The satellite assembly building.

10 m underground, from which controllers watch the launch by TV and periscope and relay information to the main national tracking centre in Xian.

Adjacent are the Huxi Xincun range control centre, assembly buildings, blockhouses and electricity station. The main processing building is 140 m long with an area of 4,587 m^2 with a 90 m by 8 m assembly hall and a 24 m by 8 m fuelling hall. Equipment can be moved around by a crane able to lift 16 tonnes. Adjacent are 25 test rooms for checking out parts of a spacecraft. Beside them are a solid rocket motor checkout and processing hall, 24 m by 12 m with crane, storage and test facilities. The halls guarantee clean room standards of 100,000 class (one dust particle in 100,000 or less), temperatures of 20.5° C and humidity in the 35% to 55% range.

Fuels are stored in underground bunkers. There are barracks for the militia who assist in the launchings (Russian rockets are also launched by troops) and four-storey Soviet-style flats for other workers involved in the maintenance of the site. Willow and white poplar trees are planted around the buildings and walkways to provide wind breaks and colour. Launches from Jiuquan curve over to the south-east – otherwise they will fly over Mongolia to the west, or Russia to the north. Visitors can watch launches from an observation site 4 km to the east of the pad. Sven Grahn, the first western visitor to see a Chinese launch, recalled: 'tables and chairs were arranged directly on the sand and there were loudspeakers on telephone poles to relay the countdown in Chinese'.

Originally, Jiuquan had a surface to air missile site to the south-east, though whether this was for training or to guard the facility or both is not clear. The first American intelligence assessments took the view that Jiuquan was at least inspired by Soviet planning, for the layout had many similarities to the Russian range at Kapustin Yar, the original test centre near the Volga river.

The second substantial expansion of Jiuquan began in 1992, following the start of the manned space programme, project 921. This led to the construction of the new site to the south east, beyond the coal-fired power station. New paving was put down and in 1997 construction began of the vertical vehicle assembly building and the new, steel launch tower.[2] The new construction area was to the east of the river, south east of the old tracking station. The south site comprises:

- Technical centre – vehicle processing building, transit building, non-hazardous operations building, hazardous operation building, solid rocket motor building
- Crawler and tarmacadmed road to the pad
- Two new launch pads

The tower at Jiuquan. (Courtesy Sven Grahn.)

- Umbilical tower
- Launch control centre

Traditionally, Chinese rockets were assembled on the ground in a horizontal position and then reassembled vertically on the pad by cranes. Now, with the vehicle processing building, it is possible to do all the assembly vertically indoors and roll out a ready-to-go rocket to the pad, with fuelling the only major task remaining before countdown. At the new pad, the turnaround period is three days, which means that a new rocket could be got ready for a mission within 72hr of the previous launching.

The vehicle processing building is the equivalent of the Vehicle Assembly Building at Cape Canaveral. The Chinese vehicle processing building is 81.6 m high, 27 m wide and 28 m long, with a 13-floor platform, cranes able to lift 17, 30 and 50 tonnes, with two high bays and two vertical processing halls. Although designed initially to handle the Long March 2F *Shenjian* (it could prepare two for launch at one time), it was also built with the Long March 5 in mind. Engineers can access the CZ-2F from nine different levels of walkways. The door of the building is 74 m tall, 8 m wide at the top and 14 m wide at the bottom and weighs 350 tonnes. Not being able to afford a steel framework, the Chinese constructed the building from reinforced concrete.

Adjacent are the horizontal transit building, 78 m long by 24 m, used to test out the launch vehicle; transit room; non-hazardous operations building for spacecraft checking in clean room conditions, and a hazardous operations building where fuels are loaded before launch. Beside the vehicle processing building is the launch control centre, equipped with a main control room and two smaller control facilities, facing the launch pad 1,500 m distant. The normal criteria for launch are temperatures of $-10°$ C to $+40°$C, humidity less than 98%, winds of less than 10 m/sec, visibility of 20 km and no lightning or thunder within 40 km. It's a huge construction project, sitting on the brown desert with low hills visible through the haze in the distance.

To get to the pad, the assembled Long March 2F travels on a crawler 24 m long, 21 m wide, 8 m high, weighing 750 tonnes. Travelling at 20 cm/sec, it takes the crawler 40 mins to travel from the vehicle processing building to the launch pad 1,500 m distant. There, the launcher and spacecraft are grappled by the umbilical tower, an 11-floor fixed steel structure 75m tall, with floors for fuelling, electrical connexions, fire-fighting equipment and an elevator. Underneath is an underground equipment room.

A second pad branches just off to the left. Originally, it was thought that this would be used for launches, much later, of the Long March 5, but it made its début a mere three weeks after Yang Liwei blasted off nearby. For its first mission, the second pad was used by a Long March 2D to put into orbit the FSW 3-1 recoverable satellite (also known as Jian Bing 4). The FSW was brought down to the pad by a new, 91 m tall launch tower used to test, integrate and fuel the assembled complex, equipped with no less than 40 testing workshops. The introduction of yet another new facility at such a pace was dramatic evidence of the modern state of Chinese launching infrastructure.

Jiuquan: the new pads. (Courtesy Reuters.)

By the time of the first manned mission in 2003, ordinary people could have the kind of bird's eye view of the launch pad that had traditionally been the protected privilege of Pentagon analysts. The small, independent commercial American satellite Ikonos, whose startlingly clear pictures could be ordered over the world wide web, snapped pictures of the vehicle processing building with its support buildings clustered round about and a long black slash across the terrain like a runway marking the route down to the pads to the south-east.

Visitors to Jiuquan fly in to Dingxin airport, 97 km to the south. This was built to a Soviet design over 1958–62. Getting to the launch centre requires a 90 min train journey through desert featuring bushes and small trees. At one point sand dunes encroach onto the railway: a detachment of soldiers is assigned nearby, their principal job being to clear the dunes when they drift onto the track. Staff at Jiuquan must, to a certain extent, learn to be self-reliant. Some of them keep pigs. The reservoir for the launch site, which is replenished during the rainy season, is used for breeding fish. Air conditions there can be extraordinary clear: one observer brought out his particle detector at the launch site and found that the level of particles was one in a million – the standard of clean room conditions![3]

It is cold to work in Jiuquan in mid-winter. Personnel there receive a subsidy for their winter clothing. The catering record shows that for the new year celebrations in 2002, they got through 3,600 kg of fresh vegetables, 5,000 kg of fish and 6,300 kg of pork, beef and goat. As the population of Jiuquan swelled during preparations for the manned spaceflight programme, steps were begun to improve the environment and appearance of the town and its environs. This took the form of fresh bush and wood plantation, irrigation, wind breaks, measures against pests and the construction of three city parks, called the natural park, the railway park and the sculpture park. For recreation, a swimming pool and football stadium were built. A book shop was put in and department stores, bank, post office, and even a karaoke club.

XI CHANG

When seeking a launch site closer to the equator, the Chinese considered a number of candidates. Altogether, 80 sites were surveyed in 1972, shortlisted to 16 before Xi Chang was selected. The final decision appears to have been made by Zhou Enlai personally. Xi Chang was constructed in the course of 1978 to 1982. The first launch rehearsal was conducted in 1983, with the centre opening in January 1984. The geographic coordinates are 28.2°N, 102.02°E.[4]

Xi Chang is located 1,826 m high in mist-shrouded mountains, near Mt Lijang in

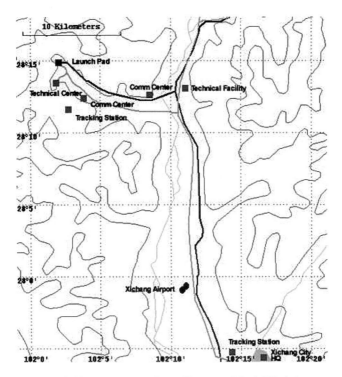

Xi Chang: an overview. (Courtesy Mark Wade.)

the direction of Burma. It must be one of the most scenic launch sites in the world. Nearby are rice paddies and grazing buffalo. To the north lie mountains and giant panda reserves and to the south lie lakes. The launch site is near Xi Chang city and 270 km from Chengdu by rail and road. Xi Chang is on the old south-western silk road which started at Chengdu and headed through Xi Chang into Burma (250 km distant) and India. It is a main passing point for migratory birds. Average temperature is a pleasant 17°C, with 320 days of sunshine a year. The only problem time is midsummer, when there is often heavy rain and at worst, the danger of flooding.

Western commentators have expressed surprise at the selection of a launch site so far inland, in difficult terrain, with poor communications facilities in a relatively populated rural area. The Chinese subsequently explained that during the tense seventies, an inland site was preferred, because coastal sites, though more southerly, were vulnerable to attack by China's enemies.

Xi Chang city has a combined civil and military airport – one now able to take jumbo jets – and is the headquarters of the launch site organizations. To reach Xi Chang launch site by land, one follows the Kunming-Chengdu railway northward along the valley floor of the Anning river until a single branch line turns west into the launch site valley, which one enters by passing a Soviet-style people's memorial. Visitors arriving there drive along roads cluttered by bicycles, water buffalo and farm workers carrying chicken and vegetables to market. One then passes a communication centre, technical centre and command and control centre. To the right of the railway is the first launch pad, for the Long March 3, served by a large 900-tonne 77 m tall gantry which has 11 work levels and a crane. A cement flame trench has been constructed to take away the flames of the rocket on take-off.

The clean room.

To the left of the railway, 1,000 m away, is a second launch pad, constructed subsequently for the Long March 2E, 3A and 3B. It was built in the course of 14 months and first used for the CZ-2E in 1990. This second pad has a huge 4,580-tonne service tower, 97 m tall, with 17 work levels. Just 80 mins before launch, the tower moves back to a distance of 130m. Exposing the satellite to dust and humidity, Chinese satellites were originally stacked on the pad, the shroud then being clamped on top, but Xi Chang now has an air-conditioned clean room on the upper gantry work level.

Also to the left of the railway line lie buildings for storing launchers, the various stages and payloads, though they are finally assembled vertically on the pad by crane. The launch towers are protected by 100 m high lightning rods. Around the gantries are fuelling lines – one set to keep the liquid hydrogen third stage topped up; a second to provide helium which pressurizes the fuel tanks; another for storable fuels. Liquid hydrogen is topped up in the third stage until just 3 mins before lift-off.

Launches out of Xi Chang take a curving trajectory to the south-east, flying over southern Taiwan, north over the Philippines and toward the equator. The ascent is tracked from either side by ground stations from Yibin and Guiyang, Nanning. Satellites are still just over China when they reach the edge of space.

The countdown is carried out in a blockhouse close to the pad, but the overall operation and the subsequent flight are monitored from the launch control centre in a deep gully. The launch control centre comprises a large gymnasium-size room with walls of consoles and a large, 4 m by 5.3 m visual display at the front. The centre is about 3,000 m from the pad, but is not hardened against explosions or falling débris. There is an observation room, able to take 500 people at a time. Laser theodolites, set in domes, track rockets as they ascend to orbit.

A launch campaign in Xi Chang takes 40 days. The rocket is first delivered by rail into a transit hall measuring 30.5m by 14m, before being brought into a much larger assembly room of 91.5 m by 27.5 m. Payloads are checked out in a clean room measuring 42 m × 18 m called the non-hazardous operations building where temperatures and humidity are kept within tight limits. The stages and payloads are then transferred to the hazardous operations and fuelling building where the fuelled solid-rocket stages are installed. Final checks take place in a last checkout and preparation building. The site also has an X-ray facility to check any equipment against cracks. The rocket stages are trolleyed to the pad, one by one, before being assembled vertically.

Xi Chang began to promote itself abroad for investment purposes, calling itself China's 'space city' and even the imitative 'China's Houston'. An American company won the contract to build a 200,000kW hydropower station. Xi Chang director is Tang Xianming.

TAIYUAN

Taiyuan launch centre was originally developed as a launch site for the Dong Feng missile. The base is set in gently rolling hills in Kelan county south-west of Beijing, 1,500 m above sea level. Its precise coordinates are 38.8°N, 111.5°E and it was built

Taiyuan 207

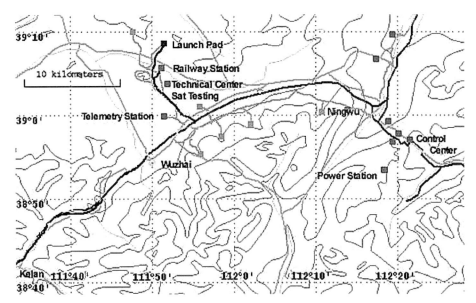

Taiyuan: an overview. (Courtesy Mark Wade.)

Taiyuan launch centre.

on loamy yellow rocks outside the city of Taiyuan itself (38.4°N and 115°E have also been published). The first photographs of the site were not released until 1990: it is much less well-known outside China than Jiuquan or Xi Chang.

Construction was authorized by Mao Zedong, Liu Shaoqi and Zhou Enlai in March 1966. The early construction crews had a difficult time, surviving on Chinese cabbage and potato, but they succeeded in their main task of building a high launching platform of 11 stories and a height of 76.9m. The first Dong Feng was launched from there on 8th December 1968. Taiyuan was used for missile tests in the seventies (e.g. for the submarine-launched ballistic missile) but was not brought into the space programme until the early 1980s. The first static rocket tests were carried out there in September 1984. The best weather for launches from there comes between May and September.

Taiyuan was eventually introduced as a national launch site in 1988, with the first launch of the Feng Yun weather satellite on the Long March 4A. It has a single pad for rocket launches. Initially, only the Long March 4 was used here, but the site then became a busy place when it supported the seven Long March 2C-SD launches of Iridium comsats over 1997–9. Its principal use since then has been for the Long March 4B to put into orbit the Feng Yung metsats and CBERS and Zi Yuan resources satellites.

HAINAN

Hainan is a large but poorly known island to the south-east of China and has a maritime border with Vietnam. It rose briefly to prominence when an American EP-3 spyplane was forced down there in early 2001 and its crew interned.

Construction of a small launch site near Haikou, on the west coast of Hainan launch site began in 1986 and it was commissioned in 1988. The site was intended to serve as a launching platform for sounding rockets able to reach between 120 km and 300 km. The site consists of a launch pad, underground control centre, tracking and payload retrieval systems. Location is 19°N, 109.5°E. It was used for the launch of four Weaver Girl 1 sounding rockets in 1988 and a single Weaver Girl 3 launch in 1991.

In 2000, there were reports that a different part of Hainan might be developed into a full-scale launching centre. CALT disclosed that negotiations had been underway between the central government in Beijing and the regional government of Hainan for a full-scale launch centre near the city of Sanya. Elsewhere the location was given as Wenchang, on the north-east of the island, 70 km from Haikou. The launch site would be 3 km from the coast, with tracking stations on two small islands, 35 km and 70 km to the east.

Further south than Xi Chang, near the Gulf of Tonkin, a new Hainan site could offer a 7.4% payload advantage (in turn, 18.5% more than Jiuquan), the delivery of rocket stages by sea rather than over land and launches out over the ocean, conferring a considerable safety advantage. The cost of development was estimated at €600 m. According to CALT, the cost could be kept down if the new launch site were built in stages, attracted capital investment funds from Japan and Hong Kong and linked to high-tech industrial and tourism services. A launch site for the CZ-2E and CZ-3 family was proposed, supported by a Jiuquan-style vehicle processing building, with an adjacent tourist centre and industrial park. If this were to be successful, it could well have a detrimental effect on Xi Chang. What progress has been made since 2000 is uncertain. The political leadership of Hainan lobbied strongly for the project and by 2003 seem to have convinced most of the key figures in the Chinese space programme.[5]

So much for the launch sites. Next we turn to China's families of launchers.

LAUNCHERS

China has developed two families of launchers – the Long March (LM), known as the Chang Zheng, or CZ, and the Feng Bao (FB, or Storm). The Long March family is divided into four series – 1, 2, 3 and 4. The Feng Bao launcher was used from 1971 to 1981 and is no longer in service. A considerable amount is now known in the west about Chinese launchers, both because they are commercially promoted in the west and due to the work of analysts such as Clark.[6]

At first, the numbering pattern for Chinese launchers can be quite confusing. Although the Long March 3 series (the 3, 3A and 3B) is designed for geostationary orbit, one of the 2 series (2E) is also used for these missions. Thankfully, the Chinese are visually helpful in enabling us to identify rocket launchers, for their white – painted rockets invariably have the launcher type painted in big red letters in large English script on the side after the Chinese pictograms for 'China' and 'Hangtian', the latter meaning 'space' or 'cosmos' in Chinese.

Although, to an outsider, all rockets, being rocket-shaped, appear to have the same means of propulsion, in fact there are many important distinctions between them. First, rockets may use either solid fuel or liquid fuel. Solid fuel rockets operate on the same principle as fireworks. A gray sludge-like chemical is poured into a rocket container. When the nozzle is fired, the stage burns to exhaustion. Solid rockets are very powerful. Their main disadvantage is that they cannot be turned off – they simply burn out. They are less precise and less safe.

Liquid fuel rockets are more complex. They have two tanks – a fuel tank and an oxidizer. Both are pressurized and fuel is injected, at great pressure, into a rocket engine where is ignited. On liquid fuel engines, the level of thrust may be varied (throttled) and the engine may be turned off and restarted. This system is more complex but more versatile and, from a manned spaceflight perspective, safer. Liquid fuel rockets may be divided into three sub-categories, according to the type of fuel used. Most Russian and American civil rockets have used kerosene (a form of paraffin) as a fuel. These are powerful fuels, but they degrade if they are kept in a rocket for more than a few hours at a time. If a launching is missed, the fuels have to be drained and reloaded, a tedious and time-consuming process. From the 1960s, Russian and American military rockets began to use storable propellants, generally based around nitric acid or nitrogen tetroxide and UDMH (unsymetrical dimethyl hydrazine). The advantage of storable propellants is that they can be kept at room temperature in rockets for long periods before they are fired, a necessity when military rockets must be kept in a constant state of readiness. The disadvantage is that such fuels are highly toxic, presenting hazards for launch crews and horrific consequences in an explosion. In 1960, a Soviet R-16 missile exploded at Baikonour cosmodrome. 97 engineers, supervisors and rocket troops died in the ensuing fireball, but the level of casualties was made much worse by the toxic nature of the exploding fuel. It remains the worst rocket disaster in history. Finally, there is the use of liquid hydrogen as a fuel. Liquid hydrogen is enormously powerful, but has to be kept at extremely low temperatures.

Most modern Chinese rockets use storable propellants for their main stages and solid rocket boosters for their small upper stages. The Chinese introduced a hydrogen-fuelled upper stage with the Long March 3 in 1984. The main centres for the development of Chinese rockets are the China Academy of Launch Vehicle Technology (CALT) in Beijing; the Shanghai Academy of Space Technology (SAST) and the Liquid Rocket Engine co, in Shaanxi.

Chinese launchers and their design bureaux

Launcher	First stage	Second stage	Third stage
Long March 1	CALT	CALT	Solid Engine Research Institute
Feng Bao 1	SAST	SAST	
Long March 2C	CALT	CALT	
Long March 2E, D, F	SAST	SAST	
Long March 3 series	SAST	SAST	CALT
Long March 4 series	SAST	SAST	SAST

Adapted from Clark, *Chinese Launch Vehicles*, 8

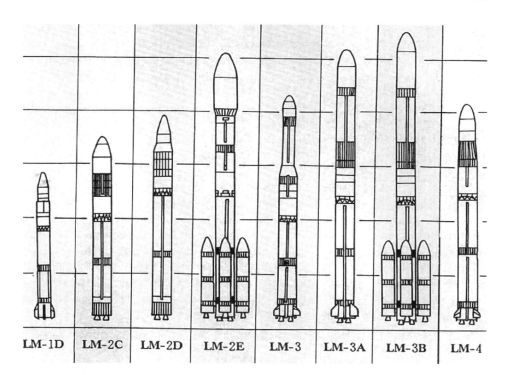

Long March evolution.

China's leading rocket designers

Long March 1	Ren Xinmin
Long March 2	Tu Shoue
Long March 3	Xie Guangxuan
Long March 4	Sun Jingliang
Feng Bao	Shi Jinmao

Now follows a description of each of China's main launchers, with technical details and their launch histories. As with many aspects of the Chinese space programme, this compilation is a hazardous exercise. Precise technical details of Chinese rockets vary slightly from one publication to another. Designators vary even more, especially when it comes to rocket engines. For example, the YF-20 engine when clustered as a first stage engine is called the YF-21; when used as a second stage engine is called the YF-22 but when linked to YF-23 vernier engines is called the YF-24!

Long March 1 (CZ-1): launcher of China's first two satellites
The Long March 1 launched China's first two satellites into orbit – Dong Fang Hong in 1970 and Shi Jian 1 the following year. The Long March 1 is essentially a three-stage version of the Dong Feng 4 medium-range missile developed over 1965-70, a weapon planned to hit targets as far away as the mid-Pacific. On top of the DF-4, a small third stage solid-rocket motor stage was fitted to get the two satellites into orbit.

Three subsequent versions of the Long March 1 have been proposed. The CZ-1C and the CZ-1M were proposed in the mid-1980s as improved versions of the Long March 1, each with more powerful third stages, but neither was developed nor flown. A smaller version, the CZ-1D, was offered commercially from the mid-1990s, but has yet to find customers. The Long March 1 series still has an honoured place in China's astronautical history: launcher of China's first two satellites and a 100% reliability record.

Long March 1 (CZ-1)

Dimensions	29.45 m long, 2.25 m diameter		
Weight	81.6 tonnes		
Capability	300 kg to 440-km orbit at 70°		
	First stage	*Second stage*	*Third stage*
Dimensions	17.835 m × 2.25 m	5.35 m × 2.25 m	4 m × 0.77 m
Weight	64.1 tonnes	14.85 tonnes	
Engines	Four YF-1A	YF-3	GF-02
Fuels	UDMH, nitric acid	UDMH, nitrogen tetroxide	Solid
Thrust	104 tonnes	30 tonnes	3 tonnes
Burn time	140 sec	120 sec	

Flight record

24 Apr 1970	Dong Fang Hong 1 (first Chinese satellite)
3 Mar 1971	Shi Jian (first Chinese scientific satellite)

Long March 2 (CZ-2): launcher of China's recoverable satellites

The Long March 2 series must be subdivided into six:

- Long March 2A, which made one failed launch in 1974 and was then improved;
- Long March 2C, which introduced the FSW recoverable satellite programme in 1975;
- Long March 2D, which introduced heavier, recoverable FSW satellites from 1992;
- Long March 2E, a radically more powerful version, used for lofting communications satellites to 24 hr orbit;
- Long March 2F, used for the Shenzhou.
- Long March 2EA, to be used to launch a small space station.

(The Long March 2B was a cancelled design for a version to carry a small payload to 24 hr orbit).

Long March 2C launch.

The Long March 2A made one flight, in November 1974, which ended disastrously. This was the first attempt to place a recoverable FSW satellite into orbit. The series of extensive improvements to the Long March design which followed saw the new rocket being renamed the Long March 2C (other versions say the 2A programme continued and give an effective introduction date for the 2C as 1982). The Long March 2C is one of the most successful of China's rockets. From 1975 to 1993, it flew 14 times, each launching being successful. One satellite failed to return, but this was not the fault of the launcher.

During a typical mission, the Long March 2C rocket begins to pitch over into its flight trajectory 10 sec after liftoff. Staging takes place at exactly 2mins: it is a hot staging, the second stage firing first, explosive bolts detonating the now-expired first stage which falls away a second later. 20 explosive bolts fire to separate and release the fairing over the payload at 230 sec and the second stage engine completes its burn at 305 sec. The payload is released at 569 sec. Telemetry relays back as many as 300 different parameters during launch.

The Long March 2C series might have ended in 1993, had it not been for the American Motorola company which booked the Long March 2C for 11 launches of its Iridium global telecommunications satellite (22 satellites altogether). The 2C was adapted with a longer second stage (2 m longer) and what is called a 'smart dispenser' (SD), designed to spring the small comsats into orbit. The Taiyuan site was used for these flights, which flew into a new, higher orbit of 700 km at 58°. This launcher is referred to as the Long March 2C-SD. A test of the SD was made on 1 September 1997, following which seven successful launches took place before Iridium filed for bankruptcy. A further refinement, the CTS, was used for the Chinese-European Doublestar project in 2003.

Long March 2 (CZ-2A, 2C)

Details of 2C version

Dimensions	32.57 m long, 3.35 m diameter
Weight	192.15 tonnes
Capability	2,500–2,800 kg to 170–300-km orbit at 56–67°.

	First stage	Second stage
Dimensions	20.52 m × 3.35 m	7.5 m × 3.35 m
Weight	151.55 tonnes	38.5 tonnes
Engines	4 × YF-21	YF-24
Fuels	UDMH, nitrogen tetroxide	UDMH, nitrogen tetroxide
Thrust	284 tonnes	73.2 tonnes
Burn time	120 sec	130 sec

Flight record (2A, 2C)

5 Nov 1974	FSW 0 (failed after 20 sec)
26 Nov 1975	FSW 0-1 (first Chinese recovery)
7 Dec 1976	FSW 0-2
27 Jan 1978	FSW 0-3

214   **Launch centres, rockets and their engines**

9 Sep 1982	FSW 0-4
19 Aug 1983	FSW 0-5
12 Sep 1984	FSW 0-6
21 Oct 1985	FSW 0-7
6 Oct 1986	FSW 0-8
9 Sep 1987	FSW 0-9
9 Sep 1987	FSW 1-1
5 Aug 1988	FSW 1-2
5 Oct 1990	FSW 1-3
6 Oct 1992	FSW 1-4 (with Swedish satellite Freja)
8 Oct 1993	FSW 1-5 (recovery failed)

Flight record (2C-SD version)

1 Sep 1997	Iridium demonstration
8 Dec 1997	Iridium 42, 44
25 Mar 1998	Iridium 51, 61
2 May 1998	Iridium 69, 71
19 Aug 1998	Iridium 3, 76
20 Dec 1998	Iridium 11A, 20A
11 Jun 1999	Iridium 14A, 21A

Flight record (2C-CTS version)

29 Dec 2003	Tan Ce 1

Long March 2D (CZ-2D)

The Long March 2D was introduced in 1992 to carry the heavier, third generation of FSW recoverable spacecraft, the FSW 2. The payload of the Long March 2D was 3,400 kg to low Earth orbit, about 400 kg more than the Long March 2C. The launcher was slightly heavier, with improved performance in a number of areas. Three launches were made, all successful and all from Jiuquan. Liftoff mass is 236 tonnes.

Long March 2D (CZ-2D)

Dimensions	38.3 m long, 3.35 m diameter		
Weight	236 tonnes		
Capability	3,400 kg to 200 km		
	First stage		*Second stage*
Dimensions	24.92 m × 3.35 m		7.92 m × 3.35 m
Weight	187.7 tonnes		36.2 tonnes
Engines	4 × YF-21B		YF-24B
Fuels	UDMH, nitrogen tetroxide		UDMH, nitrogen tetroxide
Thrust	302 tonnes		80 tonnes
Burn time	154 sec		94.4 sec

Long March 2 launch.

Flight record

9 Aug 1992	FSW 2-1
3 Jul 1994	FSW 2-2
20 Oct 1996	FSW 2-3
3 Nov 2003	FSW 3-1/Jian Bing 4

Long March 2E (and 2EA): China's heaviest commercial launcher

Although the Long March 3 launcher is primarily associated with the missions to geosynchronous orbit which began in 1984, a version of the Long March 2 was adapted for flights to 24hr orbit. This is the Long March 2E (CZ-2E). Approval of the CZ-2E project was made at a State Council conference chaired by premier Li Peng in 1988. Inception to production took only 18 months, the launcher making its début on 16th July 1990 under the guidance of Liu Jiyuan.

The Long March 2E was a stretched version of the Long March 2, with more powerful engines and strap-on rockets. To adapt the original Long March 2 for flights to geosynchronous orbit, the Chinese fitted four liquid-fuel strap-on boosters, the first use of strap-ons in the programme. The use of strap-on boosters is not unusual in rocketry – they have been used on the famous American launcher, the Delta and Europe's Ariane. What is unusual is that China uses *liquid* fuel strap-ons, unlike other launchers which largely use *solid* rocket motors to provide additional thrust. For the CZ-2E, the Chinese added four strap-on boosters with YF-20B engines. At lift off, eight engines light up simultaneously – the four core engines of the first stage and the four strap-ons. The noise is tremendous, reaching 142 decibels.

The Long March 2E was first delivered to the launch centre in April 1990. After software modifications, it was on the pad by June 1990. Humid weather caused condensation on the rocket and there were several delays. Until the appearance of the Long March 2F for the Shenzhou, it was China's heaviest launcher, weighing 463 tonnes on the pad and remains the heaviest commercial rocket (though not the most powerful: that distinction belonged to the 3B version).

In a typical launch profile, the Long March 2E drops its four strap-on rockets when they burn out at 2 min 5 sec. Staging takes place at 2 min 40 sec, the second, top stage entering orbit at 10mins. At this point, the solid rocket motor is expected to fire the payload into geostationary transfer orbit. Two such motors have been used – the Chinese-built perigee kick motor, the EPKM and the American Star 63 on American satellite payloads. The CZ-2E was first launched in July 1990, when it lifted the Pakistan satellite Badr into orbit. However, the attempt to fire the EPKM to put a model comsat into geostationary transfer orbit failed, the EPKM firing the wrong way and bringing the model out of orbit. The subsequent use of the EPKM on Asiasat and Echostar was entirely successful. The use of the American Star 63F kick motor has been controversial. Although the Star 63F performed as advertised on its first and third mission (the Australian Optus B-1 and B-3), the second and fourth missions were lost (Optus B-2 and Apstar 2).

The Long March 2E was only twice used again after the Apstar 2 explosion though, presumably, it was available. However, it was to have a new lease of life as the launcher for China's manned spaceship, the Shenzhou. The Long March 2F was an improved version of the 2E, with the same two stages and four strap-ons, though with a quite different set of payload fairings and, of course, an escape tower.

A much-upgraded version of the Long March 2E, called the Long March 2EA, has been in preparation since 1998. This would also be based on Jiuquan, not Xi Chang and lift cargoes of 14 tonnes into low Earth orbit, the size of a Kvant-class Russian small space station module. Design of the Long March 2EA is reported to have been frozen in late 2000 but its current state of development is uncertain. The 2EA is a bigger, fatter version of the 2E, with strap-ons double the length and a much larger fairing. Launch would be ear-shattering, with twelve YF-20B engines roaring together to produce a massive thrust of 906 tonnes.

Long March 2EA design. (Courtesy Mark Wade.)

Long March 2E (CZ-2E)

Dimensions	49.7 m long, 3.35 m diameter			
Weight	463.7 tonnes			
Capability	8.8 tonnes to low Earth orbit, 3.4 tonnes to geostationary orbit			
	First stage	*Strap-ons (4)*	*Second stage*	*Third stage*
Dimensions	23.7 m × 3.35m	6.017 m × 2.25 m	15.523 m × 3.35 m	3.07 m × 1.7 m
Weight	195.7 tonnes	41 tonnes	93.5 tonnes	5.9 tonnes
Engines	4 × YF-21B	4 × YF-20B	YF-24B	SPTM-17
Fuels	UDMH, nitrogen tetroxide	UDMH, nitrogen tetroxide	UDMH, nitrogen tetroxide	Solid
Thrust	300 tonnes	300 tonnes	93.5 tonnes	21 tonnes
Burn time	160 sec	125 sec	300 sec	75 sec

Flight record

16 Jul 1990	Aussat demonstration model, Badr (partial success)
13 Aug 1992	Optus B-1
21 Dec 1992	Optus B-2 (fail)
27 Aug 1994	Optus B-3
25 Jan 1995	Apstar 2 (exploded after 70 sec)
28 Nov 1995	Asiasat 2
28 Dec 1995	Echostar 1

Long March 2EA (CZ-2EA)

Dimensions	53.628 m long, 3.35 m diameter
Weight	695 tonnes
Capability	14–16 tonnes to low Earth orbit

	First stage	*Strap-ons (4)*	*Second stage*
Dimensions	23.7 m × 3.35 m	6.017 m × 2.25 m	15.523 m × 3.35 m
Weight	195.7 tonnes	41 tonnes	93.5 tonnes
Engines	4 × YF-21B	4 × YF-20B	YF-24B
Fuels	UDMH, nitrogen tetroxide	UDMH, nitrogen tetroxide	UDMH, nitrogen tetroxide
Thrust	300 tonnes	300 tonnes	93.5 tonnes
Burn time	160 sec	125 sec	300 sec

Long March 2F (*Shenjian*)

The Long March 2F is an adaptation for the Long March 2E for manned flight. A total of 55 engineering changes were documented in the redesign of the Long March 2E to make it capable of manned flight. President Jiang Zemin bestowed on it his own name, the *Shenjian*, or magic arrow, in 2002. The principal difference is the addition of an escape tower based on the Russian design for the Soyuz spacecraft. In the event of a mishap either on the pad or in the first 160 sec of flight, the tower fires, pulling the Shenzhou rapidly high and clear of the rogue rocket. Once the thrust is exhausted (after only a few seconds), the cabin drops out of the bottom of the tower. This is a tricky manoeuvre, for the three Shenzhou modules must then separate very quickly, giving the descent cabin time to get free, deploy its parachute and fill it with air. Four retardant panels are deployed on the tower to slow its fall and avert the danger of it tangling with the cabin. All this must be done in seconds. The escape tower is probably the aspect of the 2F design taken most directly from the Russians. Director of the CZ-2F programme is Huang Chunping of the China Academy of Launch Technology (CALT).

Long March 2F. (Courtesy Mark Wade.)

Long March 2F (Shenjian)

Height	58.34 m
Capability	7,600 kg to low Earth orbit
Weight	479.8 tonnes

	First stage	*Strap-ons (4)*	*Second stage*
Dimensions	23.7 m × 3.35 m	6 × 2.25 m	15 m × 3.35 m
Engines	4 × YF-20B	4 × YF-20	One YF-22
Fuels	UDMH & nitrogen tetroxide	UDMH & nitrogen tetroxide	UDMH & nitrogen tetroxide

220 **Launch centres, rockets and their engines**

Flight record

19 Nov 1999	Shenzhou 1
10 Jan 2001	Shenzhou 2
25 Mar 2002	Shenzhou 3
30 Dec 2002	Shenzhou 4
15 Oct 2003	Shenzhou 5

Long March 3

The Long March 3 was introduced in order to give China the capability to fly to geosynchronous orbit. The Long March 3 adapts the Long March 2 as its first two stages, but adds an upper stage, while later versions add powerful strap-ons to give the rocket an extra kick at take-off.

The Long March 3 was introduced in January 1984, although the satellite launched was left stranded in low Earth orbit. Since then, the rocket has been used for both domestic and foreign communications satellite launches. In June 1997, the Long March 3 successfully launched China's first geostationary meteorological satellite, the Feng Yun 2. The upper stage is broadly comparable to the American Centaur. The first two stages of the Long March 3 are manufactured in the Xin Zhong Hua factory in Shanghai. About a thousand people work there on rocket production, including 400 engineers. The centre can handle up to six Long Marches at a time.

The Long March 3 has been launched 13 times. The launcher has clearly experienced problems with its third, hydrogen-fuelled stage, for it has failed on three occasions, even though the precise causes for the failures seem to have varied. All these failures occurred on domestic comsat missions. The Long March 3 is likely to continue in business, placing Feng Yun 2 and similar payloads into orbit.

Long March 3 (CZ-3)

Dimensions	44.86 m long, 3.35 m diameter
Weight	205 tonnes
Capability	1.4 tonnes to geostationary orbit

	First stage	*Second stage*	*Third stage*
Dimensions	20.219 m × 3.35 m	9.707 m × 3.35 m	7.484 m × 2.25 m
Weight	149.45 tonnes	39.7 tonnes	10.5 tonnes
Engines	4 × YF-21	4 × YF-24	YF-73
Fuels	UDMH, nitrogen tetroxide	UDMH, nitrogen tetroxide	Liquid hydrogen, liquid oxygen
Thrust	284 tonnes	73.2 tonnes	4.5 tonnes
Burn time	130 sec	130 sec	13 min, 20 sec

Flight record

29 Jan 1984	Shiyan Weixing (3rd stage failure)
8 Apr 1984	Shiyan Tongbu Tongxin Weixing (first success to 24hr orbit)

1 Feb 1986   Shiyong Tongbu Tongxin Weixing 1 (first operational comsat)
7 Mar 1988   Shiyong Tongbu Tongxin Weixing 2
22 Dec 1988   Shiyong Tongbu Tongxin Weixing 3
4 Feb 1990   Shiyong Tongbu Tongxin Weixing 4
7 Apr 1990   Asiasat 1 (formerly Westar 6)
28 Dec 1991   Shiyong Tongbu Tongxin Weixing 5 (third stage failure)
21 Jul 1994   Apstar 1
4 Jul 1996   Apstar 1A
18 Aug 1996   Zhongxin 7 (third stage failure)
10 Jun 1997   Feng Yun 2-1 (first geostationary metsat)
25 Jun 2000   Feng Yun 2-2

Long March 3A: double the performance
The Long March 3A was the first variant of the Long March 3, being introduced ten years later. Compared to the Long March 3, it offered substantially improved

Long March 3A launch.

performance and was able to place about twice the weight in geosynchronous orbit. It had a stretched first stage and bigger third stage. The third stage was entirely redesigned and carried two YF-75 engines (rather than one on the Long March 3). Ten small engines were fitted to the third stage in an attempt to settle the propellant before its second ignition. The Long March 3A had a new, advanced digital computer system. It has made eight launches, all successful, three over 1994-7, three in 2000, when it lofted Feng Huo 1 and the two Beidou navigation satellites, and two in 2003. It is slated to fly China's first moon probe, the *Chang e*, in 2006.

The third stage hydrogen engine was developed by Zeng Guang Shang, Wang Heng, Shen Weigou and their team. It took them five years, working in an open assembly plant which in the course of the seasons let in cold north-west winter winds and the scorching sun of summer. The welding of the engine chamber produced particular difficulty, designers breaking into tears in their frustration. A team of seven women designers, led by Fang Rongchu, was assigned to the electronic control system of the Long March 3A, the system passing its test in a 480hr soak in vacuum and thermal chambers. There was an innovation on the Long March 3A which put Beidou 3 into orbit in summer 2003, for it carried the first laser-based inertial measurement unit.

Long March 3A (CZ-3A)

Dimensions	52.52 m long, 3.35 m diameter		
Weight	241 tonnes		
Capability	2.3 tonnes to geostationary orbit		
	First stage	*Second stage*	*Third stage*
Dimensions	23.075 m × 3.35 m	11.256 m × 3.35 m	7.484 m × 3 m
Weight	182.83 tonnes	34.963 tonnes	21.257 tonnes
Engines	4 × YF-21B	4 × YF-24B	2 × YF-75
Fuels	UDMH, nitrogen tetroxide	UDMH, nitrogen tetroxide	Liquid hydrogen, liquid oxygen
Thrust	296.16 tonnes	73.2 tonnes	8 tonnes
Engine SI*	2,550 m/sec	2,910.5 m/sec	4,286 m/sec
Burn time	146 sec	110 sec	8 min
Diameter	3.35 m	3.35 m	3 m

*Specific impulse

Flight record

8 Feb 1994	Shi Jian 4, KF-1
29 Nov 1994	Zhongxing 6
8 May 1997	Dong Fang Hong 3-2/Zhongxing 8 or 6B
16 Jan 2000	Feng Huo 1
31 Oct 2000	Beidou 1
21 Dec 2000	Beidou 2
25 May 2003	Beidou 3
15 Nov 2003	Zhongxing 20

Long March 3B: China's most powerful launcher

The Long March 3B is the most powerful rocket in the Chinese armoury of unmanned spaceflight, equivalent to Russia's Proton and Europe's top version of the Ariane 4. In essence, it adds the four strap-on rockets from the CZ-2E to the CZ-3A to get a payload of 4.8 tonnes to geosynchronous orbit. The 3B took a number of

Long March 3B.

systems directly from the 3A, such as engines, electronics, guidance and computer controls. This weight is comparable with the best rockets operated by Russia, Europe and the United States. It has larger propellant tanks and better computer systems than its predecessors. Although not as heavy as the Long March 2E (425 tonnes compared to 463 tonnes), it is slightly more powerful, generating 5,923kN of thrust (compared to 5,919kN for the CZ-2E). To accommodate larger satellites, a larger nose fairing was developed in 2001/2, 4.2 m in diameter. Chief designer is Long Xuehao.

The Long March 3B has made five flights. The purpose of its maiden flight was to carry into orbit the Intelsat 708, but it crashed only 2 sec into its mission. No test flight had been carried out, largely because the CZ-3B had so many features in common with previous Long March launchers that it was not considered necessary. This judgement may have been a mistake. Repairing the problems took 18 months, the first successful mission being the launch into 24 hr orbit of the Philippine satellite Mabuhay in August 1997. After Apstar 2R, two more successful missions followed in 1998, so the St Valentine's Day Massacre seems to have been an 'unlucky first'.

Long March 3B (CZ-3B)

Dimensions	54.838 m long, 3.35 m diameter			
Weight	425 tonnes			
Capability	4.8 tonnes to geostationary orbit			
	First stage	*Strap-ons (4)*	*Second stage*	*Third stage*
Dimensions	23.075 m × 3.35 m	6.017 m × 2.25 m	13.75 m × 3.35 m	3.07 × 1.7
Weight	180.3 tonnes	41.2 tonnes × 4	55.6 tonnes	21.7 tonnes
Engines	4 × YF-21B	4 × YF-20B	YF-24B	2 × YF-75
Fuels	UDMH, nitrogen tetroxide	UDMH, nitrogen tetroxide	UDMH, nitrogen tetroxide	hydrogen and oxygen
Thrust	302 tonnes	305 tonnes	73.2 tonnes	16 tonnes
Burn time	146 sec	125 sec	185 sec	75sec

Flight record

14 Feb 1996	Intelstat 8 (fail after 2 sec)
20 Aug 1997	Agila
16 Oct 1997	Apstar 2R
30 May 1998	Chinastar 1
18 Jul 1998	Sinosat 1

Capacities of the Long March series to geosynchronous orbit

Long March 2E	3.4 tonnes
Long March 3	1.4 tonnes
Long March 3A	2.3 tonnes
Long March 3B	4.8 tonnes

Long March 4 (CZ-4)
The Long March 4 was developed in the 1980s in order to fly meteorological satellites (the Feng Yun 1 series) into polar orbit from the new launch site of Taiyuan. The rocket is built in the same plant that designed and constructed the Feng Bao in Shanghai: the project provided much-needed work for the factory when the Fang Bao programme terminated in 1982. In effect, the Long March 4 is a stretched version of the Long March 2, using the same first stage, but with a totally new third stage and engines (YF-40). For the CZ-4, Chinese rocket designers stretched the CZ-2C first stage by 4m and the second stage 3m.

Long March 4 night launch.

A further improved version, the Long March 4B, was introduced in May 1999, with the launch of Feng Yun 1-3, which also brought to orbit the small scientific satellite Shi Jian 5. More applications missions followed: CBERS 1 and 2, the two Zi Yuan 2 and Feng Yun 1-4/Haiyang 1. The main difference in the Long March 4B is that it has a more powerful, restartable third stage, longer than the present third stage. Its thrust level is 3% greater and can burn for a third time longer. Its capacity is 4.2 tonnes to low Earth orbit or 2.8 tonnes to polar orbit. The 4B is slightly taller on the pad, 44.1 m tall compared to 41.9m.

The Shanghai Institute for Space Technology has designed a successor version to the Long March 4B, called the Long March 4B-8S. The '8S' part denotes the intention to fit, to the bottom stage, eight solid rocket motors to give a substantial kick to the rocket at liftoff, similar to the American Delta (the first such Chinese use of *solid*-fuel strap-ons). This would raise the weight of the rocket at liftoff to 270 tonnes and enable it to send 2.6 tonnes to polar, sun synchronous orbit.[7]

Long March 4 (CZ-4)

Details of 4B version

Dimensions	45.6 m long, 3.35 m diameter
Weight	248.5 tonnes
Capability	2.8 tonnes to Sun-synchronous polar orbit at 900 km, 3.8 tonnes to low Earth orbit

	First stage	*Second stage*	*Third stage*
Dimensions	24.66 m × 3.35 m	10.407 m × 3.35 m	1.92 m × 2.9 m
Weight	192.2 tonnes	40.05 tonnes	15.15 tonnes
Engines	4 × YF-21B	4 × YF-24B	YF-40
Fuels	UDMH, nitrogen tetroxide	UDMH, nitrogen tetroxide	UDMH, nitrogen tetroxide
Thrust	302.8 tonnes	73.6 tonnes	10.2 tonnes
Burn time	156 sec	127 sec	5 min 21 sec

Flight record

6 Sep 1988	Feng Yun 1-1
3 Sep 1990	Feng Yun 1-2 with Qi Qui 1 and Qi Qui 2 balloons
19 May 1999	Feng Yun 1-3
	Shi Jian 5
14 Oct 1999	CBERS 1
1 Sep 2000	Zi Yuan 2A
15 May 2002	Feng Yun 1-4, Haiyang 1
27 Oct 2002	Zi Yuan 2B
21 Oct 2003	CBERS 2/Chuangxin

Feng Bao (1974-81)

The Feng Bao was conceived at around the time as the Long March 1 and, like the CZ-1, benefited from the experience of the DF-5 missile. The rocket was developed in record time. Despite heroic efforts by the design and production teams, the Feng Bao turned out to be China's least successful rocket. Its first and second missions failed, in September 1973 and July 1974. The Feng Bao then had three successes in a row, putting into orbit the three Ji Shu Shiyan Weixing (JSSW) technology payloads over 1975–6. The sixth launching, carrying what would have been the final JSSW, also failed. It was originally called the Feng Bao 1, but in the absence of a successor is here called the Feng Bao.

Following this disappointing early history, a number of modifications were made and these were tested in the course of sub-orbital missions (15th September 1977 and

13th April 1978). This gave the Chinese the confidence to proceed to a 3-in-1 satellite launch on 28th July 1979. Unhappily, the second stage vernier engine, which had caused the earlier failures, malfunctioned yet again and the mission was lost. The reason for this approach was that the main second stage engine just lacked the capacity to get the payload into orbit and in effect the manoeuvring engine was used to fire a further 223 sec to achieve the necessary velocity (other countries have used similar techniques). The 3-in-1 mission eventually took place as Shi Jian 2,2A and 2B in September 1981.

Because of its spotty launch record, coupled with the fact that the rocket had to rely increasingly on equipment taken from the Long March 2, it was felt that its continued existence was superfluous. The rocket had never performed as well as its contemporary, the Long March 2. Although the design specification had been to lift two tonnes to 200 km orbit, the best it could manage was 1.5 tonnes to 190 km. Its engines developed less thrust and had to fire longer than those of the Long March 2.

The association of Shanghai with the Mao Zedong was of little help once the great helmsman had passed away in 1976. In September 1981, after the 3-in-1 launch, the staff in Shanghai were transferred to the Long March 3, whose development was then reaching a critical stage and where they were more urgently needed.[8]

Feng Bao

Dimensions	32.57 m long, 3.35 m diameter
Weight	192.679 tonnes
Capability	1.5 tonnes to 190 km

	First stage	*Second stage*
Dimensions	23.819 m × 3.35 m	8.6 m × 3.35 m
Engines	Four FY-21	Five FY-23
Fuels	UDMH, nitric acid	UDMH, N_2O_4
Thrust	280 tonnes	300 tonnes
Burn time	128 sec	117 sec (main), 323 sec (vernier)

Note: The 'FY' engine designator was used by the Chinese for the Feng Bao, reversing the 'YF' terminology used on the Long March.

Flight record

18 Sep 1973	Ji Shu Shiyan Weixing (fail)
14 Jul 1974	Ji Shu Shiyan Weixing (fail)
26 Jul 1975	Ji Shu Shiyan Weixing 1
16 Dec 1975	Ji Shu Shiyan Weixing 2
30 Aug 1976	Ji Shu Shiyan Weixing 3
10 Nov 1976	Ji Shu Shiyan Weixing (fail)
28 Jul 1979	Shi Jian (fail)
19 Sep 1980	Shi Jian 2, 2A, 2B 3-in-1 science mission

SOUNDING ROCKETS

The role of the T-7 in preparing the way for the first Earth satellite has already been described. Overall, though, our knowledge of the sounding rocket programme in China is fragmentary and incomplete. There is no full list of all the tests carried out since the programme began in the late 1950s. Histories have some tantalizing references, such as that small sounding rockets were tested from flat ships on inland lakes in 1964.[9] Here is a story that has yet to be written.

He Ping 2

Following the T-7 in the 1960s, two meteorological rockets were developed and used, the He Ping ('peace') 2 and the He Ping 6. The He Ping 2 two-stage solid fuel meteorological rocket was developed for the armed forces. Designer Song Zhongbao began work on the project in 1965 under the auspices of the China Academy of Sciences Applied Geophysics Research Institute. He Ping 2 was 6.645m tall, weighed 331 kg and was able to reach 72 km. Its first flight took place in 1967 and serial production began the following year. He Ping used a new launch site, an undisclosed location in Heilongjiang in the north-west of the country. Forty nine He Ping 2s were launched there from January 1970 to February 1973.

He Ping 2. (Courtesy China Astronautics Publishing House.)

He Ping 6

Its successor, the He Ping 6, was much smaller, being, weighing only 60 kg, capable of firing a 2.8 kg package of instruments to 80 km. Considerable efficiencies were achieved by miniaturization and by the development of a high-thrust engine. Made by the Space Physics Research Institute, the He Ping 6 deployed a 1.6 m reflecting ball to act as a tracking target during its descent. He Ping 6 used a mobile launcher with guide rails. The first flight of He Ping was from Jiuquan at the end of 1971 and it reached 75 km. Four batches of tests were carried out between 1972 and 1975, but the results were disappointing due to poor quality equipment resulting from the cultural revolution. The final set of tests, made in Yunan in December 1979, were completely successful, sending nine rockets to an altitude of up to 68 km to 90 km. A new sounding rocket, code-named project 761 was ordered in 1977, but its outcome is unclear.

Zhinui – Weaver Girl

Sounding rocket tests resumed with four Zhinui launches in 1988. The first Zhinui (Weaver Girl) launch took place in the morning of 19th December 1988 in the newly commissioned launching pad near Haikou, capital of China's southernmost island province of Hainan. 2 min 10 sec into the mission the first stage dropped away and later fell into Beibuwan Bay. The rocket reached such an altitude that the instruments took 2hrs to float back to Earth, landing some 64 km distant from the launch site. A further three Zhinui 1 tests were made over the following six days to conclude the series. The main purpose of the flights appears to have been to collect information on the structure of the atmosphere.[10]

The Weaver Girl 3 programme was begun by the Academy of Sciences in 1988. The purpose was to measure the elements and density of the atmosphere to an altitude of 120 km. The Weaver Girl rocket itself was 4.87 m tall, weighed 285 kg and had a payload of 45 kg. It took 3 min 24 sec to reach the highest point of its ballistic arc. On 22 January 1991, China began a new series of sounding rocket tests, sending the Weaver Girl 3 sounding rocket to an altitude of 120 km.

ROCKET ENGINES

Sergei Korolev, Russia's great chief designer, once remarked that at the heart of a successful space programme lay a sound rocket engine. It may be no coincidence that rocket engine design was a high priority in their programme, the Russians continuously leading rocket engine designs from the earliest day to the present time.

In Russia, rocket engines were designated RD – (*raketa dgvatel*), or rocket engine, from – 1 onward. China has followed a similar system, using the designator YF-, or *yeti fadong* (liquid type engine). Data on Chinese rocket engines are much less satisfactory than the Russian ones. In some cases, only poor-quality photographs are available. The Chinese have not marketed their rocket engines in the west, as the Russians have (with considerable success). Chinese reticence is surprising, granted the success of their engines and their development in a relatively short time period (the Russians started as far back as 1921, with the Gas Dynamics Laboratory in Leningrad).

China has developed a very small number of rocket engine types, but with many variants. In essence, there are four types – the YF-1 to YF-3 series; the YF-20 to YF-24 series; the YF-40 series; and the YF-73 and YF-75 series. These rocket engines have been adapted and modified to serve the entire range of the Long March and Feng Bao families. Hundreds of rocket engines have been manufactured in the YF-20 family, since they are the basis for the main stages of the Long March 2 and 3. In addition, China has developed a small number of solid rocket motors and minor engines.

For their lower stages, the Chinese have used toxic fuels – UDMH and nitric acid at the beginning and then UDMH and nitrogen tetroxide. For the upper stages of the Long March 3, hydrogen is used as a fuel, with oxygen. For its new generation of launchers, China will use liquid kerosene – fuelled lower stages and hydrogen –

The YF-2A engine. (Courtesy China Astronautics Publishing House.)

fuelled upper stages. The debate on which is the most appropriate fuel to use was a dominant and divisive one in the Soviet space programme in the 1960s, some bureaux favouring kerosene (e.g. Korolev), others toxic fuels (e.g. Chelomei, Yangel). To what degree and in what way this has been a live issue in China is not known.

YF-1 and YF-3 series
China's first liquid fuel engine was the YF-1 which was the motor used for the Dong Feng 3 and Dong Feng 4 ballistic missile from the 1960s. Using UDMH as fuel and nitric acid as oxidizer, the YF-1 had a thrust of 28 tonnes. For the Long March 1, four YF-1s were clustered together to provide a thrust of 102 tonnes (this configuration was called the YF-1A).

The other early rocket motor was the YF-3, which was the engine for the second stage of the Dong Feng 4 and then the Long March 1. It also used UDMH. Design started in 1965, the first engine being completed 14 months later and the first test run being carried out four months after that. The YF-3 was designed to operate from an altitude of 60 km. For the Long March 1, a solid fuel upper stage was added, the GF 02.

Designers of the YF-1 and YF-3 engine

Ren Xinmin
Mo Tso-hsin
Zhang Guitian

YF-20 series
The YF-20 engine and its variants have been used for the Long March 2, 3 and 4 rockets, being officially introduced on the Long March 2C in 1975. Chief designer was Li Boyong. The YF-20 has a thrust of 70 tonnes and uses UDMH with nitrogen tetroxide as oxidizer. Design began in 1965, exhaustive tests being run over 1966-8. By the following year, the Chinese had clustered four YF-20s together to provide a lift-off thrust of 280 tonnes (in this configuration it was called the YF-21). An improved version, the YF-20B, with 7% more thrust, was developed for the Long March 2E, 3A, 3B and 4 (when clustered, they are called the YF-21B). The YF-20B was introduced on the Long March 2D in 1992. A single YF-20B engine is used on each strap-on booster for the Long March 2E and 3B. These are big engines, weighing nearly 3 tonnes each (2,850 kg).

The YF-22 engine is a modification of the YF-20 and is used for second stage rockets. The engine is designed to light at altitude. The YF-22 was introduced on the Long March 2C second stage in 1975. An improved version, the YF-22B, with slightly extra thrust, was introduced on the Long March 2E second stage in 1990. The fuel used on both is UDMH with nitrogen tetroxide. Later versions were called the YF-23 and YF-24 series, several with A and B sub-designators.

YF-40 series
The YF-40 is the third stage engine used on the Long March 4 rocket introduced in 1988. Third stage engines are relatively small in size and thrust compared to the first and second stage, but they have longer burn times, in the order of 320 sec. These are small engines, 166 kg in weight, 1.2 m long and 65 cms diameter.

Liquid hydrogen third stage engines: YF-73 and YF-75
When the Long March 3 flew in 1984, China became the third country in the world to tame liquid hydrogen – fuelled upper stages. It was preceded by the United States (Centaur) and Europe (Ariane) and followed by Japan, the Soviet Union and India. A powerful upper stage is highly desirable in order to put large comsats into geosynchronous orbit. A complication is that the third stage must be restartable – firing once to enter Earth orbit, a second time about 50 mins later for the transfer to geosynchronous orbit.

Design of a third stage, restartable hydrogen-fuelled engine, the YF-73, began in 1965, the initial hardware tests being carried out in 1971. Testing of the restarting capability began in 1979 and by the time it was ready for flight over a hundred tests had been carried out. Despite this, development of the upper stage continued to prove problematical, for it failed to restart on its first flight in January 1984. This problem must have been promptly identified and remedied, for the next mission, four months later, went perfectly. The thrust of the liquid hydrogen third stage is 4.5 tonnes, with a burn time of 13.3 mins. It is still in use with the Long March 3. An improved version, the YF-75, was introduced with the Long March 3A in 1994 and was since used by the 3B. The YF-75 weighs 550 kg, is 2.8 m tall and 3 m in diameter and has a thrust of 8 tonnes. Restarting problems have, disappointingly, recurred from time to time (though such problems are not entirely absent from the American or Russian programmes either).

Minor engines
The second stage of the Long March 2 carries vernier engines to provide additional thrust and more stability at the bottom of the rocket. The engine motors concerned are the YF-23 (Long March 2C from 1975) and the YF-23B (Long March 2E from 1990). These low thrust engines are used not just for steering but also assist to get the second stage into orbit – this was the phase that proved the undoing of the Feng Bao.

Solid rocket motor engines: GF and PKM series.
China has developed two families of solid rocket motor engines: the GF series and the PKM (perigee kick motor). Considering first the GF series, the two-stage Long March 1 rocket lacked, on its own, sufficient thrust to reach orbit. Accordingly, a small upper stage was required. In developing a solid rocket upper stage, the Chinese relied on solid rocket motors previously used on sounding rockets. The GF 2 was designed in 1967 at the Solid Fuel Engine Research Academy, the chief designer being Yang Nansheng and first tested the following year. The GF-02 exploded on its first test 26th January 1968, but was subsequently mastered, 19 firings (including five altitude tests) having been carried out before its historic first assignment in April 1970. For the Long March 1D, the new light launcher, a new third solid rocket motor engine has been developed, the GF-36. This is a 729 kg engine which has already been used as an apogee engine for the Feng Yun 2 series of geostationary metsats. The GF-14, 23 and 23A solid rocket engines have been used as retrofire engines for the FSW 0, 1 and 2 series respectively. The GF-23A engine, for example, weighs 265 kg.

With the beginning of flights to 24hr orbit in 1984, a new generation of solid fuel rockets was required to carry out the manoeuvres necessary to ensure that communications satellites accurately reached their final orbital destinations. The PKM was developed as a kick motor to complete the transfer of comsats to geostationary orbit. The PKM comes in two versions – the basic (PKM) and the EPKM (used on the Long March 2E). PKM stands, somewhat crudely, for perigee kick motor, though it has also gone under the title of FG-46. It is built by the Haxi Chemical and Machinery Company. It is 1.7 m in diameter and 2.5 m long. It weighs 5,978 kg of which 5,444 kg is propellant. On two Long March 2E missions, the Americans insisted on using their own solid rocket fuel kick motor, the Star 63F. Both these missions were lost a minute after liftoff. The GF-15 solid rocket motor (500 kg) was developed as the apogee motor for the Dong Fang Hong 2 comsats and the 15B for the Dong Fang Hong 2A.

New engines

Development of the new fleet of Chinese launchers for the 21st century, like the Long March 5 series, will require the construction of a range of new engines running on kerosene and hydrogen fuels. Progress has clearly been made, for the first reports of tests of new kerosene-fuelled engines came through in 2000.[11] The first technical details of two entirely new engine designs emerged three years later (see chapter 11).

RELIABILITY

Not long after the St Valentine's Day Massacre, at the 1996 International Astronautical Congress in Beijing, the President of the Chinese Academy of Launching Technology (CALT) declared that the primary problem for the Chinese space programme was trying to ensure the reliability of its launchers. At that time, Chinese launchers had become uninsurable. The main commercial launching base, Xi Chang, which had known a flurry of activity only a few years earlier, was idle and no rockets were to be seen.

The development of the space shuttle by the United States in the 1980s began to create the public notion that somehow spaceflight could be made as reliable and safe as flying in an aeroplane. The loss of the *Challenger* was a frightening reminder that this was not and could not be the case. It is highly unlikely that conventional rockets will ever achieve aeroplane-like reliability. From time to time, the major rocket programmes of all the major space powers suffer periodic reminders as to just how complex and dangerous modern rocketry remains.

So, just how reliable are Chinese rockets? The following table lists the total number of known launches and failures by launcher type. The following tables list the record of Chinese launch failures and stranded satellites since the start of the programme.

Launch failures

18 Sep 1973	Feng Bao	Ji Shu Shiyan Weixing	Second stage failure
14 Jul 1974	Feng Bao	Ji Shu Shiyan Weixing	Second stage failure
5 Nov 1974	Long March 2A	Fanhui Shi Weixing	First stage failure
10 Nov 1975	Feng Bao	Ji Shu Shiyan Weixing	Second stage failure
28 Jul 1979	Feng Bao	Shi Jian	Second stage failure
21 Dec 1992	Long March 2E	Optus B-2	Satellite broke up
25 Jan 1995	Long March 2E	Apstar 2	Exploded 70 sec
14 Feb 1996	Long March 3B	Intelsat 708	Failed at 2 sec

Stranded satellites

29 Jan 1984	Long March 3	Shiyan Weixing	Loss of thrust 3 sec
28 Dec 1991	Long March 3	Shiyong Tongbu Tongxin Weixing 5	3rd stage failure
(29 Nov 1994	Long March 3A	Dong Fang Hong 3-1	Failure of satellite motor)
18 Aug 1996	Long March 3	Zhongxing 7	Transfer failure

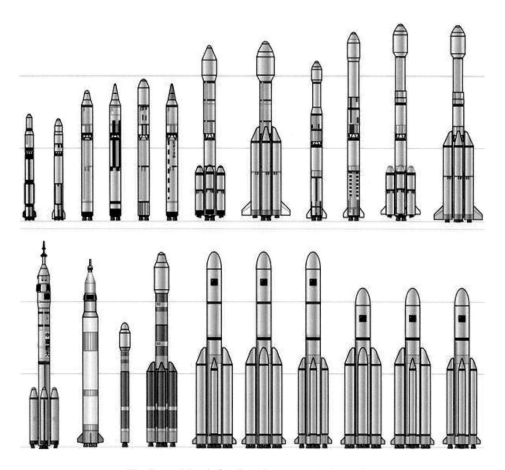

The Long March family. (Courtesy Mark Wade.)

Thus, over the period from 1970 to 2003, China had eight launch failures out of 83 attempts (giving an overall 90% reliability rate); and there were four further instances where the insertion into geosynchronous orbit was either wholly or partially unsuccessful.

As may be seen, five of the launch failures took place during the first ten years of the programme, when incidents of this kind were most likely. The two Long March 2E failures of 1992 and 1995 were, as we know, contentious, with blame being passed between the launching company, the satellite maker and the kick motor manufacturer. Three rockets were lost on their maiden flights – the Long March 2, the Feng Bao and the Long March 3B. Maiden flight losses generally account for a third of all first-time launchings worldwide, so this outcome is well within international norms.

Are some Chinese rockets more reliable than others? The following table gives a breakdown of the reliability of each launcher in the Chinese fleet.

Chinese launcher reliability rates

Launcher	Bureau	First launch	Launches	Failed	Reliability
Long March 1	CALT	24 Apr 1970	2	0	100%
Long March 2A	CALT	5 Nov 1974	1	1	0%
Long March 2C	CALT	26 Nov 1975	14	0	100%
Long March 2C-SD	CALT	1 Sep 1997	7	0	100%
Long March 2C-CTS	CALT	29 Dec 2003	1	1	100%
Long March 2D	CALT	9 Aug 1992	4	0	100%
Long March 2F	CALT	19 Nov 1999	5	0	100%
Long March 3	CALT/SAST	29 Jan 1984	13	3	77%
Long March 2E	CALT	16 Jul 1990	7	2	71%
Long March 3A	CALT	8 Feb 1994	8	0	100%
Long March 3B	CALT	14 Feb 1996	5	1	80%
Long March 4 (A & B)	SAST	6 Sep 1988	8	0	100%
Feng Bao	SAST	18 Sep 1973	8	4	50%

CALT: Chinese Academy of Launcher Technology
SAST: Shanghai Academy of Space Technology

Some are clearly more reliable than others. Some rockets are extremely reliable. The Long March 1, 2C, 2C-SD, 2D, 2F, 3A and 4 series have 100% reliability rates. The problem launchers were the poorly performing and abandoned Feng Bao and Long March 2A (discontinued), the Long March 3 (three final stage failures), the Long March 2E (two exploded) and the Long March 3B (maiden flight failure).

One of the most trying aspects was the string of recent upper stage failures associated with the Long March 3 in which several satellites were stranded in incorrect orbits. Although failures to reach geostationary orbit can afflict the space programmes of other nations from time to time, the number of such failures has been higher than the international norm and more than what can be explained away by random failure. What has been most frustrating for the Chinese is that the causes appear to have been different and unrelated on each fresh occasion. Doubly frustrating is that these problems have emerged in a programme that has always had a strong commitment to quality control and testing. The Chinese do not claim to be world leaders in rocket and spaceflight technology, nor that their equipment is the most advanced in the world, for they know it is not; but they will not accept that it is inherently below professional standards. Because Chinese space budgets are restricted, the programme can afford exploding rockets and satellites breaking down much less than others. Each launch costs at least ¥10m, leaving aside the value of the payload. Accordingly, there is a strong emphasis on quality control and rigourous ground testing, considerable resources being so invested, as chapter 5's description of the extensive space infrastructure indicated. The Chinese have introduced a 'testing pyramid' of checking individual components, combined parts and each system as a whole.

236 Launch centres, rockets and their engines

LongMarch 4B leaving Taiyuan. (Courtesy *Aerospace China*.)

Most rockets of the developed space powers now have very high reliability rates: for example, those of the American Delta and the European Ariane are in the order of 98%. Others are not far behind (e.g. Titan). Most well-established rocket programmes now experience only rare failures (e.g. Russia's Soyuz, Cosmos, Proton). If we make a simplistic comparison with them, then China's rockets come out poorly. However, context is important. Most of these rockets achieved their current high reliability rates after tens, even hundreds of missions over decades in which the bugs have been systematically ironed out (Russia's Soyuz and its many antecedents have made no less than 1,658 missions). Their bugs were ironed out in the 1950s and 1960s. The rockets of the developed nations experienced their worst failure rates many years ago, at the start of programmes that now have long flying histories. With much fewer launches, only 80 altogether, China has not had the luxury of development periods with high early failure rates. If one compares developed countries' rockets with Chinese records since their programme started, the Chinese will always come out second best. A more relevant comparison is the number of Chinese failures *in recent years*, now that China has had the opportunity to eliminate the main problems. Since 1996, its last failure, China had over thirty straight successes in a row – very much in line with the best of the best elsewhere.

ASSESSMENT AND CONCLUSIONS

China has an impressive family of launchers and variants and is able to put a range of payloads into several types of orbits, from small to large payloads into low Earth orbit; to specialized payloads into polar and sun-synchronous orbits; and large communications satellites into geosynchronous orbit. Such versatility is the result of the imaginative adaptation of the Long March to fulfil a variety of possible tasks. In a mirror image of the way in which the Long March has been adapted, Chinese rocket engine technology shows the ability to adapt a limited range of models to suit a variety of launchers. Chinese rocket engines have demonstrated a high rate of reliability.

From 1996 to 2003, China had 34 straight launch successes in a row. Detailed examination of the statistics of Chinese launchers shows that they are not inherently unreliable, as some ill-informed western commentators have suggested. Some individual launcher types have exemplary records, 100% reliability. The Chinese have faced a persistent problem of incorrect injection into geosynchronous orbit, one which, barring random failures, may now be overcome. The Chinese experience highlights the much tougher and more costly learning curve for new space nations and that some individual launchers have particular problems which must be ironed out. The Chinese realized from the start that they, of all countries, cannot afford failures and from the beginning invested huge efforts in quality control and testing. Western observers of the Chinese space programme have commented that while the technological capacity of the programme may lack depth, standards of manufacture and quality control have been very high.[12] This investment may at last have paid off.

Another striking feature of the Chinese space programme is the youth of its engineers. When 38-year old Yang Liwei circled the Earth in October 2003, many of the people who designed his spaceship and controlled his mission were younger than him. 80% of the engineers were under 40 and some were even under 30. Shenzhou designer Qi Faren, a 70-year old man who when he was 37 witnessed the launching of the Dong Fang Hong, welcomed the large number of well-educated professionals and managers in the programme. Shenzhou programme designer Wang Yongzhi likewise pointed to the emergence of a large group of young specialists as the key to a successful long-term programme.[13]

China maintains three main cosmodromes. Whilst small in scale compared to Cape Canaveral, Vandenberg, Baikonour or Plesetsk, the three launch sites are able to carry out a flexible range of missions. Whilst some of the launch sites may have been modest in some of their infrastructure, they operate to the highest standards of launch preparation. With the arrival of the manned programme, Jiuquan has been modernized, the new south site having some of the most state-of-the-art launching facilities in the world. They must be the envy of their neighbours and especially their former mentor, the Russians.

REFERENCES

1 The original description of these launch sites was provided in Kenneth Gatland (Ed): *The Illustrated Encyclopaedia of Space Technology*, 2nd ed., Salamander, London & New York, 1989.
2 Sven Grahn: *Launch from the desert – memories of Jiuquan launch centre.* http://

www.users.wineasy.se/svengrahn/histin/). The details of the early Jiuquan launch site are taken from National Photographic Interpretation Centre: *Shuang Chengtzu Missile Centre, China*. Photographic interpretation report, March 1962.
3 Christian Lardier: Première visite au champ de tir de Xi Chang. *Air & Cosmos*, 25 octobre 1996.
4 Chinese official denies Xi Change launch site to be decommissioned. BBC reports, 25th August 1999.
5 Hainan spaceport in detail. *Go Taikonauts!* 12th December 1999. http://www.geocities.com/CapeCanaveral/Launchpad/1921/news; Zong He: NPC deputies pushing for new satellite launch centre. *Aerospace China*, vol 4, #2, summer 2003.
6 Phillip S Clark: *Chinese launch vehicles*. Molniya Space Consultancy, 1996.
7 Théo Pirard: SAST prépare une LM-4 amelioré. *Air & Cosmos*, 16 avril 1999.
8 For a history of this rocket, see Phil Clark: The Feng Bao 1 launch vehicle programme. *Journal of the British Interplanetary Society*, vol 55, #7/8, 2002. Fresh details of the Long March 2E are provided in Phillip S Clark: *The development of China's piloted space programme – from sounding rockets to Shenzhou 5*. Monograph, Molniya Space Consultancy, 2003.
9 Théo Pirard: Chinese secrets orbiting the Earth. *Spaceflight*, 19, October 1977.
10 The most detailed accounts of these missions may be found in Phillip S Clark: *Chinese space activity, 1987–8*. Astro Info Service Publications, West Midlands, 1989.
11 David Whitehouse: Taikonauts 'ready for 2001'. *BBC News on-line*, 3rd July 2000, http://www.news.bbc.co.uk/hi/english/sci/tech
12 For a detailed treatment of these issues in the earlier period of the programme, see Jim Harford & Wilbur Pritchard: *China space report*. New York, American Institute of Aeronautics & Astronautics, 1980, an account of their inspection of Chinese space facilities.
13 Young space élite emerges from manned space programme. *People's Daily*, 27th October 2003.

9

Shuguang's false dawn: project 714

China's present manned space programme dates to 1992. Long before then, there had been fleeting rumours that China was planning to send astronauts into space. Not until 2002 did we learn that China's manned space programme actually dated back to the early 1970s as project 714, also known as Dawn (Shuguang).[1]

ORIGINS OF A CHINESE MANNED SPACE PROGRAMME: PROJECT 714

Like everyone else, the Chinese were greatly impressed with Yuri Gagarin's historic flight into space on 12 April 1961. The Academy of Sciences was spurred into holding a series of symposia starting that summer. Tsien Hsue Shen was one of the organizers. Twelve meetings were held between then and 1964. Their purpose was to keep in touch with developments abroad and discuss how a manned and deep space exploration programme could best be organized in the distant future. Tsien's book *An introduction to interplanetary flight*, the basis for instruction of all engineers in the space programme, included a chapter on manned space flight. So the idea of a manned flight was there from the very beginning.

The March 1966 conference on the space programme was notable for laying down the broad lines of future space development, especially the artificial satellite project (project 651) and the proposals for a recoverable satellite (project 911). We now know that there was also a closed session where the idea of a manned spacecraft was discussed at the Jingxi Hotel, parallel with the main space conference. The National Defence Science Committee formed a three-strong committee to develop the concept. They comprised Cai Qiao, Vice President of the Military Medical Sciences Academy of the People's Liberation Army, Bei Shizhang, Manager of the Institute of Biophysics of the Chinese Academy of Sciences; and Shen Qizhen, Chairman of the Chinese Academy of Medical Sciences.

The committee spent 20 days working out the aims, objectives and methods of a Chinese manned flight, after which it filed a 20-page report. There was much discussion about whether animal flight should precede manned flights. The Academy of Sciences then threw the space programme open for wider discussion in the course

240 Shuguang's false dawn: project 714

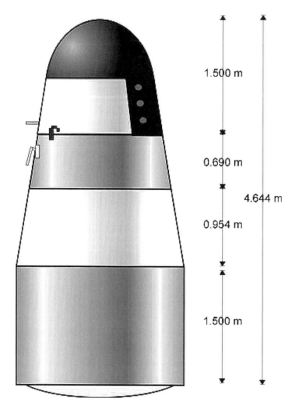

Shuguang: Was the FSW cabin the basis of its design? (Courtesy Mark Wade.)

of meetings held over 11–25 May 1966. Zhao Jiuzhang provided the overview, Qian Ji the plans for a retrievable satellite and Jia Shuguang (Academy of Military Medicine) discussed the concept of a manned spacecraft. Only if the recoverable satellite project went well would the manned project go ahead. The manned spacecraft would be based on the planned recoverable satellite, project 911. The manned programme was assigned the name of *Shuguang*, or Dawn.

In April 1968 the government took a decisive step by setting up the Institute of Space Medicine in north-west Beijing (originally it was called the Space Medicine Project Research Institute and it has also been identified as the Research Centre into Physiological Reactions in Space). The institute was formed out of the Institute of Biophysics of the Academy of Sciences and the Military Work Physiology Research Institute of the Military Medical Science Academy. They took over the buildings of the Agricultural College, whose staff had been relocated, as part of the cultural revolution, into the countryside. The new institute was made a military body, though it is not clear whether this was because of its part-military origins or for their own protection during the cultural revolution. Tsien Hsue Shen was made the first assistant director, which must have been a promotion for him since he had been demoted to canteen worker during the cultural revolution. The centre was equipped with acceleration chairs, pressure chambers, centrifuges and revolving chairs. It even investigated whether traditional Chinese breathing exercises, qi gong, could be of help to astronauts in space.[2] The institute was to remain permanently in existence, despite the subsequent ups and downs of the manned programme. Its continued operation was one of the main reasons why Chinese denials concerning a manned space programme were never entirely convincing. Later, the centre was renamed the Institute of Medical Engineering in Space and it did indeed become the main training centre for Yang Liwei and his colleagues. It is possible that officialdom never managed to issue the order to close it down.

In his position as deputy director of the Institute of Space Medicine, Tsien Hsue Shen asked the National Defence Science Committee and Air Force to take the first steps towards final design of the spacecraft and identification of the astronaut candidates. Guo Rumao, Director of the Fourth Air Force Research Institute, Senior Colonel for Medical Affairs since 1955, was made responsible for the biomedical aspects of astronaut selection and training. He was a veteran from Mao's revolutionary war. He started by examining how the Soviet Union and United States had recruited their first batch of cosmonauts and astronauts. Both had recruited from their top fighter and test pilots, the Russians going for younger, less experienced pilots. Guo Rumao saw no reason to depart from their approach, opting for active fighter pilots with a perfect medical record, rated on their psychological stability and ability to act calmly under pressure. Minimum and maximum age, weight and height limits were set. They must be physically perfect. They must have secondary schooling and some technical training, a range of flying experience (e.g. all-weather). Inevitably granted the background of the cultural revolution, they were expected to have the right family background and to have expressed 'correct revolutionary thoughts'.

Selection began on 5th October 1970. The selection committee worked with the air force medical units to examine candidates according to political reliability, family background, physical condition and service record. The following month, the first of a thousand pilots were sent to the new institute for space medicine in Beijing for screening. While there, so as to preserve secrecy, they were housed in an air force hospital and forbidden to have contact with people on the outside. They were put through centrifuge tests, which some failed. Like their Russian counterparts, they were not initially told the real purpose of the tests, though they guessed soon enough, especially when they were flown on weightless trajectories in specially adapted aircraft. When they were shown films about Soviet manned spaceflight, they knew for certain.

FIRST ASTRONAUTS

There was a 10-step procedure to eliminate the group down to a final group of 19 astronauts. As the group got smaller and smaller, the members were ferried to their tests individually, lest several be lost together in bad bus accident. The final selection was made on 15th March 1971. China thus became the third country in the world to select an astronaut squad. The process was so secret that no one, apart from those immediately involved, knew about this at the time, nor for another 30 years.

Not all their names are known. Some are. Fighter pilot Lu Xiangxiao was a MiG pilot who had shot down unmanned American spycraft over China and had engaged American bombers over Vietnam. Wang Zhiyue had also shot down a high-altitude American drone. Dong Xiaohai had shot down three drones, one at the great altitude of 18 km. Fang Guojun was a veteran of the People's Liberation Army from 1949, a volunteer for the Korean war and was trained to fly MiGs by Russian advisors. Yu Guilin was a member of the original group, but suffered a lung injury during centrifuge tests and had to leave the group.

The final squad of 19 was called 'project 714', after the year and month that confirmed their selection (April 1971) and the term seems to have been eventually applied to the whole project. Wu Faxian was the air force commander responsible for the project. Wu Fuxian appointed Xue Lun commander of the astronaut squad, the opposite number of the famous General Kamanin, who headed the Soviet cosmonaut squad in its early years. Xue Lun had fought the Americans in Korea and had proved himself subsequently by successfully tackling high accident rates in the airforce arising from pilots unable to control stalling and spinning planes.

Project 714 was assigned 500 support workers, from supervisors to trainers and guards. Wu Fuxian was told that the first flight would take place at the end of 1973 and that he had two years to get the men ready to fly. The 19 astronauts reported for duty in Beijing in June 1971. Full-up training began that November. Instructors were brought in from the universities in such subjects as physics, sciences, rocketry and English. A British-built Trident aircraft was obtained from China's civilian airline, CAAC, for weightlessness training.

The all-important political director of the astronaut squad was Li Zhengjun, the man who carried out all the background family checks into the members of the squad. The astronauts would be expected to carry out a lot of training in high-performance fighters: here, Liu Shuzhi was responsible. Coordinator of training was Zhou Yongli. Guo Rumao had overall responsibility for biomedical training. Major general Yang Guoyu and state planning committee director Yu Qiuli were responsible for obtaining supplies for the project.

The search began for a permanent training base. Xue Lun and Xu Peigin began their quest in August 1971, starting in Xi Chang. It was while they were doing this that project 714 became caught up in a national political crisis. On 13th September 1971, Minister of Defence Lin Bao died when his jet crashed in Mongolia after what was seen at the time as a failed coup attempt. Li Zhengjun and Xu Peigen were at the time at Mount Zijinshan Observatory, Nanjing, to discuss astronaut training, presumably in stellar navigation. By sheer chance, Lin Bao's plotters had used the same code-number as project 714 and in the paranoid atmosphere, they too came under suspicion. A thousand military were dismissed at the time. The project still continued, but under a cloud.

Because the project was a secret one, they found it difficult to commandeer resources. The initial equipment of the squad comprised only one car and one telephone. Budgets were underestimated and they had difficulty getting flying time from the air force. Despite these problems, a wood and cardboard spacecraft mock-up was built. Space food was prepared for toothpaste tubes.

China's first group of astronauts

Dong Xiaohai
Fang Guojun
Lu Xianxiao
Wang Zhiyue
Yu Guilin

DISBANDED

Yu Qiuli eventually went to Mao himself to ask for the resources to enable the project to continued. Mao rounded on him, told him that the project should have been able to take care of itself and expressed the view that terrestrial needs must be put first. In spring 1972 the astronauts returned to their air force units. The project may definitively be said to have come to an end on 13th May 1972 when the last staff member of project 714 closed up the office, turned out the lights and returned to his unit. Not a word about the project leaked out of China. Much later, Chinese histories adjudged the project to be premature and impossible due to weak economic foundations and a low level of technical, industrial, manufacturing and processing development.

What would the spacecraft have been like? The origin of project 714 alongside the recoverable satellite project suggests that the spacecraft might have been a derivation of the FSW recoverable satellite. In the Russian case, the film and camera payload of the Zenit cabin were taken out and replaced by the ejector seat and controls used by Yuri Gagarin and his successors – and called Vostok. The Chinese could similarly have squeezed their astronaut into the FSW for Dawn 1. The cabin would have been about 4.6m tall, with a diameter of 2.2 m. It would have been a tight fit for the single astronaut, but no more so than for the American Mercury cabins flown by Alan Shepard, John Glenn and the other early astronauts. According to western analysis of a postulated manned version of the FSW, the Chinese astronaut would have had a rough ride: at least 6G on launch and 8G on reentry. Once in orbit, the astronaut would have orientated himself with Earth and Sun sensors, assisted by a flight control control computer, manoeuvring with a cold gas thruster system. Power would have come from 1,300 amp hour 27 volt silver-zinc batteries. Missions could have lasted up to five days. For retrofire, the astronaut would have fired the 212 kg solid fuel 31kN 680 mm diameter, 896 mm long motor for 18.5 sec, giving a delta V of 340 m/s, providing sharp downward thrust and a diving, ballistic reentry. FSW had a weight of 1,800 kg, making it larger than the John Glenn's American Mercury cabin of 1,360 kg.

Even without a manned space programme, the institute of space medicine continued its work. It actually expanded to 60 technical staff who carried out work in space medicine, suits, food and equipment. Western acquaintance with the facility

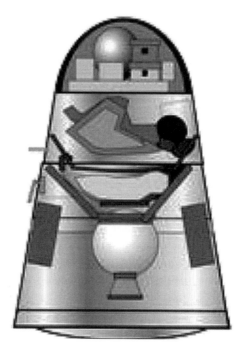

Possible Shuguang design. (Courtesy Mark Wade.)

began with a visit by shuttle astronaut Gordon Fullerton of the STS-3 crew in December 1982. By then, it had carried out 455 successful experiments to predict how humans would react to space travel. Eventually foreign scientists from America, Germany, Russia, England, Japan, Sweden, Canada, Portugal and Pakistan visited.

By way of a medical postscript to the project, as part of a medical test, the institute for space medicine contacted all the members of the astronaut group 30 years later. All were still in good health and none had developed illnesses, such as cancer. Most now held high ranks in the air force. They had chosen well.

RUMOURS AND REFUTATIONS

Ironically, the idea of an early Chinese manned spaceflight now began to circulate in the west – but well after the closure of the 714 programme. The first FSW mission was made in 1975 and western analysts reckoned it could be paving the way for a manned spaceflight. Why else would China be interested in recovering such large cabins at such an early stage in its programme? Confirming these suspicions, the Chinese newspaper *Guang Ming Ribao* described the FSW as the first step toward a manned flight, lending credence to these rumours. In March 1979, posters appeared in Beijing demanding to know why space programmes had come to a halt during the period of the Gang of Four, demanding progress be resumed. Their writers must have known something the rest of us did not.

Rumours of planned manned flights persisted. In February 1978 an article in *Navigation Knowledge*, a Chinese technical journal, wrote that China was working on the problem of manned spaceflight. On 10th April 1978, Fang Wi, the deputy prime minister for science and technology said that China had plans, over the next eight years, to launch science and applications satellites, a national communications satellite and a manned space mission and orbital laboratory. In November 1978 the head of the Chinese Space Agency, Jen Hsinmin, confirmed that China was working on a manned space capsule and a 'Skylab' space station. Rough drawings appeared sketching a manned cabin atop a Feng Bao booster, with an escape tower above.

On 11th January 1980 *Wen Hui Bao* reported a visit to Chinese 'astronaut trainees' at the space medicine institute. Photographs appeared of the astronauts in training with their space food.[3] Pressure suited astronauts were shown in pressure chamber tests. Other trainees were shown at the controls of a space shuttle-like spaceplane cockpit and what appeared to be astronauts in spacesuits undergoing training in simulators, 25 m drop chairs, centrifuges, altitude chambers, vibration systems, spinning gondolas and aircraft (to simulate zero gravity). Candidate astronauts trained in a mockup spacecraft which could simulate high altitudes. It even had star views outside the windows to test the ability of the trainees to recognize the constellations. It is still not clear if these were air force pilots doing some training at the institute of space medicine – or, conceivably, older photographs taken from 1971–2. Similar articles appeared in another Chinese paper, *Jiefang Ribao*. Much earlier, the Russians had published photos of pressure-suited trainees in the late 1950s, provoking a similar flurry of rumours and the misidentification of an entire cosmonaut squad.

How do we interpret these continued rumours? It is possible that some of those involved in the manned programme had not entirely given up on the Shuguang project and were trying to find a way to restore it to the programme. They must have been encouraged by the successful first flights of the FSW series. Whatever was going on, it seems that the new leadership of Deng Xiao Ping decided to put a stop to these ambitions. In January 1981, the general secretary of the China Space Research Society and chief engineer in CALT, Wang Zhuanshan, was quoted as saying that China had postponed a manned spaceflight for at least ten years due both to economic considerations and a reappraisal of Chinese space aims and objectives. Fundamental economic development was given priority. Another official, Shen Yuang, President of the Aeronautical and Astronautical Institute in Beijing, confirmed in June 1981 that manned flight was no longer a priority project.

Even as these options closed off, others opened. 1978, the first non-Russians and non-Americans began to fly in space. The first, a Czech, was flown by the USSR up to the Salyut 6 manned orbital station. Other Soviet block countries followed. With the introduction of the shuttle in 1981, a range of other nationalities began to fly in orbit. President Reagan, in the course of a visit to China in spring 1984, offered the Chinese a seat on the American space shuttle. That autumn, *People's Daily* reported that a group of trainee astronauts would be selected. Chinese payload specialists were even slated to visit the Manned Spaceflight Centre in Houston Texas in January 1986 for an orientation tour – but then, sadly, *Challenger* was lost that month, the shuttle was grounded for over two years, guest programmes were suspended indefinitely and the American offer had to be withdrawn. Mikhail Gorbachev also offered the Chinese a seat into space, to the Mir space station, but this did not progress far either. It seemed that China preferred to settle for the longer, tougher and more demanding task of becoming the third country to put up its own astronauts under its own steam. By way of a footnote to the period, American astronaut Dr Taylor Wang became the first Chinese-*born* person to fly in space (mission 51B, *Challenger*). Shannon Lucid, America's longest-flying woman astronaut to the Mir space station, had been born in Shanghai where her parents worked at the time.

FRESH PLANS

But this was a story that just would not go away. In the late 1980s, further photographs of fresh training candidate astronauts appeared. These were reinforced by the fact that the Long March 2E was in development, a booster capable of putting nine tonnes into low Earth orbit, more than enough for a manned spacecraft (and heavier than the Russian three-man Soyuz). *China Daily*, carrying a story from the *People's Daily*, reported in September 1986 that China had made a simulated space cabin, that selection of an astronaut group was under way and that a life support system had been fully tested.[4] In 1990, Wang Shuanglin of the Department of Oceanic Affairs stated that China was studying a manned spacecraft. In summer 1992, a Hong Kong paper reported that a manned spaceflight centre was being built near Jiuquan to support a manned space programme. In August, reports were carried that China would launch astronauts by the year 2000. In reality, the key date was 1989, for in that year there had been a feasibility study of the prospect of manned flight under the Project 863 technology development programme.

In 1992, at the International Astronautical Federation congress, a Chinese paper called *A modular space transportation system* was presented by Xiandong Bao which outlined a new launcher system able, in different variants, to life a range of payloads. One, using liquid oxygen and kerosene fuel, could loft 11 tonnes into low Earth orbit, more than adequate for a small manned spacecraft. China's future launch system would move away from its current toxic fuels to the use of liquid oxygen with kerosene for the bottom stage and hydrogen for the upper. He showed a diagram of a design for a manned spacecraft presented by the Shanghai Bureau. The design looked like the Russian Soyuz but had a much larger bell-shaped reentry cabin, a cylindrical orbital module and a forward thrust package on the front. Two years later, the China Aerospace Corporation (CASC) issued a brochure of a modular Chinese space station, showing four Soyuz class modules with solar arrays clustered around a central hub. The picture was reminiscent of early 1970s Russian designs. Clearly, something was stirring.

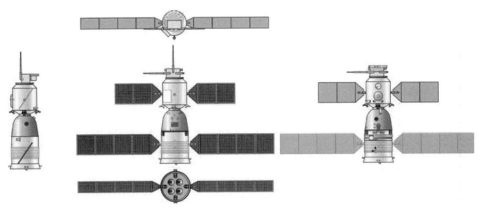

Project 921: the original design. (Courtesy Mark Wade.)

DECISION

It now appears that the formal, definitive decision to begin a manned space programme was taken by the government following a proposal of the State Science and Technology Commission in April 1992 and that it was confirmed that September. The decision, we now know, was the outcome of a series of feasibility studies carried out over several years in the late 1980s under project 863, though we have no details of these studies. The eventual proposal won the full endorsement of President Jiang Zemin, who pressed for a positive decision.[5] Once adopted, it acquired the title of project 921, after the year, 1992.

The 921 decision set the year 2000 as the target date – over-ambitious in the event – with the introduction of the modular rocket system by 2020. However, despite the decision in principle, signing off a suitable design was some distance away. In October 1993, the Shanghai Bureau shipped the blueprints of its design to Beijing, presumably hoping that this would be the system chosen.

It seems that the government took the decision to approve the development of the manned spacecraft, but using versions of existing Long March rockets, rather than waiting for the arrival of the first of the new fleet of Xiandong Bao's launchers. In a further effort to keep costs down, they would get design and hardware advice from their former allies in the former Soviet Union. None of this was announced publicly at the time.

Project 921 necessarily involved the building of a huge infrastructure – new launch pads at Jiuquan being the first requirement, along with assembly buildings. A new mission control centre must be built, tracking systems reinforced and a training centre for the astronauts was constructed. Responsible for the infrastructure was 33-year old Yuan Jiajun who was instructed to build the Beijing Space Technology Research Development and Test Centre. This was subdivided into ten laboratories which must be built to world class standards. His main challenge was construction of a new KM vacuum chamber, able to provide a simulated space environment.

RENEWAL OF CONTACT WITH MOSCOW

The next move took the Chinese back to their long-estranged partners in Moscow. Links were reopened when the chief of staff of the People's Liberation Army, Chi Haotian, visited Star Town, the cosmonaut training centre in Moscow, in early 1993. Formal agreement on cooperation was signed in Moscow on 25th March 1994 between the Russian Space Agency (RKA) and the Chinese National Space Administration (another agreement was formalized 25th April 1996). That September, President Jiang Zemin visited the Russian flight control centre, the TsUP, in Kaliningrad (now called Korolev), spoke to the cosmonauts then on board Mir and visited a Russian rocket factory in Ekaterinburg in the Urals.

In March 1995, a group of Chinese space experts visited Moscow for the first time since the great split in 1960. They expressed an interest in buying equipment which could be used for manned space flight, such as environmental control systems, docking and emergency systems. They bought some RD-0120 rocket engines, the advanced engines which the Russians used on the *Energiya/Buran* space shuttle. They tried to buy the engine used on the first stage, the powerful RD-170, but the Russians declined to negotiate. Russian experts made a return visit to China later that year. Following this, the Chinese bought an entire spacecraft life support system. The Chinese also bought a docking module and the Kurs rendezvous system, used to dock supply craft to the Mir space station. They did buy a full Soyuz capsule, but it was a stripped down shell, without any equipment or electronics. It is reported that in order to help the Chinese refine their designs of a manned space raft, they gave them access to the unflown Zarya design. Not to be confused with the name of the Functional Control Block of the International Space Station, Zarya was a 6-person spaceship to replace the Soyuz, to be launched on the top of the Zenit rocket. It was the last work of chief designer Valentin Glushko and he completed the blueprints shortly before his death in 1989. They bought a *Sokol* spacesuit: this is the suit worn by Soyuz cosmonauts during ascent and landing, but not for extravehicular activity (another account says they bought only samples of the suit materials).[6] The Russians had used this suit since Soyuz 12 in 1973.

On 20th August 1996, a third group of Chinese (estimates of numbers ranged from 20 to 50), led by Shen Jungjun, arrived in Star Town. Although Star Town had traditionally been very much off limits to westerners, this had ceased to be the case when the collaborative missions with the Americans got under way in early 1995. Indeed, American specialists and astronauts were ubiquitous. Although the Russians were unusually coy about the presence of the Chinese in Star Town, they could not deny that they were there, for they could not exactly be hidden! From this period on, the presence of 15 to 20 Chinese specialists was the norm.

So what were the Chinese doing in Star Town? There were many conflicting interpretations of the Chinese presence in Star Town. Two of the Chinese were named in December 1996 – Wu Tse and Li Tsinlung, being variously described as cosmonauts, cosmonaut trainees and instructors. The most reliable interpretation of their role is that they were cosmonaut instructors but would not necessarily fly into orbit themselves. After a year, they would return to China and train in the first cadre

Li Tsinlong. (Courtesy Neil Da Costa.)

of Chinese astronauts. Apparently, the Chinese paid €1.17 m for the training, medical advice and supervision. They returned on 19th November 1997. There were some subsequent reports of more Chinese cosmonauts training there later but this does not appear to be the case.

Recruitment began for China's first astronaut squad in 1996. It seems that the competition was thrown open to all air force pilots with more than a certain number of flying hours, meaning a large recruitment pool of between 1,000 and 1,500 people. The selection process took two years. Medical tests were the focus of the selection and according to the People's Liberation Army, some of the tests carried out 'bordered on the unreasonable'. A final selection of 12 was made in 1998, with the two instructors later added, giving a first squad of 14 men. Deputy commander of the group was Qin Wenbo. Requirements were:

- Height: up to 175 cm (160–170 cm preferred).
- Weight: up to 80 kg.
- Age: 20–45 (25–36 preferred).
- Flying time: 1,000+ hrs.
- Qualifications: a degree in engineering, biology, physics or mathematics.
- Languages: at least one foreign language.

In the event, the average age was 30 years, weight 50 kg and height 157 cm. Yang Liwei, when he flew, was above average, being 168 cm tall, 63 kg in weight and 38 years old by the time of his mission. Where would they train? From 1988, there had been unconfirmed reports of an astronaut training centre located in a walled village in the western suburbs of Beijing and described as well up to the standards of facilities available in Moscow and Houston. The centrifuge was the biggest in the world with a rotary arm of 30 m. The training centre had a spinning chair which whirled people up and down, left-and-right, around and around, in dizzying combinations; and an isolation, thermal and vacuum chamber from which the air was sucked out and where astronauts learned to live in an air-free environment for several days, testing their psychological fitness to the limit and subjecting them to a range of temperature and humidity regimes. For gravity tests, the astronauts were put in a cylindrical tower 10 m tall and then shot up at great speed, to simulate the stresses of launching. To test the other end of the mission, they were dropped in a fast lift in a four-storey high building. There was plenty of theory to learn too. When they arrived, the astronauts were handed a 600-page manual *Manned spaceflight engineering*, covering everything from flight dynamics to cosmic rays and navigation systems.

The astronauts trained there five days a week in a secluded area protected by military guards. They returned home to their families each weekend. They had ordinary apartments to the standards of a cadre division commander. During the week, they had their own transport and police escort for visits outside the training centre, but at the weekend they were expected to get around like anyone else by

Wu Tse. (Courtesy Neil Da Costa.)

Training in Moscow. (Courtesy Neil Da Costa.)

bicycle. As was the case with many in the Russian cosmonaut squad, most of their wives also worked in the training centre or in the space industry.

One outstanding question: remained: what to call China's spacemen? Several terms were in circulation at this stage, in addition to the traditional Russian and American ones of 'cosmonaut' and 'astronaut'. The most popular term in China was yuhangyuan, first used in the late 1950s, the word applied in the Chinese media to the cosmonauts and astronauts of the space superpowers. This would be the term most used by everyday Chinese. In the end, this became the official agreed term and is the one used in this book.

A second term in use was 'hangtianyuan', which is the professional or academic Chinese term for astronaut introduced in the 1970s. The Chinese term for 'space' or 'cosmos' is 'hangtian' and 'yuan' means something or someone who goes there. The symbol for 'hangtian' is normally painted on the Long March rocket. Also in use is a third term 'taikongren', the term most familiar to overseas Chinese and people in Hong Kong and Taiwan. It began to be used there, in southern China and some other parts of the People's Republic from the 1980s. An anglicized version of 'taikongren' is 'taikonaut', which has the merit of symmetry with 'cosmonaut' and 'astronaut'.[7] This was favoured by the western media and even gained ground in China in 2003.

China's second group of yuhangyuan

Yang Liwei	Liu Buoming
Nie Haisheng	Liu Wang
Zhai Zhigang	Zhao Chuandong
Wu Tse	Chen Quan

Li Tsinlong   Fei Junlong
Deng Qingming   Pan Zhanchun
Jing Haipen   Zhang Xiaoguan

The story of Yang Liwei is already known. What about the others? Nie Haisheng was born deep inside China, in Yangdang, Zaoyang, Hubei in 1964. His family was so poor and found such difficulty paying for his education that he once had to give his teacher a rabbit in lieu of money. He borrowed textbooks, unable to afford any for himself, learning them by heart before returning them. Only two students from Yangdang primary school made it to secondary school in his year and he had to work herding cattle through his holidays to pay his fees. He persuaded a visiting commissar to give him a chance in the air force. There he excelled and was decorated for trying to save his plane when it spun out of control as a result of a compressor failure. He would not bail out till the last possible moment, but eventually did so, parachuting unconscious into a rice field. Zhai Zhigang came from Longjiang, Qiqihaer, Heilongjiang. His domestic circumstances were also difficult, for his father was invalided and his mother illiterate although determined to get her children an education. Like Nie Haisheng, the air force was his route to a career. His colleagues reckon that he won selection in the first three because of his sanguine behaviour: 'he always wore a smile, whatever happened', they said. His hobbies are fixing gadgets, calligraphy and dancing.

MISSION CONTROL

As for the new mission control centre approved in 1992, this was built in Yenshan (swallow mountain) district, 40 km north from Beijing's centre, not far from one of the emperor's summer palaces. Called the Beijing Aerospace Command and Control Centre (BACCC), it opened March 1996. BACCC has five walls of consoles, 75 in all, with a huge wall-to-wall screen at the front, with clocks, images of the worldwide tracking system and television relays from the launch centre, its gleaming and futuristic appearance confirmed by up to four presentations of three-dimension displays at the front. Its appearance was not unlike that of mission control in Moscow, the TsUP, used to control the Mir space station and then the International Space Station. In between missions, the controllers spend time honing their skills in simulations. When they are not doing this, the screen puts up a graphic of a Long March taking off against a background of pagodas and distant mountains. Computers and high-speed links connect BACCC to China's national ground control system and the *Yuan Wang* comships.

Director is Sui Qisheng. The main feature of the people who work there is their youth: the average age is well under 30. None of the departmental directors completed their graduate education earlier than 1984 while the heads of sections are all post 1995 graduates. Typical monthly pay is low and a fraction of what these bright graduates could get in the private sector – only about €2,000 a year. By way of compensation, working in manned mission control is prestigious. Morale is high and

15 hr days are done cheerfully and enthusiastically. The centre provides extensive training programmes and fast tracks the best into onward career opportunities in science, technology and the military. The controllers are encouraged to take masters courses in Qinghua University and Beijing University of Aeronautics and Astronautics and to study abroad in Britain, France and the United States. On the subject of youth, an outcome of the success of the Shenzhou 5 mission was that many of the chief engineers got national medals – but in the 'youth' section!

PROJECT 921 IN THE OPEN

The nature of Chinese intentions was publicly – though incompletely – revealed at last at the 1996 Beijing International Astronautical Federation conference.[8] The vice-administrator of the China National Space Administration, Wang Liheng, announced that China intended to make a breakthrough to manned space flight before the end of the century. It was confirmed that a manned space programme had been decided in 1992 and was code-named project 921. A schedule was even reported, one which was to prove far too ambitious and not fully accurate either. An unmanned prototype would fly in 1998, with a manned flight by two astronauts in a Gemini-class cabin before October 1999, just in time to mark the 50th anniversary of the revolution. The launcher would be either the Long March 3, 3A or 3B. China's ultimate intention was to have its own manned space station by 2020.

At the Paris Air Show in 1997, the Chinese delegation confirmed that work was proceeding on the Chinese manned space flight. Work had already started on the new Long March pad at Jiuquan. The main effort at present was concentrating on the adaptations necessary for the launcher, which was now correctly identified as based on the Long March 2E. Considerable efforts would be made to improve its electronics, guidance and quality control.

Appointed chief designer was a person then unknown outside China (and probably little inside China either). Qi Faren, born 1933, represented the main design team from the China Academy of Space Technology, CAST, assisted by the Shanghai Academy of Space Technology, SAST. He had graduated from the Beijing Institute of Aeronautics and Astronautics in the historic year of 1957 and thirteen years later was involved in the building of China's first satellite, Dong Fang Hong. He was appointed general designer and leader of Project 921 in 1992, with a thousand scientists and engineers under his command. Shin Jingmaio led the contribution from the Shanghai Academy. Wang Yongzhi was identified as the person originally in charge of the overall programme. Guidance engineer was the greying aerospace engineer Chen Zugui.

ASSESSMENT AND CONCLUSIONS

Although we did not know this until the new century, manned spaceflight had been an objective of the Chinese space programme from an early stage. It is now evident that Tsien Hsue Shen and his colleagues forever had their sights set on higher ambitions: a manned flight into space. Tsien Hsue Shen had, after all, outlined ideas of manned flight in the United States in the 1940s. The symposia held after Yuri Gagarin's flight, his publication *An introduction to interplanetary flight* and the setting up of the Research Centre into Physiological Reactions in Space confirm these ambitions.

The *Shuguang*, Dawn project of 1966–73 was a brave venture in the light of Chinese technology levels at the time, but the chaotic political environment of the period was the main reason for its undoing. Even during the period of rectification in the 1980s, the Chinese never gave up their ambition for a manned flight. The continued existence of the medical space research institute meant that a tradition of knowledge was kept alive.

In the more certain political and financial leadership of the early 1990s, the Jiang Zemin leadership took the view that the time had now come to revive the project. Unlike the Soviet – American space race of the 1950s, Chinese space leaders did not have a blank cheque and they had to make some judicious, cost-effective choices about their methods, hardware and equipment. Existing rockets would be adapted for the manned space programme, rather than await a promised fleet of new launchers, although planning for them would still go ahead. It was decided to work closely with Russia, an option impossible in the 1970s. The Chinese sought and obtained help from the Russians in limited but carefully chosen and distinct ways, principally in hardware, design and training. These decisions on rockets and cooperation saved time and money, without compromising the essentially indigenous nature of the manned project. The cabin which emerged clearly owed much to the Soyuz by way of inspiration.

First, though, they must put the building blocks in place: mission control, new test facilities, a second team of yuhangyuan. This they did carefully and methodically. The Chinese had a steady but limited budget, one which had to be spent carefully, systematically and sparingly. As was the norm with the Chinese space programme, the money, effort and energy were concentrated on the ground segment, ensuring that all equipment was exhaustively planned, designed and tested before being committed to launch. But what about the spaceship itself? This is the focus of the next chapter, *The Shenzhou precursor missions*.

REFERENCES

1 The original story was first recounted as 'Send Chinese yuhanghuans into space!' in *Times Literature* in 2001, republished in *Beijing literature* of November 2001 (put on line at sina.com). The first western account was published by Mark Wade on his *Encyclopaedia Astronautica* astronautix website on 5th May 2002 (www.astronautix.com/craft/shuguang1.htm) and the narrative here relies in many ways on his story. For an earlier account of the early development of China's manned spaceflight programme, see EP Grondine: Chinese manned space programme: behind closed doors, also posted on *Encyclopaedia Astronautica*, then at http://www.friends-partners.org/-mwade/

articles. See also Zhou Xiaofei: Development and prospects of China's manned spaceflight project. *Aerospace China*, vol 4, #3, autumn 2003. An independent appraisal of the Shuguang story is provided by Phillip S Clark in *The development of China's pilot space programme – from sounding rockets to Shenzhou 5*. Monograph, Molniya Space Consultancy, 2003.
2 Yu Qingtian: Keeping fit in space. *China Pictorial*, November 1988.
3 Chinese astronauts train in simulators. *Aviation Week & Space Technology*, 26 January 1981.
4 Portrait d'un vaisseau fantome, in Philippe Coué: La Longue marche des cosmonautes chinois. *Ciel et éspace*, février 2000; For a contemporary exploration of China's space capabilities at this time, see Gerald Borrowman: China's long march to orbit. *Spaceflight*, 25, 5, May 1983.
5 Tian Zhaoyun & Xi Qixin: Launching Shenzhou 3 is a preview of sending astronauts into space. *Beijing Liaowang*, 1st April 2002, translated, text supplied by Jim Oberg.
6 James Oberg: Taikonauts prepare for liftoff. *Spectrum*, 18th December 2001. The evolution of the Chinese manned space programme is also traced in the following: Phillip S Clark: In business and advancing fast – Chinese space activity. *Spaceflight*, 30, February 1987; Philippe Coué: Des cosmonautes chinois bientot en orbite. *Air & Cosmos*, 25 septembre 1998; BBC reports, 10th March 1999; G Lynwood May: China advances in space. *Spaceflight*, 30, November 1988; for a discussion of the spacesuit, see Wei Ling: New design for Chinese spacesuit revealed. *Spacedaily*, 9th March 2001. http://www.spacedaily.com/news.
7 I am grateful for Chen Lan of the *Go Taikonauts!* website for this clarification (www.geocities.com.Cape Canaveral/Launchpad/1921).
8 The principal western report on this aspect of the conference was filed by Craig Covault: Chinese manned flight set for 1999 liftoff. *Aviation Week & Space Technology*, 21 October 1996.

10

The Shenzhou precursor missions

By 1999, many of the new building blocks were being put in place that would underpin the new Chinese manned space programme. Little of this was in evidence outside China itself. Indeed, several people still doubted whether the Chinese had a manned space programme at all. Where was the evidence? they asked.

ROLL-OUT

The Chinese manned space programme literally rolled out for the first time on 9th June 1999. Pictures were published on the internet of a brand-new rocket which looked like a Long March 2E at the bottom but with the top resembling the Russian Soyuz. Engineers mingled around the base of a large crawler used to move the white-and-red rocket across to its new launch site in Jiuquan. In the foreground, on the right, stood a 12-storey blue-grey steel assembly tower, with six swing arms reaching out at the side to grapple the rocket. Even more astonishing, in the left background, was something that looked remarkably like the Vehicle Assembly Building at Cape Canaveral, the largest structure in the world. The right front door of the 80 m tall structure was rolled half-down, to indicate the new rocket had emerged from the white and blue-grey structure. As was the normal custom, the Chinese had helpfully marked the rocket with the national flag on the top and 'Long March 2F' in Chinese and Roman script vertically down the side. Later, the rocket was to acquire a special name, the *Shenjian*, or 'magic arrow'. The new Long March, never before seen, had four strap-on rockets on the bottom. At the top was, just as on the Russian Soyuz, an escape tower and four aerodynamic flanges to be used to drop the cabin out of the nosecone during an emergency. Closer examination of the 10 m pin-shaped tower revealed four sets of ten engines. Apparently, different combinations of the engines could be used for escape below 39 km (the first three sets), 39 km to 110 km (the second and third set) and to whisk the tower away if still unused (the small top set). Escape at low altitude would be memorable experience, pulling 20G.

The photographs caused a sensation among China spacewatchers on the internet. No one had expected these pictures to appear before a first launch. Still less did

Long March 2F.

anyone expect such clear photographs, well scaled by the accompanying personnel, never mind the launch tower and the huge vehicle assembly building in the background. The Chinese did not publish the pictures, which, it transpired, had been taken by a Dutch engineer visiting the launch site in preparation for a scientific mission – and who had seen the rollout by chance (another version was that the pictures came from a Mongolian construction company). As ever, American intelligence agencies remained typically inscrutable, for there was no way that their spycraft would not have noticed the rollout which took place in clear weather. Later, the National Reconnaissance Office admitted that it had been keeping an eye on new building at Jiuquan since the mid-1990s. A full-scale mockup of the new launcher had first been rolled out in May 1998.[1]

For China space watchers, these unexpected pictures of such startling clarity was too good to be true. Within hours, some experts had denounced them as forgeries.[2] Others, reared in the hard school of Soviet forgeries from times past, were also suspicious. With computer aided design, digital cameras and photographic manipulation, it was possible to make a convincing forgery. The scale of the building was not right and shadows did not fall in the right way, critics hinted. The story faded. The Chinese never commented on the photographs and if they had any intention of using the new Long March, they kept it to themselves. The 50th anniversary of the revolution passed without any intimation of a manned spaceflight. Maybe they had been a forgery after all.

SHENZHOU 1 FLIES!

Forgery or not, the new Long March 2F rocket took to the skies at 10.30 pm GMT on 19th November 1999, a date which will always be remembered in the history of Chinese astronautics. Eight YF-20 engines fired together to lift the satellite into the dark skies. Night turned to day as the flames billowed under the Long March 2F at Jiuquan launch centre. The gleaming white rocket, the red flag with five stars on two sides, headed skyward, the pin-shaped escape tower shooting free after 160sec. Night-time had been chosen in order to track the later stages of the ascent to orbit through China's clear winter skies.

Although there had been no pre-launch announcement from China, there was now one way whereby space-watchers could predict upcoming Chinese manned launch attempts, a technique learned from Soviet times. Because China lacked tracking bases in other countries, it was impossible to keep up round-the-clock contact with its spacecraft – not something that mattered on unmanned flights, but a question of vital importance during a crewed mission. Accordingly, China had fitted out a fleet of four communications ships, or comships, the *Yuan Wang*. In the period before the mission, these deployed as follows:

- *Yuan Wang 1*: North-eastern China (to track entry into orbit)/north Pacific
- *Yuan Wang 2*: North-east of New Zealand/southern ocean
- *Yuan Wang 3*: Coast of Namibia (to prepare for retrofire)/south Atlantic and
- *Yuan Wang 4*: South-west Australia/Indian ocean.

During Soviet times, the arrival of big Russian comships on station was a sure sign of planned manned, lunar, Mars or Venus missions. Now, whenever rumours flew of an upcoming Chinese mission, the experts always asked: *But where are the tracking ships?*

The new spaceship was called Shenzhou. On the cabin was painted the Chinese symbol for Shenzhou, meaning 'magical heavenly vessel', inscribed on its side by no less a person than the President of China, Jiang Zemin. Shenzhou settled down into an orbit of 197–325 km, inclination 42.6°, orbital period 89.74 mins. Signals came back from orbit through no less than seven different wave bands.[3] The main solar panels were deployed to provide electrical energy for the spacecraft. There was sustained applause in Beijing mission control when the many mission controllers got confirmation of the good news. Back at the launch pad, the task of the rocket team

Shenzhou night launch. (Courtesy *Aerospace China*.)

was done and they set off little firecrackers to celebrate on the hot, sooty, still smouldering launch pad.

On board Shenzhou were several kilos of biological experiments (10 vegetables and 30 medicinal herbs), plants, a dummy and some commemorative souvenirs. The vegetables were 10 g each of seeds of melons, tomatoes, peas, radish, rape, green peppers, maize and barley. The dummy, or mannequin, was 1.7 m tall, dressed in a silver-grey spacesuit and was laid at 15° to the horizontal, with his knees tucked up, much as a real astronaut would be. Sensors in the cabin recorded the mannequin's journey, in particular the temperature, humidity and oxygen level. The last two unmanned Soviet missions before Yuri Gagarin in March 1961 called Korabl Sputnik 4 and 5 had carried dummies – creating a problem when unforewarned villagers found the cabin on its return, convinced themselves that the dummy was a dead cosmonaut and insisted on giving him the last rites. A dummy had been in development for many years, according to the man responsible for the environmental control system, Su Shuangning. Everything was done to ensure that the 14-part dummy mimicked human behaviour as closely as possible: the same weight, the same heat, consuming 600 litres of oxygen a day and generating 12 megajoules of energy. Also on board were the flags of China, Hong Kong, Macau, first day stampcovers, a red banner signed by project participants and 1001 commemorative gold plaques. Shenzhou was not an operational model. The front solar panels did not deploy and the life support system was not active.

Shenzhou made no manoeuvres while in orbit. After 20 hr 20 mins hours circling the Earth, in the course of which it made 14 orbits, retrorockets fired over the coast off west Africa (35.2°S, 0°E) and the cabin began a long searing descent through the flames of the upper atmosphere. The retrofire command was sent up by the *Yuan Wang 3* comship – not an easy task, for the weather had taken a turn for the worst in the past 20 hours and the ship was now pounding into 10 m tall waves, with white water washing over the sides. Shenzhou swung in a giant arc over western Asia. Reentering at 80 km, a hot plasma shell formed around the cabin at 40 km. Radar picked up the descending cabin and controllers relayed directions to three circling helicopters. Shenzhou came down gently under its parachute just to the east of where it had been launched, at 7.41 pm the next day, after 21 hr 11 mins aloft. Parachutes deployed at 30 km, still high in the thin atmosphere. As it came into land, four small solid-fuel retrorockets fired at a distance to 1.5 m to break the fall for the final distance. The landing took place in darkness at 41°N, 105°E, 110 km from Wuhai, 52 minutes after retrofire, about 415 km east of the launch site. The cabin came down 12 km from the point predicted and the heat shield was found 5 km away. The area selected was steppe grassland – so selected because of its flatness.

The front cabin, the orbital module, which had been separated just before the firing of the retrorockets, stayed for eight days longer in orbit until late on 27th November when it lost altitude to 122 km and decayed.

AFTERMATH

Once back on the ground, the Shenzhou was brought back to Beijing. The cabin was personally opened by President Jiang Zemin. Also present were Chinese vice-president Hu Jintao, members of the military Commission Zhang Wannian and Chi Haotian, programme head on the military side Cao Guangchuan, with CAST President Xu Fuxian and chief designer Qi Faren.

In the public meeting which followed, military officers made speeches as red banners hung in the background and a worn, browned and blackened cabin sat behind the podium, on a platform. A thin layering of paint had already peeled off. Chief designer Qi Faren, along with military officers, was photographed peering into the mental ring at the top of the cabin. The tracking fleet returned home, *Yuan Wang 1* and *2* to Huangshan port on the Yangtse, *Yuan Wang* 4 to Hainan and and the last, *Yuan Wang 3*, coming back to its home port of Nanjing on 31st December. The Chinese flag flown on Shenzhou was erected over Tienanmen square on 1st January to mark the new millennium.

The mission received enthusiastic coverage in the Chinese press. Apparently, the word had gone out from on high not to over-publicize the mission, but this message was received too late. Journalists in the Chinese papers reflected popular interest and gave it extensive publicity. Pictures shown by Central China Television showed

Shenzhou in the new vehicle assembly building in Jiuquan, surrounded by white-coated technicians, the cabin being hoisted by cranes and attached to the Long March 2F. The facilities were brand-new, spotless and futuristic. They must have caused pangs of envy from Russians trying to maintain their crumbling space infrastructure and impressed the American technical press.[4] Mainstream western press coverage was patchy and minimal, some of it focussed on how China had copied Russian designs to achieve a modest success.[5]

Now it was possible for students of the Chinese space programme to make a proper assessment of the Shenzhou. Their task was made easier when China ran a full-scale exhibition of the Shenzhou in Hong Kong the following August.[6] Shenzhou later went on show at the first World Space Week – a United Nations designation – in the Military Museum of China in Beijing in October 2000 and at the 3rd Zhuhai air show in southern China the following month. A 1/40th scale model was even produced. The cabin, it was announced, was built by the China Academy of Space Technology, CAST, with, as partners, the Shanghai Academy of Spaceflight Technology, SAST, the latter concentrating on the descent cabin, solar power systems and radio systems. The heat shield was developed by the Institute of Space Engineering in Beijing and the avionics by the North West Polytechnical Institute in Xian.[7] Shenzhou was a limited, basic test to see if the spacecraft could be safely launched, orbited and recovered. No extras were built into the mission and all complications were avoided.

The links between Shenzhou and Soyuz were the subject of much commentary and speculation, some of it ill-informed. That there was cooperation between China and Russia on the project was never denied, China purchasing equipment, expertise and advice on a commercial basis. The precise nature of these exchanges was never publicly identified in the protocols agreed between the two countries, so in a sense the Chinese have themselves to blame for not publishing the technical details. The protocols do involve some fairly frequent exchanges between scientists, managers and directors, right up to prime ministerial level.[8] The improved nature of the Sino-Russian relationship was exemplified on 12th April 2001, the 40th anniversary of Yuri Gagarin's flight into space, when leading Chinese scientists in Beijing joined in the worldwide series of parties and commemorative events to mark the occasion. The Russian ambassador was there, as was cosmonaut Valeri Tokarev who had just flown to the International Space Station on the American space shuttle *Discovery* on mission STS-96 in May–June 1999.

So what did we know of the Shenzhou? Like Soyuz, Shenzhou comprises a service or propulsion module, descent cabin and orbital module. The service module contains four reentry rockets, manoeuvring engines, two solar panels and radiators to discharge heat. The headlamp-shaped, sometimes called beehive-shaped descent module had room for three, possibly four, crew members. The orbital module, at the front, has two solar panels, manoeuvring engines and space for a scientific package on the front. For all its similarities with the Soyuz, there were differences:

- It was larger: 2.8 m in diameter, and 8.8 m long (Soyuz was 7.5 m).
- It was heavier: 7.8 tonnes (Soyuz was 7.2 tonnes).

Aftermath 263

Shenzhou: the spaceship.

- It had solar panels not only on the service module (24 m^2), but also on the orbital module (although the latter were not deployed on Shenzhou 1). These were reckoned to deliver up to three times more power than the Soyuz.
- The orbital module was heavier: 2 tonnes. It could be left in orbit for independent flight, and had four groups of four manoeuvring engines.
- The descent module was slightly larger: 2.5 m in diameter (Soyuz was 2.17 m), and 2 m long (Soyuz was 1.9 m).
- The propulsion service module was 70 cm longer.

The Chinese themselves made comparisons between the Shenzhou and the Soyuz. The Shenzhou was, they say, larger, roomier, better, with an additional forward system and a different docking system.[9] Overall internal volume is 13% larger. For docking, Shenzhou was expected to use an androgynous petal-style docking system, rather than the probe-and-drogue of the Soyuz. The following is a comparative table. As may be seen, Shenzhou was clearly influenced by the Soyuz design, but to describe it as a 'copy' would be both inaccurate and unfair. The Chinese became quite sensitive to allegations of copying and at press conferences stressed that Shenzhou was a 'Made in China' project ('Made in China' stated emphatically in English).

Comparison of Shenzhou and Soyuz

	Shenzhou	Soyuz*
Complete spacecraft		
Weight	7.8 tonnes	7.25 tonnes
Length	8.8 m	7.48 m
Diameter	2.8 m	2.72 m
Propulsion module		
Weight	3 tonnes	2.95 tonnes
Propellant	1.1 tonnes	900 kg
Diameter	2.5 m	2.5 m
Base	2.8 m	2.72 m
Solar panels	Two, 24 m^2	Two
Descent module		
Weight	3.2 tonnes	3 tonnes
Length	2.059 m	1.9 m
Diameter	2.5 m	2.17 m
Orbital module		
Weight	2.0 tonnes	1.3 tonnes
Length	3.4 m	2.98 m
Diameter	2.8 m	2.25 m
Solar panels	Two, 12 m^2	None

* The Soyuz TM version, which operated from 1986 to 2002. The original Soyuz was 6,500 kg.

The Shenzhou escape tower is 15.1 m long, 3.8 m in diameter and weighs 11.26 tonnes. It can be fired anytime from 15 min before launch to 160 sec after liftoff. In the American space programme, the lever the astronaut uses to activate the escape system is called the 'chicken switch'. Assuming the launch goes normally, the escape shroud is normally fired clean free in any case and not dragging the capsule with it. The escape system can be activated by the yuhangyuan, mission control or if the automatic guidance system detects that the rocket is heading badly off course.

The escape tower might be called upon to work, but at the other end of the mission, the parachute of the descent module must always work. Here the Chinese made the largest ever parachute for a returning manned spacecraft. The landing sequence would triggered as the descent module reached subsonic speed 15 km above the ground. First, the pilot chute comes out for 16 sec to slow the module from 180 metres a second to 80 m/sec. Next, the deceleration chute comes out,

The escape system.

slowing the cabin to 40 m/sec, bringing out the main parachute. This is a huge canopy, 80 m tall, 30 m across, with an area of 1,200 m^2, 20% broader than the Soyuz parachute, held by a hundred 25 mm diameter cords, each able to bear a weight of 300kg. Once it billows out, it slows the Shenzhou to its descent speed of 15 m/sec. The parachute is made of 1,900 pieces of thin strong fabric able to withstand high loads and temperatures of up to 400°C. Should something go horribly wrong, like the parachute twist or Roman candle, then a reserve parachute can be ejected. This is much smaller, 63% the size of the main chute, 760 m^2. But assuming all is well, the final action takes place as the cabin comes in to land. 1 m above the ground, a gamma detector senses the touchdown and fires solid fuel retrorockets to cushion the final descent to 1 m/sec, simultaneously severing the parachute so that it will not drag the cabin in a high wind.

The orbital module was sufficiently large for basic comforts to be provided for the orbiting yuhangyuan. They would be able to sleep in sleeping bags mounted on the wall. A sealed plastic tent was provided so that they could shower (the Russians had provided showers on Salyut and Mir, but never in the Soyuz ferry).

SHENZHOU 2

Although forthcoming about the flight of the first Shenzhou after it happened, Chinese space officials were extremely vague about the timetable for the next, or future missions. In July 2000, there were some indications that a second mission would follow in October. The month came and went without comment, although, as we knew later, the second capsule had arrived at Jiuquan in September and the launch had originally been planned for several weeks thereafter. In early December, there were rumours that the second mission would last seven days and would take place on 5th January. The day before, word was passed around that there would be a delay of up to five days. So the experts asked: where was the tracking fleet? The *People's Liberation Army Daily* told its readers on 3rd January that the *Yuan Wang 2* was now in the Pacific Ocean. Later, the full list of deployments was announced:

- *Yuan Wang 1*   Bo Hai, off the north-east coast of China.
- *Yuan Wang 2*   South Pacific.
- *Yuan Wang 3*   South Atlantic, off Namibia.
- *Yuan Wang 4*   Indian Ocean, off western Australia.

The northern wintertime, southern hemisphere summertime was chosen for these launches because the weather was normally calm for the comships. Not this time, for as it waited *Yuan Wang 4* had severe swells, with seas up to 10 m tall. The *Yuan Wangs* relayed their messages straight back to mission control in Beijing, using the code word 'Changjiang' (Yangtse river).

This time the rumours were accurate enough. The countdown began 8hr before liftoff and, following the firing of a green pistol flare by the base commander, the pad was evacuated 30 mins before take off. People stood calmly, waiting. It was after

Shenzhou: a side view. (Courtesy Mark Wade.)

Shenzhou deploying its panels. (Courtesy *Aerospace China.*)

midnight local time and the stiff north wind coming in from Mongolia had reduced temperatures to to −13° C. In fact, the choice of a night time launch was deliberate, designed to ensure the best possible tracking against a cold, clear night sky. Shenzhou 2 duly rose from wintertime Jiuquan at 5 pm GMT on 9th January, 1 am local time on the 10th. In the bunker, the controllers watched closely to make sure that the red marker of the ascending vehicle matched the path drawn on the wall screen. Shenzhou was the first launch from any country in 2001, China achieving the distinction of making the last flight of the second millennium (Beidou 2) and the first of the third (Shenzhou 2) (accepting of course the astronomical millennium rather than the popular one!). Its international designator was satellite 2001-001. Shenzhou 2 entered orbit as it passed over the Chinese coast at 5.10 pm in a path which circled the Earth every 91.1 minutes at 197 by 336 km, 42.58°. A ground observer in Houston, Texas spotted Shenzhou in binoculars six hours later. It was a magnitude +2 to +3.5 and could just be seen with the naked eye.

But how ambitious would be the flight of the second Shenzhou? 12 experiments were carried on the orbital module, 15 in the descent module and 37 in a scientific unit both inside and attached to the orbital module on the outside (equivalent Russian add-on packages are called nauka modules, nauk being the Russian word for 'science') – 64 experiments in all. The Chinese announced that animals were on board, along with a cargo of plants, seeds and snails. The exact nature of the animals was not revealed – one newspaper quoted a dog, a monkey and a rabbit, another rats (some wit volunteered 'a panda'). Post-landing announcements gave the cargoes as six mice, fruit flies and small aquatic animals. The specimens were chosen by the Institute of Medical Space Engineering. There were three containers with 20,000

Shenzhou: a view from beneath. (Courtesy Mark Wade.)

plant grains and seeds including tomato, cucumber, cabbage, Chinese cabbage, wheat, potato, corn, apple, pear, asparagus, carrot and fungus. Several science packages were designed to monitor cosmic and gamma radiation and the space environment. The experiment to detect gamma ray bursts was developed by Zhang Heqi and 20 colleagues at the Zijnshan Astronomical Observatory and aimed to detect the distance of each burst and its properties. Life sciences payload manager was Dr Liu Yongding.

No one had expected the Chinese to develop such an elaborate scientific programme for the early Shenzhou missions. The early Russian and American manned precursor flights had carried few scientific payloads (the Russian Korabl Sputnik *2* was the notable exception). Instead, the Centre for Space Science and Applied Research in the Academy of Sciences set up a dedicated Payload Operation and Application Centre to develop scientific payloads for flying on the Shenzhou.[10] Other experiments dealt with life sciences, astronomy, physics, materials sciences, semiconductors, oxide crystals, crystal growth of protein and biological macromolecules in zero gravity and the effects of the space environment on cells and microorganisms. Ten biological experiments were flown, including micro-organisms, plants, aquatic organisms and invertebrates. There was a multi-chamber crystal growth furnace for semiconductors, oxidized monocrystals and metallic alloys, photographed by camera. There were 25 life science experiments, selected from 87 proposals to the Academy of Sciences.

Shenzhou relayed television from the descent cabin: in the course of time, this would send back pictures of the first yuhangyuan on board. Shenzhou 2 carried, unlike its predecessor, the full environmental control system to provide air and proper temperatures for a crew. There was also a political message: a banner supporting Beijing's candidacy for the Olympic games in 2008 and a flag of Macau, the Portuguese colony recently returned to China. Shenzhou 2 was 100 kg lighter than its predecessor, due to the removal of wiring carried on its predecessor but no longer deemed necessary.

Second, would Shenzhou test its manoeuvring ability? The question was soon answered. At 1.23 pm on the 10th, 20 hr 20 mins into its mission and off the coast of Namibia, Shenzhou 2 raised its low point to adjust its orbit to a more circular path of 329–334 km. On the 12th January at 12.19pm, there was a small manoeuvre to reestablish the orbit from decay. At 10.34 am on 15th January, Shenzhou 2 adjusted its course over the Arabian Sea to an apogee of 345 km, so as to get on track for reentry the following day. The first two manoeuvres took place over the coast of south west Africa, commanded by *Yuan Wang 3* and the third may have been directed from China itself.

The orbital module was cast free at 10.23 am on 16th January as it was passing over 42.5°S, 64.7°W. Retrofire duly took place ten minutes later over 34.2°S, 7.3°W, off the coast of south west Africa and over *Yuan Wang 3*. In Beijing, this is what they heard:

Changjian 3 reporting, attitude correction!
Attitude correction completed!
Orbital and descent modules separated!
Successful separation!
Attitude correction!
Ignition!
Shut-down!

Shenzhou passed over Tanzania, Somalia, the coastline of Saudi Arabia and Pakistan, eventually passing over the Jiuquan launch site to come down over inner Mongolia at 11.22 am GMT. In the recovery zone, darkness had fallen. The recovery team, equipped with four helicopters and six recovery vehicles, was ready in perishingly cold conditions, with temperatures tumbling to $-30°C$. Far to the west, the fireball of Shenzhou 2's reentry was spotted as the spacecraft went into the blackout zone. There were cheers when the first radar station picked up the cabin high in the atmosphere. The drogue parachute came out under 20 km and then the 1,200 m^2 main parachute. As the cabin touched down, the circling helicopters saw the brown and orange flash of the landing rockets in the dark and headed toward the spot. It was a bitter evening during one of the coldest winters for many years.

Total flight time was 6 days, 18 hr 21 mins and the cabin had made 108 orbits and travelled 5.4 m km. The official announcement of the landing was flashed soon thereafter, stating that the cabin had landed smoothly, had been quickly recovered and that the mission had been a complete success. President Jiang Zemin promptly telephoned Cao Gangchuan, in charge of the recovery, to congratulate him. Once the

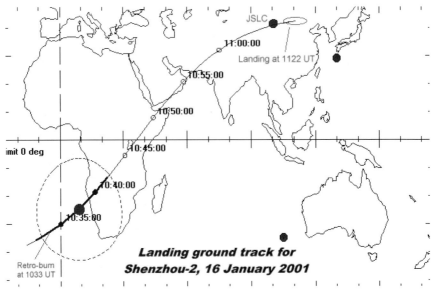

Shenzhou 2's landing track. (Courtesy Sven Grahn.)

cabin was back, the project team which had been encamped in Jiuquan since 23rd October returned home to Beijing and Shanghai. Scientists whose experiments in materials, life, astrophysical and space environmental sciences said that their experiments on Shenzhou 2 had made key breakthroughs, though they were not more specific. Life sciences experts said that the centrifuge of the biological incubator had functioned properly.

SHENZHOU 2 POSTSCRIPT

But where was the descent module? There was no triumphant parading of the returned cabin in Beijing. No pictures were even released of it landing (difficult presumably, since it was dark in local time). Officially, it was to be shipped to Beijing 'shortly'. In no time western commentators were speculating that 'something had gone wrong'. There were some reports that the cabin had been damaged at the final stage of landing because one of the parachute cords had broken free. To add to the confusion, the manager of the life studies experiments in the cabin, Dr Liu Yongding, ridiculed some press reports about the experiments flown on the mission. Even if nothing was wrong, the least one could note was that the return of the cabin had been treated very differently this time, for no apparent reason. Commentators further speculated a number of possibilities as to what might have gone wrong, from a hard landing to being lost in snowdrifts.[11] Some time later, the Chinese stopped denying that there had been a hard landing, resulting from a broken parachute connexion.

After the return, Chinese spaced officials announced that several further tests

Shenzhou's descent. (Courtesy *Aerospace China*.)

would be needed before manned flights could be contemplated. Further progress continued to be made with preparations for manned flight during the spring and in March, pictures were published of two astronauts kitted out in their *Sokol*-style spacesuits. The Chinese stressed that they had developed and manufactured their own spacesuits themselves at a cost of about €10 m. In the event of a depressurization of the Shenzhou while in orbit, systems would pump air into the cabin for 10 mins, but trainee yuhangyuans had been able to put their *Sokol* suits fully on in 3 mins.

There was a second postscript for the mission, for the two-tonne orbital module continued its flight long after the descent cabin was recovered. It was bigger than its Soyuz equivalent and, unlike Soyuz, carried its own solar panels and manoeuvring engines. These were used promptly on 17th January to raise the altitude of the module by 60 km and thus prolong its orbital lifetime. More surprises followed. The module manoeuvred again on 20th February, raising its orbit from 375–391 km to 389–403 km and again on 15th March from 382–390 km to 394–405 km. The ability to manoeuvre clearly required an autonomous flight capacity, navigation and control systems as well as engines, fuel and orientation systems. The Russians had never used the Soyuz orbital module in this way and it had always been discarded as débris.

The Chinese were far from reticent about the new assignment for the orbital module and hailed the experiment as a means of getting considerable scientific value added from an engineering test. The extended mission for the orbital module was officially announced on the day the descent cabin returned. There were four categories of instruments on board Shenzhou 2:

272 The Shenzhou precursor missions

- Space medicine and life sciences: tests of life support system, with mice; effects of the space environment on insect eggs, larvae and seeds; experiments with genetically modified tomaties, potatoes and Chinese cabbage;
- Biotechnology: protein crystal growth, macromolecular biology;
- Materials science: semiconductors, crystal and metal alloys;
- Earth and space sciences: atmospheric density, astronphysics and solar physics.[12]

The 64 experiments on the orbiting module included a gamma burst detector, a wide-spectrum instrument to detect solar flare high-energy electromagnetic waves in space, hard X-ray detector and supersoft ray detector. The space environment instruments were an atmospheric composition detector and an atmospheric density detector. Another experiment would determine the density of atomic oxygen, with a view to selecting the best orbiting altitude for the spaceship and the best type of protective material. These experiments would last several months. According to the Chinese, sensors examined the orbital environment to obtain key information on its composition, particle densities and radiation characteristics. The three high-energy radiation detectors in the astrophysics experiment obtained complete light curves and energy spectra of high temporal resolution of several gamma-ray bursts, allowing astronomers to trace the evolution of high-energy radiation and its structure. The instruments detected a hundred solar flares and thirty gamma ray bursts. The space environment experiment gave scientists a detailed mass spectral map of the atmospheric composition and density data, with a distribution map of the diurnal variation of the global atmospheric density. Data were transmitted whenever the orbital module made an overpass of a ground station in China. This was China's first big astronomical payload, making up for the loss of the canceled Tianwen Weixing mission (planned 1976, canceled 1984).

After the March manoeuvre, the module's orbit was allowed to decay naturally. By mid-August it was down to 209 km, 88.9 mins. The module eventually burned up on 24th August after 260 days (decay point was 33.1°S, 260.4°E, in the Pacific west of Chile).

Shenzhou reaches orbit. (Courtesy Aeropsace China.)

SHENZHOU 3: THE LONG WAIT

The Chinese originally planned to follow Shenzhou 2 with their third launch in August 2001. In the event, the third mission did not fly for another seven months. So what went wrong? The long delays in getting Shenzhou 3 airborne were an instructive example of the deliberative, cautious approach of the Chinese to their manned space programme. More than ever, they were determined not to rush their fences and only fly once they were totally happy with their equipment.

Shenzhou 3 was rolled out to the launch pad in Jiuquan in late July 2001. We know this from space imaging available from commercial spaceborne imaging agencies using small satellites with very high ground resolution – though American military satellites would probably overfly the launch site frequently enough as well. Not only did the space imaging satellites pick up the new Long March 2F on the pad, but they also saw that the Chinese were building a second pad close by, set at a slightly different inclination.[13] What about the tracking ships? Confirmation of an

Long March 2F in the vehicle processing building at Jiuquan.
(Courtesy *Aerospace China*.)

imminent launch came when the Hong Kong newspaper *Wen Wei Po* reported that three of the *Yuan Wang* tracking ships had completed refurbishment in Shanghai and one had already left.

The launch crews were not happy with their checkout of the Long March 2F and were unable to make the necessary modifications *in situ*. Product quality was at fault, indicated one report. The problem must have been deep inside the rocket. In accordance with Soviet practice, they sent it back on the railway line to Beijing for modifications. The rocket was removed on 12th October and sent back by rail. The rocket was back in Jiuquan again by early November. The tracking fleet slipped out to sea again on 16th December for a 10-day rehearsal of its tracking routines in the East China Sea, handing control over from one ship to another. They had a tough time, for they were hit with 5 m high waves, 60 km/hr winds and the crews were badly seasick.

A fresh attempt to count down the vehicle was made over the new year and 8th January was even given as the launch date. The launching teams even forewent their 1st January holiday to ensure the count made progress. But no launch took place: this time, apparently, the avionics were at fault. Internal systems had to be taken out and fully replaced. It is not known exactly what work had to be re-done (one report suggested it was the environmental control system for the cabin), but it was extensive and forced a further three-month delay. The deputy mission director Yuan Jiajun might have had something to do with this, for his idea of weekend fun was a coordinating meeting to review quality control.

The lengthy delays were not at first acknowledged by the Chinese. Eventually, in March 2002, academician Zhuang Fenggan gave an interview to *People's Daily* in Beijing, defending the delays as being a necessary part of quality control. Of course we wanted to launch last year, he said, but when the rocket returned to the pad the systems still weren't right and some had not been properly corrected. Ground control teams spent their time rehearsing the launch procedures and executing simulated countdowns. As they waited in the winter cold, the area was hit by some of the worst sandstorms in the region for many years.

Eventually, by the end of March, things were ready. The trusty *Wen Wei Po* newspaper leaked the warning that launch would take place within days. At Jiuquan, technicians were placing the three dummy yuyangyuans into the cabin. Mission director for the new flight was 39-year-old Yuan Jiajun.

JIANG, YUHANGYUAN AT LAUNCH SITE

Shenzhou 3 was eventually launched at night, 22.15 Beijing time (14.14 GMT) on 25th March. Its pillar of flame lit up the gantry alongside and sent orange smoke spewing up the side of the site. The escape system was operated in live conditions for the first time. In ten minutes, Shenzhou 3 had reached orbit, one slightly different from the previous Shenzhou, at 41.4° rather than 41.6°, with an altitude of 195–336 km, 89.84 mins. Confirmation that orbit had been achieved was greeted with applause in mission control. Sometime between 7hr and 9hr 15 mins

Shenzhou enters orbit. (Courtesy *Aerospace China*.)

after launch, Shenzhou manoeuvred to its standard orbit of 335 km, 91.216 mins, one which brought it exactly over Jiuquan every 31 circuits.

Shenzhou's launch was important for who attended. First, on hand was Chinese president Jiang Zemin, an indication of the importance of the mission to the Chinese leadership, in military uniform. With him was vice-premier Wu Bangguo. After the launch he toured the site to shake hands with members of the launch teams, some of them in office suits, some in blue technician overalls with, in true American style, peak caps and sporting mission patches on their shoulder arms.

Equally significant was the presence, for the first time, of China's yuhangyuan squad of 12 (with two instructors). The purpose of their visit was, besides watching the launch, to test the procedures for leaving the cabin in an on-the-pad emergency. In the event of having to leave the cabin quickly, they would exit the Shenzhou in 5 sec each, run to a tunnel, descend eight floors on a slide and shelter in a bunker. Some of their personal souvenirs were flown on the Shenzhou 3. Villas had been built for the young yuhangyuan, much like the wooden chalets occupied at Baikonour by Yuri Gagarin and his successors before their missions.

Shenzhou 3 carried dummies (two or three, depending on one's sources) with simulated blood pressure, pulses and breathing. Voice recordings were transmitted to and from the cabin. A full monitoring system was installed. Zhuang Fenggan was asked why the Chinese did not fly monkeys or another animals. They would break loose and interfere with the controls, he told *People's Daily*, possibly remembering a Russian monkey who allegedly took over control on Cosmos 1887. There would also be protests for animal lovers, he added. Also on board was a streamer done by 30 calligraphers called 'soaring Chinese dragon' based on a traditional painting.

Reports were issued during the mission from the Beijing Aerospace Command

276 The Shenzhou precursor missions

A dummy yuhangyuan. (Courtesy *Aerospace China*.)

and Control Centre. The dummies were still 'alive and well' and voice recordings had been sent to and from the cabin. The half-way point of the mission was signalled several days later, indicating that the mission was not intended to be longer than the week of Shenzhou 2. More television pictures were relayed of the dummies in the cabin while another shot showed Earth through the porthole. It was announced that 44 experiments were on board, 13 in the descent cabin and 31 on the orbital module.

On 29th March at 18:15 Beijing time (10.15 GMT) on the 61st orbit, the apogee was raised slightly while Shenzhou 3 was directly over the *Yuan Wang 3* comship off Africa in a 8 sec burn to trim it for reentry. The orbit was raised from 331–336 km, 91.2 mins to 335–342 km. A final trim took place on the 31st, adjusting the orbit from 330–337 km to 330–340 km. On the following day, 1st April, 15.52 the orbital module was separated retrofire took place at 16.02 (Beijing time). Shenzhou crossed the equator for the last time at 16.14 at 34°E. In mission control in Beijing, the path of the incoming spaceship was marked up on the 48 m^2 liquid crystal display screen. 29-year old mission controller Shen Jiansong called out each crucial stage as it happened, from retrofire through to parachute deployment and then touchdown in Chinese Mongolia. Stormy applause broke out. The landing came at 1651 after a 162hr mission in which it had flown 108 times around the Earth, covering a distance of 5.4 m km. It was an hour and a half before sunset.

MONGOLIAN LANDFALL

Would the Chinese release pictures of the return cabin? Indications that some of the criticism arising from Shenzhou 2 had hit the mark came when pictures of the Shenzhou 3 cabin were posted on the internet within minutes! A rescuer was pictured rushing forward toward the cabin which had alighted on the grassy steppe brush, with a Mil-8 helicopter in the background. Late afternoon sunlight flooded into the cabin as they opened the hatches to take out the dummies. In fact, Chinese media coverage was more extensive than ever, with quality colour pictures of the mission being made available on the internet from the start. The Shenzhou 3 cabin was the first one to come down in daylight.

Later, reports were given of the success of the recovered experiments. The descent module contained an experimental microchip, an incubator, seeds, seedlings and a vaccine experiment. The seeds were taken from plums, vines and alfalfa. The seedlings project was masterminded by Academy of Sciences genetics professor Liu Min. It was the first time that China had orbited seedlings (as distinct from seeds). This time grape, raspberry and orchid had been chosen. On their return, they grew at five to seven times the normal rate, he reported. The grape seedlings would later be attached to adult grapevines. 38 varieties of seeds were supplied by the Tian Xiang Ecoagriculture Company in Sichuan, including rice, wheat, vegetables and traditional medicinal herbs. Nine Blacklion chicken eggs flew aboard the Shenzhou in an experiment developed by chicken researcher Yang Anning. 30 days after their return to Earth, the first three hatched out and the results would be analysed for programmes to breed more successful chicken varieties. The descent cabin also carried protein crystallization and space cell culture experiments.

As was the case with the previous mission, the orbital module then began its own solo career, scheduled to last six months. On 1st April, the day of the landing, its engines fired to raise its orbit to 354–257 km, 91.64 mins. This had decayed back to the original altitude by 24th April, so early the following morning, a burn put the craft back up, this time to 382–388 km. This was a slightly lower altitude than its predecessor module. Up to 100 experiments were carried on the module, including China's first moderate resolution imaging spectrograph. The module carried 44 scientific experiments. The medium resolution imaging spectrometer had a 34-aperture lens, making it the second most advanced in the world. The instruments included a solar constant monitor, solar ultraviolet spectral irradiance monitor, atmospheric density detector, atmospheric composition monitor, Earth radiation budget measurement system, cirrhus sounder, multi-chamber crystallization furnace, spave protein crystal experiment, cell bioreactor and solid matter tracking detector.

On 13th June, the orbit had decayed, so a manoeuvre by the engines raised the module's orbit from 91.79 minutes, 356–369 km to 92.15 minutes, 375–385 km. It was flying over the Jiuquan launch site every 32 circuits. Later, details were released of the hundreds of images collected by the spectrometer, including the sea around north China and forest fires in north America. Atmospheric composition and solar constant charts were published. American analysts later made a distinctly military interpretation of the payload, believing it was used for electronic intelligance. It is possible that the module carried both electronic direction finders to detect and localize radars while the 550 m aperture camera could be used for visual military observations.[14] When the Shenzhou 3 orbital module had completed over a thousand

Shenzhou with sensors. Some believe these could be military in nature. (Courtesy Sven Grahn.)

revolutions around the Earth, a status report was issued by Karshi ground tracking station. Karshi is an important station in the Chinese network. Being the most westerly station in China, it was normally the first to pick up Chinese spaceships as they cross the country from west to east. Chief engineer there is Tao Feng: he reported that his station had picked up signals from 112 passes of the module and that all was well on board. The orbital module eventually completed its mission on 10th October. By then, it had circled the Earth 2,821 times on a 232 day independent mission.

With the success of Shenzhou 3, the Chinese began to speak more openly about the prospects for the first manned mission. There were features on the (still-unnamed) yuhangyuan in training. *China Space News* magazine interviewed the experts preparing the space food for the first manned mission. The crew would have 20 items to choose from, ranging from canned fish and meat to rice, curry and prawns, accompanied by dried strawberries, apples, bananas and peaches. To drink they would have a choice of orange juice, red tea, green tea or iced tea. The food and drink would be available either as solid blocks or else for mixing with water in a dispenser. Responding to rising public interest, national science week (May 2002) featured a full-scale mockup of Shenzhou and eight of the experiments returned to Earth from Shenzhou 3. As the prospect of manned flight drew near, excitement began to mount in China. Several daring youths stepped forward, volunteering to take the place of the dummies known to be scheduled for the next mission.

LAST TRIAL: SHENZHOU 4

With the return of Shenzhou 3, Chinese space experts let it be known that only one more flight would be necessary before a manned spaceflight. During the summer, 52 experiments weighing 300 kg were selected to fly on the Shenzhou 4 mission, some having taken part in earlier missions, such as the atmospheric composition detector. The Shenzhou 4 experiments included a digital microwave remote sensing instrument, a cell electrofusion unit, peony seeds, experiments for life and materials sciences and ion and proton detectors. Others covered Earth observations, the monitoring of the space environment and microgravity fluid physics. The Institute for Plant Physiology and Econology in the Shanghai Institute for Biological Sciences had devised a complex set of experiments for fusing cells in animals (mice) and plants (tobacco), so sensitive that they could not be loaded on board until eight hours before take-off. Among the plants and seeds to be flown were vegetables, grain, flowers, medicinal herbs, pinellia tuber and goldthread. Peony seeds from Luoyang, Henan would be flown and subsequently exhibited at its next spring show.

In advance of the mission, the 508 Research Institute carried out five air drops of the descent module over the Gobi desert in July, pushing the cabin out of the back of a transport plane at 11,000 m to verify the parachute and landing systems. Tests were also carried out of water landings, in case the cabin made a premature or inaccurate return to Earth (splashdowns were an essential aspect of Russian training for the Soyuz and one such mission, Soyuz 23, actually came down on water, though not intentionally). The desert tests involved a range of scenarios such as deploying the main parachute without the drogue first and the intentional failure of the main parachute to see if the reserve would open. It is possible that the careful attention to these systems was related to the Shenzhou 2 hard landing. Or maybe their Russian colleagues reminded them of their first Soyuz flight when the failure to do such exhaustive tests had cost the life of solo Soyuz 1 pilot Vladimir Komarov.

In early December, it was made known that the new rocket was on its pad and

Shenzhou 4: final tests. (Courtesy *Aerospace China*.)

Shenzhou heads for orbit. (Courtesy *Aerospace China*.)

that the launch would take place between 25th December and 10th January, once again during the coldest part of the year, when night skies would be clearest and the southern ocean calmest. It was announced in advance that the mission would fly the same 108 orbit pattern of its two immediate predecessors. The tracking ships were on station, waiting.

With none of the delays that held back the launching the previous year, Shenzhou 4 soared into the cold night skies of northern China at 00.40 am on 30th December. Present were Shenzhou designer Qi Faren, yuhangyuan commander Su Shuangning, Long March 2F designer Liu Zhusheng, director of experiments Gu Yidong, launch centre director Zhou Jianping, tracking system chief Sun Baosheng and commander of the recovery forces Zhao Jung. Also present were the twelve men of the yuhangyuan squad, each of whom hoped he would be chosen to fly the next one. Several days before the launch, the yuhangyuan had each entered the cabin on the pad to test entry and exit procedures. The scene echoed the events of 25th March 1961, when Yuri Gagarin and his five colleagues of the final training group went to Baikonour to watch the launch of Korabl Sputnik 5, the final dress rehearsal before the mission of Vostok 1.

The orbit was spot on: 331 by 337 km, 91.2 minutes, 42.41°, tweaked to the perfect orbit on the fifth circuit. The whole mission seemed to go effortlessly. The tracking centre had little to report, apart from the fact that everything was normal. Two manoeuvres were made to raise the orbit, on the 31st December and the 3rd January. On each occasion, when the orbit dropped to 91.088 minutes, engine firings pushed the Shenzhou back up to 91.102 minutes and there was a further set of two thruster firings on the 4th January. The big excitement of the mission was on the 5th January, when the *Straits Times* of Singapore identified one 'Chen Long' as the commander of the first manned spaceship. A picture was published of two

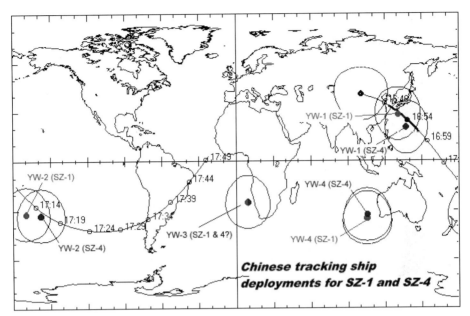

Tracking Shenzhou in orbit. (Courtesy Sven Grahn.)

yuhangyuan in their spacesuits in what could have been the vehicle assembly building, but a second crew member was not identified. During the mission, no space official was prepared to dampen media comment that the next flight would be the real thing.

Western observers were ready for the de-orbit burn on the morning of the 5th January – indeed, they ran a competition as to who could most accurately guess the actual landing time. Shenzhou 4 blasted its retrorockets over Africa, making a giant curved descent over the horn of Africa, Arabia and Pakistan. In the cold and slightly foggy recovery region, Mil helicopters, transfer cabins and recovery vehicles with direction finders on their roof moved in. The cabin came to rest in the dark in the middle of the 60 km by 36 km landing zone, targeted 40 km from Hohhot, capital of inner Mongolia. Teams in orange suits rushed forward to the silvery descent cabin, which lay on its side, retrieving the two dummy yuhangyuan inside. The descent cabin returned to Beijing three days after it touched down. The experimental section was opened and the yuhangyuan climbed inside to examine its condition.

As was the case with previous missions, Shenzhou 4 left an orbital module behind. The module manoeuvred first to 353–358 km and then a day later up to 382–388 km. On board were eleven experiments, including an upper atmosphere detector, high energy radiation and low energy radiation detector, biological module and microgravity fluid tester. An experimental microwave remote sensing system was used to watch the oceans, winds, waves, vegetation and crops and was able to do so through clouds. The orbital module carried a laser microwave altimeter, able to measure the altitude of the module from the ground to extreme extreme accuracy, for example to 331.25631 km. The altimeter was only 20 cm across and 800 g in weight and was installed on the bottom of the spacecraft. Over 2,000 measurements were taken over three days to test out the accuracy of the laser altimeter, which was so accurate it could measure wave heights and thus infer important information about changes in the Earth's oceans.

The orbital module continued operations until 9th February, by which time its altitude had declined to 331–346 km. The rockets on board fired to raise its orbit to 359 to 366 km. Additional corrections were carried out on 17th and 22nd April to restore the orbital altitude. The module finally decayed on 9th September 2003 after the longest orbital module mission so far. The following table lists the instruments carried:

Scientific instruments carried by Shenzhou 4

1	Earth remote sensing: Microwave radiometer, radar altimeter, radar diffuser
2	Space weather: High energy proton and heavy ion detectors; high energy electron detector and low energy particle monitor; particle track monitor; single particle detector
3	Fluid science: liquid drop migration, electrophoresis separation
4	Biology: Electrofusion, animal cell fusion, plant cell electrofusion
5	Precision orbit determination: Global positioning, S-band, laser ranging

Source: Ulivi (2004)

The Shenzhou precursor missions

Date	Spacecraft	Mission duration	Orbital module
19 Nov 1999	Shenzhou 1	21 hours	8 days
10 Jan 2001	Shenzhou 2	7 days	260 days
25 Mar 2002	Shenzhou 3	7 days	232 days
30 Dec 2002	Shenzhou 4	7 days	283 days

WHAT NEXT?

China's first manned flight was never intended as an end in itself. As far back as the time of the Shenzhou 2 mission, the manager of the manned programme, Wang Yongzhi was quoted as saying that the next steps would be:

- Longer flights and observations from orbit.
- Spacewalking.
- Rendezvous and docking.
- A semi-permanent, visited space laboratory.
- A permanent space station.

As the Shenzhou programme progressed, this sequence was emphasized by more and more Chinese scientists, engineers and spokesmen. In March 2004, Chinese space officials announced their intention of proceeding to a two-man, five-day to seven-day flight with Shenzhou 6 in summer 2005. The yuhangyuan squad was formed into seven groups of two and began training for the mission. Director of manned space engineering Zhou Xiaofei said China would dock space vehicles and then go on to build a space laboratory but, he said, 'our plans are in their early stages. This won't be like Mir or the International Space Station'. Plans were announced in 2004 to recruit another group of yuhangyuan, this time including women. Shenzhou 7 was considered the likely candidate for the first docking mission.

Pictures appeared, in a publication to mark World Space Week in October 2000, of a Shenzhou with the orbital module kitted out as an Earth observation platform, with a battery of scanning cameras underneath. The Soviet Union once flew a Soyuz on such a mission (Soyuz 22 in 1976) and the last Apollo was also equipped with an Earth resources package (Apollo 18 in 1975). Other pictures appeared showing additional scientific packages on the front of the orbital module, much like the Nauka modules flown by the USSR under the Cosmos programme.[15] Pictures of a spacewalking suit were published by *China Daily* in 2003.

The real objective of Shenzhou was and remains a long-term space station.[16] Designs of a possible Chinese space station were first circulated in the mid-1990s. They showed a small space station, about the size of the Soviet Salyut space station of the 1970s, to which a number of Shenzhou-type spacecraft or modules were docked. China had the option of first launching a small space station, around 14 tonnes in weight or proceeding directly to a larger, Salyut or Mir-class space station, over 20 tonnes when the new Long March 5 rocket became available.[17] A small space laboratory of the size of Russia's Kvant (attached to Mir in 1987) could be put up by

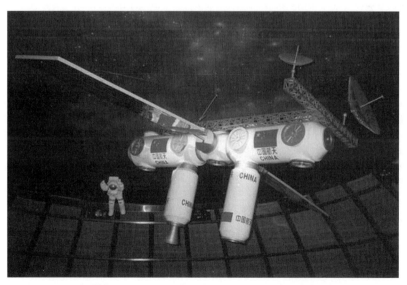

A Chinese space station. (Courtesy Mark Wade.)

a Long March 2 derivative, the Long March 2EA, capable of lifting 14 tonnes to low Earth orbit. A model of a Salyut-class space station was exhibited at an air show in Hannover, Germany in 2000. Soon thereafter, there were reports on the development of a small Chinese space station 11 m long and 4.2 m in diameter, with two compartments – a design not hugely different from the Salyut.[18] When Russian deputy prime minister Ilya Klebanov returned to Moscow after a 3-day bilateral meeting in March 2000, he confirmed that Russia was helping China to build a space station, but that it would be a Chinese-made station.[19] Reports have been made that the building of a Kvant-class space station was approved in February 1999, with the first design review held the following May.[20] The Chinese have indicated that such a station would be manned periodically at first, then permanently, exactly the pattern the USSR followed with *Salyut* and then *Mir*. In February 2003, photographs were published showing a small Kvant-size space station beside a hydrotank. The words 'space laboratory' were inscribed in Chinese on the side. The spacelab was a grey steel cylinder, painted white, with a docking unit on the front, large enough to serve as a home to at least two yuhangyuan.

Granted the costs and effort involved, several observers have been sceptical of China's desire, or intentions, to build its own space station. They have argued that China would really prefer to have access to the International Space Station (ISS), the first building block of which was launched in 1998 and which was designed to operate until at least 2015. An article in the Hong Kong *Wen Wei Po* on 28th July 2001 quoted Chinese space experts at length as saying that the costs of a Chinese space station were prohibitive. They really appeared to be pitching to join the ISS instead, suggesting that the European Space Agency could be middlemen to negotiate entry to the project. When a team from the prestigious American journal *Aviation Week & Space Technology* visited China in November 2001, it was told by the director of the China National Space Administration Luan Enje that the ISS without China was not a true international programme.[21] Several experts explained the Shenzhou programme as a means whereby China would so impress the international community that it would be allowed on board the International Space Station. This view was widely held, both in the west and in Russia.[22] In 2002, the Chinese pointed out that the Shenzhou *could* dock with the international space station if it so chose (and assuming it were invited to do so!).

They might be waiting some time, for the United States, which regarded themselves as the lead country in the ISS project, made it clear that they would not welcome China aboard.[23] The official American view was that China should be kept out of the ISS until it signed up to international agreements against arms proliferation. The Bush administration showed little preparedness to put cooperation with China on the ISS agenda, although many observers felt that the station needed all the help it could get. The level of American political hostility to China remained strong, a typical administration viewpoint being that 'we can't have Chinese spies on the international space station'.[24]

At one stage, there was discussion that China might wish to rent or buy the Russian Mir space station. China's interest in leasing Mir was apparently sounded out in February 2000. Russia apparently offered China the use of Mir for €588 m for

A Mir-class space station.

a 3-4 year period, pointing out that this was a bargain compared to the cost of developing their own station, which would cost over €1.1 bn. Press speculation notwithstanding, Russia never considered selling Mir outright – the term was 'joint utilization'. However, these discussions did not go far and the option finally closed with the deorbiting of Mir in March 2001.[25] The Chinese preferred to build their own station – but the Russians were not sent away empty handed. The two sides reached agreement for the following:

- Russia to provide technical assistance in the design of the Chinese space station.
- Russia would build a limited number of components for the station.
- Training would be provided for astronauts and ground controllers.
- Russia would transfer thirty-six specific areas of space station technology.

Russia continued to assist China in its space station development. When Russian premier Ilya Klebanov returned from visiting China in March 2000, he confirmed that Russia was providing such advice on an on-going basis. Two cosmonauts – Anatoli Berezovoi and Anatoli Filipchenko – visited China in January 2000 to provide technical advice and to brief China's trainee astronauts. A vacuum chamber, 7 m diameter and 12 m tall, was in construction to test out the first components. Progress in the development of the space station programme was reported at a conference held in the Academy of Engineering in June 2000 called *Frontiers of Engineering Technology*.[26] By autumn 2001, the closeness of Russian – Chinese cooperation was worrying the Americans, fearing that it would improve Chinese ballistic missile capabilities.[27] There was further evidence of a high level of Russian – Chinese cooperation in December 2002. When President Vladimir Putin visited President Jiang Zemin, he brought with him Russian Space Agency director Yuri

A future orbital station.

Koptev. One concrete item was that the Russians agreed to fly Chinese experiments to the Russian modules on the International Space Station – a kilo of riceseeds from Harbin Institute of Technology on a six month long astroculture experiment.

Even as they moved ahead with their space station programme, there was evidence of considerable attention being given to the detail of how the Chinese station might operate. An example was the design of a remote arm for the space station. Remote arms are important for space stations, enabling the astronauts inside

Robotics under test.

to move equipment and experiments from one exterior part of the station to another. The Russians built remote arms and girders for the Mir space station and the Canadians developed a remote manipulator system first for the space shuttle and then for the International Space Station. In China, Harbin Polytechnical University obtained funding under project 863 for the development of a remote arm for the space station. The scientists involved were Liu Hong, Professor in the university in Heilongjiang and Liang Bin, deputy director of the Beijing Robot Research Centre. The arm was unveiled in 2001. It was much smaller than the Russian or Canadian projects, being of human size, with 96 sensors, 12 motors, four fingers each with four joints and the ability to lift 10 kg. It could use screwdriver and spanner type instruments and according to its inventors could even play the piano! It was intended to test the robot first on Shenzhou. Presumably it would be fitted to the station later. Such a robot could be equally useful in helping later to lift rocks off the lunar surface for a sample return mission.

ASSESSMENT AND CONCLUSIONS

A typical feature of a major space programme is that the building of ground facilities and related equipment takes years and years before hardware actually appears in space. Project 921 was no exception. It took seven years from approval of project 921 to the launch of Shenzhou.

In devising a programme for the testing of their manned spacecraft, the Chinese could at least look back to precedent. The United States flew several Mercury capsules into orbit and sub-orbit over 1959–61. Russia made five tests of its Korabl Sputnik before committing Yuri Gagarin to orbit. These precedents suggested careful, stage-by-stage testing, each mission being slightly more ambitious than its predecessor. For the Chinese, a sobering consideration was that they were trying to launch a spaceship much more advanced than either Mercury or Korabl Sputnik. The Chinese had taken the ambitious approach of trying to launch a multi-manned, Soyuz-class spacecraft for the first manned flight.

Although the Chinese might have once entertained ambitions of making the first manned flight by the 50th anniversary of the revolution in 1999, the Chinese were not in a race. There was none of the frantic rushing that dominated the efforts of the Soviet Union and the United States to get their first person into space, as there had been over 1960–1. From a safety point of view, such a pace was undesirable in any case. The Chinese had a limited budget, one which had to be spent carefully, systematically and sparingly. As was the norm with the Chinese space programme, the money, effort and energy were concentrated on the ground segment, ensuring that all equipment was exhaustively planned, designed and tested before being committed to launch. They gave themselves the relatively generous target of 2005 for the first piloted flight. This meant that they could proceed carefully, slowly and deliberately. The seven month delay on Shenzhou 3, in which three launch campaigns were abandoned because of suspected faults, was evidence of the extremely cautious approach to quality control and safety concerns. If delays were disappointing, it was only because they realized because there was something worse than delay, namely a disaster.

Even as the Shenzhou was tested, the second stage of the manned project got under way behind the scenes in the late 1990s. Publication of space station designs with 'CASC' written on the side were not just aspirational but an integral part of the 921 project. By the time of the Shenzhou 1 flight, design of the Chinese manned space station had hardened. Reading carefully between the lines of statements by space officials, it became ever more apparent that the objective of Shenzhou was not simply to put yuhangyuan into space, but to use the cabin as a ferry for a manned space station. This was confirmed when news came out of a second technical agreement with Russia, whereby the Russian Space Agency provided design help, the building of specific components and training.

REFERENCES

1 Craig Covault: Manned programme advances Chinese space technology. *Aviation Week & Space Technology*, 29th November 1999.
2 For a discussion, see Christian Lardier: Le vaisseau habité chinois volera vers 2000. *Air & Cosmos*, 18 juin 1999.
3 For details of the orbit of satellites in the Shenzhou series (and their modules), I am grateful to Phil Clark of Molniya Space Consultancy, Hastings, England; see also Sven Grahn: *The flight of the Shenzhou* (see *Websites*, at end). For details of the radio transmission system, see Sven Grahn: *The flight of Shenzhou 5*, on www.svengrahn.pp.se
4 Craig Covault: Chinese space facilities show major build up. *Aviation Week & Space Technology*, 6th December 1999.
5 E.g. Spaceship mainly Russian. *Irish Times*, 24th November 1999. For a balanced early assessment of Soyuz, see Jim Oberg: Taikonauts prepare for liftoff. *IEEE Spectrum*, December 2001.
6 Le vaisseau Shenzhou a Hong Kong. *Air & Cosmos*, 1 septembre 2000.
7 Philippe Coué: La capsule chinoise Shenzhou en orbite. *Air & Cosmos*, 26 novembre 1999; Manned space flight in 1999. BBC reports, SWB series, 22nd April 1998.
8 New joint space project with Russia outlined, BBC reports, 1 September 1999; China Russia discuss space cooperation projects. BBC reports, 31st May 2000.
9 Qi Faren talks about Shenzhou. BBC reports, 30th November 1999.
10 Wei Long: Shenzhou 2 returns while orbital experiments continue. *Spacedaily*, 19th January 2001, http://www.spacedaily.com/news. For other descriptions of mission cargo, see La Chine: vers la maitrise des vols habités. *Air & Cosmos*, 19 janvier 2001; Official clarifies plan to launch unmanned spaceship. BBC reports, 10th March 1999; and Christian Lardier: Récupération de la capsule de Shenzhou 3. *Air & Cosmos*, 5 avril 2002.
11 Capsule de Shenzhou 2: Incertitude de récupération. *Air & Cosmos* 2 février 2001; Albert Ducrocq: Les chinois se hatent lentement. *Air & Cosmos*, 9 février 2001;

Shenzhou 2 landing mystery continues. *Spacedaily*, 24th January 2001, http://www.spacedaily.com/news; Cheng Ho: Confusion and mystery of Shenzhou 2 mission deepens. *Spacedaily*, 27th February 2001, http://www.spacedaily.com/news.

12 An extensive description of the science mission of Shenzhou 2 is given by Paolo Ulivi in *Earth and space science payload of the Shenzhou orbital modules* (currently sent for publication). See also: Shenzhou spacecraft orbiting the Earth for tests. BBC reports, 17th January 2001; Wei Long: China says Shenzhou OK: orbital module operational. *Spacedaily*, 2nd February 2001. http://www.spacedaily.com/news.

13 Phil Clark: Chinese satellites – status report at the end of 2001. *Journal of the British Interplanetary Society*, vol 55, #7/8, 2002.

14 Phil Clark: Shenzhou mission reports, unpublished reports, 2002; Manned flights a foreseeable goal for China. *Beijing Review*, 9th May 2002, Text supplied by Jim Oberg.; HX Sun & XM Chun: *Payload data management system for Chinese unmanned spacecraft Shenzhou*. International Astronautical Federation congress, Bremen, October 2003. For the military analysis, see Craig Covault: Chinese milspace ops – military technology development is a key goal of the Shenzhou programme. *Aviation Week & Space Technology*, 20th October 2003; and Arun Sahgal: China in space – military implications. *Asia Times OnLine*, 6th January 2004. For a description of the extensive civilian experiments, see Paolo Ulivi: *Earth and space science payload of the Shenzhou orbital modules* (currently sent for publication).

15 China Space News: *World Space Week, Hong Kong, 4–10th October 2000*, pp 36, 38. For a description of the science payload on Shenzhou 4, see Sun Qing: Shenzhou 4 experiments completed. *Aerospace China*, vol 4, #2, summer 2003.

16 Plan to build long-term manned space station. BBC reports, 14th June 2000.

17 Phillip S Clark: *China's Shenzhou programme reaches flight status*. Monograph, 2001.

18 La Chine mise sur les vols habités. *Air & Cosmos*, 17 novembre 2000.

19 Russian experts provide technical consultation. *Tai Yang Po*, Hong Kong, 5th March 2000, original in Chinese.

20 See Project 921-2. Mark Wade's *Encyclopaedia Astronautica*, http://www.friends-partners.org/-mwade/articles.

21 Craig Covault: China seeks ISS role, accelerates space programme. *Aviation Week & Space Technology*, 12th November 2001.

22 Yuri Koptev, director of the Russian space agency, was quoted along these lines in *China Daily* on 29th April 2001.

23 Albert Ducrocq: L'arbitre chinois – est-ce à Mir ou à ISS qu'un module chinelab ira s'accoupler? *Air & Cosmos*, 29 janvier 1999.

24 James Oberg: China takes aim at the space station. Friends & Partners in Space posting, 31st October 2001.

25 For a discussion of this episode, see China spurned Russian offer to sell Mir space station. *Spacedaily*, 28th February 2000; Russia hopes China will seek help in building its own space station; *Spacedaily*, 22nd March 2000. http://www.space-

daily.com/; and Vladimir Radyuhin: Russia may sell Mir to China. *The Hindu* (India), 9th March 2001.

26 Wei Long: China prepares plan for space station as China eyes manned mission. *Spacedaily*, 12th June 2000, http://www.spacedaily.com/news/china.

27 Steven Siceloff: Russia – China space deal makes NASA uneasy. *Florida Today*, 10th September 2001.

11

The future of China in space

China became, with its first manned spaceflight, the world's third most prominent spacefaring nation, following the original space superpowers of Russia and the United States. Many of our planet's nationalities have now been into space, but as guests of the superpowers. Only three countries have the ability to do so on their own. Now, with progress being made on its space station, China was set to achieve a form of parity with the rest of the worldwide space community. Within a few years, there would be two space manned space stations circling the Earth: an international one, led by the United States, Russia, Europe, Canada and Japan; and a Chinese one. All this had been achieved by a country where, a little over 50 years earlier, the bicycle, the tractor and the truck represented the limits of its technology, though not of its imagination. Spurred on by Zhou En Lai, managed by Tsien Hsue Shen, a space programme had emerged from the wreckage of war and revolution. Well might Zhou En Lai add to the 24th April 1970 communiqué 'We did this through our own unaided efforts'. China's space achievements were all the more remarkable for having been developed in a country isolated in many ways from the world community.

At this stage, we study the Chinese space programme in comparison to the other space programmes in the world, major and minor. There is examination and analysis of China's formal policy, as expressed in the white paper on space policy. China's ambitions in space are discussed.

THE CHINESE SPACE PROGRAMME IN COMPARATIVE INTERNATIONAL PERSPECTIVE

If we define a space power as a country or block able to put its own satellite into orbit, the world has nine space powers: Russia, the United States, France, Britain, Europe, China, Japan, India and Israel. Of these, Britain and France no longer have a national satellite launching programme, so the current relevant number is really seven. Britain cancelled its launcher programme before its first, only but successful mission; while France merged its launcher programme with the European one. The only country likely to join this list is Brazil which has made three launch campaigns, none successfully yet. These distinctions are in some ways artificial, for many other countries have space programmes, but they use other countries' launchers. Commercial rocket companies have made some of these national boundaries less and less important. For example, the Sea Launch project is developed by Boeing, owned by Russian and American companies, using Ukrainian rockets fired from a Norwegian oil rig serviced by a command ship built in Scotland and towed into the middle of the Pacific.

Nevertheless, it is valuable to set the Chinese space programme in a comparable international perspective. The following table lists the number of launches by the different space faring nations:

Rocket launches worldwide

Soviet Union/Russia	2,704
United States	1,278
Europe	153
China	75
Japan	59
(France	10)
India	14
Israel	5
(Britain	1)
Total	4,299

Notes: Figures to 1st January 2004; Britain and France no longer have an independent national launcher programme and are bracketed accordingly.

China therefore accounts for a tiny proportion of world space launches (1.6%). However, the proportions are much higher if one takes the two super-powers, Russia and the United States, out of the equation. Of the 303 launches of the *minor* powers, China then accounts for 20% of them, a fifth. China emerges as the fourth spacefaring block in the world. Except for brief periods early on and in the mid 1990s, Russia has always been the leading spacefaring nation, followed closely by the United States and, some distance behind, Europe. China has come next as the leading *Asian* power in space, ahead of Japan and India.

Looking at deep space missions (the Moon, Venus, Mars and beyond), four of the space powers have launched deep space missions: the United States, Russia, Europe and Japan, but not China. Turning to geosynchronous orbit, only six countries have launchers able to reach 24 hr orbit: the United States, Russia, Europe, China, Japan and, since 2001, India.

Placing China in the context of current world space activity, the following table lists the number of Chinese space launches compared to the rest of the world. In 2003, for example, the following is the list of launches made:

Rocket launches worldwide, 2003

Russia	24
United States	23
China	6
Europe	4
Japan	2
India	2

This finds China as the third spacefaring group in the world in terms of launches. Europe's launch rate for 2003 was unusually low and has generally been well ahead of China. However, as already noticed, China has an indigenous manned launching capacity that Europe lacks.

CHINA'S SPACE BUDGET

Estimating China's space budget has always been problematical. As was the case in the Soviet command economy, financial transfers between organizations can often be set at notional amounts. Some costs have been effectively subsidized. For example, important functions in the space programme were and still are performed with military help (for example, the rocket troops, search and recovery operations). Another consideration is that labour costs in China are exceptionally low. As a result, formal financial estimates of the Chinese space budget have tended to be on the low side in relation to their international comparators.

From the period of the four modernizations, self-accounting has been much more in evidence – indeed, the Chinese claim that they developed the Long March 2E entirely on their own company resources, without any state help. The Chinese themselves estimate that government support for space activities is worth ¥1.45bn annually, about €154 m, which is an implausibly low figure. However, this may simply be the research and development figure, for it is known to exclude launcher operations. Several authoritative western estimates have been made, some close to one another. These are in the range of €1.59 bn (America's *Aviation Week & Space Technology*) to €1.68 bn (Britain's *Flight International*). A figure of €1.64 bn represents a mid-point between the two. After the return of Shenzhou 5, the Chinese gave an estimate of ¥32bn or €3.6 bn spent on all *science* programmes each year (but did not break down the space figure within that). In 2002, the Chinese gave a figure for the cost of project 921 from inception to the completion of the first docking mission as ¥18 bn, or about €2.5 bn. Each year, this would work out at ¥1.6 bn or just under €200 m.

The following table attempts to estimate world space budgets for 2001. This is an inherently difficult exercise, for several countries do not have published space budgets, and in others it is difficult to separate national from international programmes, military from civilian. Exchange rates present a further complication, so this table must be treated with extreme caution.

World space budgets, 2001, estimated in €million

United States	35,888
Europe	5,865
ESA	2,835
national programmes	2,297
Japan	2,033
China	1,640
Russia	750
India	580

Adapted from Sevig, *European Space Directory*, seventeenth edition, 2002, Sevig Press, Paris.

Although the absolute figures given here are a helpful guide, the *relative* outcomes may be more meaningful. These figures show the United States not only as the largest space spender, but the largest by far. This has always been the case from the very beginning. Europe comes in second, a long distance behind, with Japan following much further behind in turn, but ahead of individual national programmes in Europe (the Russian figure is problematical, for its understates the programme's huge capital assets). The table places China as the fifth space spender in the world. Its low labour costs place it below the Japanese level – otherwise it would certainly be above.

Having said that, China is unlikely to remain this low on the list for long. The build up of the Chinese space programme over 2001–5 is so extensive that it can only be accomplished by a substantial increase in funding. Luan Enjie, director of the China National Space Administration, is quoted as saying that China's space budget will double during this 5-year period. Likewise, his deputy, Guo Baozhu, was quoted as saying that space spending would 'greatly exceed' the previous 5-year plan of 1996-2001.[1] The 10th five-year plan for the aerospace industry envisaged an expansion in spending from ¥1.7 bn to ¥5 bn, though that included aircraft as well as space development.

There are no absolutely clear figures available for the numbers of people working in the Chinese space programme. The best western estimates give a figure of 200,000 people directly involved in the space industry. Of these, 100,000 are technical workers, drawn from light industry, the army's technical ranks and the polytechnical schools. About 10,000 are graduate research engineers working in 460 institutes connected to the space programme. The Chinese space programme has been able to choose the top graduates coming out of engineering schools and has been able to attract the country's most talented scientists. As earlier figures on the pay rates of mission controllers have demonstrated, their qualifications are rarely reflected in their salaries.

RHYTHM OF CHINA'S SPACE PROGRAMME

There are several approaches to analysing the rhythm and characteristics of the Chinese space programme itself. The following are the basic statistics of the Chinese space programme. There are several ways of calculating such tables. This table (and the book as a whole) categorizes a launch as one in which a payload at least reaches low Earth orbit intact.

No. of launches in which payloads that reached orbit: 75
No. of satellites put in orbit: 86
Of which, commercial satellites for international customers: 16
   (To end 2003)

The next table lists the total number of successful launches made by China by year.

Annual rate of rocket launchings, 1970–2003

1970	1	1985	1	1995	2
1971	1	1986	2	1996	3
1975	3	1987	2	1997	6
1976	2	1988	4	1998	6
1978	1	1990	5	1999	4
1981	1	1991	1	2000	5
1982	1	1992	3	2001	1
1983	1	1993	1	2002	4
1984	3	1994	5	2003	6

This gives an average launch rate of just over 2 launches a year. As may be seen, the launch rates of the Chinese space programme are low and have never exceeded six in any given year. In some years, there have been no launches at all (e.g. 1989). Even in some recent years, launch rates have been quite low (for example, there was only one launch in 2001). Some people interpreted this as indicating problems, but it is more likely that they had enough applications satellites in orbit at the time to meet their needs and they had no urgent need to replace them.[2]

Turning now to the type of satellite launched, the following are the main categories. By 2004, China had launched 86 satellites. Their categories are displayed in the following table.

Classification of satellite types

Type	Number	%
FSW recoverable	18	21
Domestic communications	12	14
Scientific	9	10
Meteorological	6	7
Earth resources/oceanographic	5	6
JSSW series	3	3
International commercial	16	19
Demonstration	3	3
Manned or manned precursor	5	6
Navigation	3	3
Others/piggyback	6	7
Total	86	

Unsuccessful launches excluded; percentages rounded

The recoverable satellite series and international commercial launches have been the largest elements of the programme, each accounting for around 20% of missions, followed by geostationary domestic communications satellites (14%) and scientific (10%). The latter figure is a generous one, for it includes the two minor payloads of the Shi Jian 2 launch and the two Qi Qi Weixing balloons lofted with Feng Yun 1-1. These categories are likely to change, now that the recoverable programme has slowed and as long as international commercial opportunities remain restricted by the United States. Applications satellites (navigation, meteorology, Earth resources) are likely to continue in importance, as of course will the manned programme. A separate space science programme has played a very minor role, though everyone will have noticed the large scientific package flown on the Shenzhous, but categorized here under the manned programme.

OFFICIAL POLICY: THE WHITE PAPER

What is China's philosophy of space exploration? China's space goals have been articulated over the years in a series of government economic, defence and planning statements, documents and policy papers. The highly political, indeed polemical language in the 1970s has given way to much more pragmatic statements using frameworks and approaches familiar to students of government and public administration worldwide.

Until recently, spaceflight operated within the context of broader plans for scientific development, the most recent being the *National long and medium-term programme for science and technology development, 2000–2020,* adopted in 1996–7. The key elements of this 20-year plan were the development of comsats, metsats, satellites for remote sensing and other applications, providing international launcher services at competitive prices and a new launcher capable of putting 20 tonnes into orbit. More recently, and probably indicating its increased importance, spaceflight development became subject to a national policy statement in its own right.

On 22nd November 2000, China published a white paper on its future space programme. Readers expecting a listing of future launch schedules, dramatic reorganization or announcements of exciting new projects will have been disappointed. Like most government white papers the whole world over, the language was bureaucratic, the aspirations general and some of the statements quite bland. Positively, the 13-page white paper was economical in the use of language, logical in its presentation, short and clear. Political sloganeering and point scoring were completely absent and there was no reference to the American blockade. Like most white papers universally, the real value was in reading between the lines and in scanning the paper for nuances of ideas in train, projects hinted and new priorities articulated. Here, at the risk of over-interpretation, are the main points.

First, the white paper recited China's space achievements, articulated overarching aims and listed broad lines of development. The paper recalled how China had to struggle against 'weak infrastructure' and a 'relatively backward level of science and technology'. The three broad aims of the space programme were

Official policy: the white paper 297

Communications – a priority.

exploration, applications and the promotion of economic development. Space development was set in its broader political context and linked to economic progress, environmental protection and international cooperation. Internationally, China would make a point of working closely with the other countries of the Asia-Pacific region.

In designing its space policy, China would select a small number of key areas of development and concentrate on them, rather than try to do everything. China would build on its best abilities and concentrate on a limited number of areas and targets according to its strengths. China would combine self-reliance with international cooperation. The short-term priorities of the space programme were:

- Earth observation of the land, atmosphere and oceans.
- Weather forecasting.
- Independently-operated communications and broadcasting systems with long operating lives, high capacity and reliability.
- An independent satellite navigation system.

The long-term priorities of the space programme were:

- Manned spaceflight.
- 'To obtain a more important place in the world in space science'.
- The upgrading of existing rockets.
- The introduction of the next generation of new, low-cost, non-polluting, high-performance rockets.
- The development of a national system of remote sensing, ensuring the effective distribution of data throughout the country.
- A new generation of satellites for microgravity, materials science, life sciences, the space environment and astronomy.
- Pre-studies for the exploration of deep space, centring on exploration of the Moon.

The white paper articulated a number of what it called 'development concepts' to guide the space programme over the next number of years. These were:

- Space industry organizations were encouraged to market their products as widely as possible, domestically and internationally.
- Resources would be available for tackling key, core technological problems.

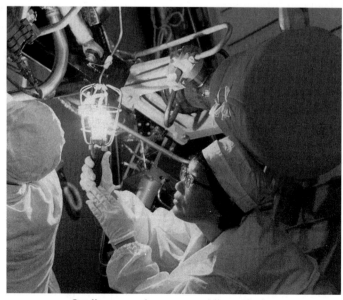

Quality control – an over-riding priority.

- The recruitment of talented people to the space industry would be encouraged, with the aim of building a contingent of young and highly qualified scientists and engineers.
- The programme would continue to emphasize quality control, risk reduction, and skilled management.

The white paper had few surprises. It confirmed the impression of a space programme that would not try to do everything but would instead concentrate on some key areas in a systematic way. The emphasis on manned flight and a new fleet of launchers was confirmed, although there was no specific mention of the planned space station. There was a renewed commitment to space applications and space science. Missions to the Moon were, for the time being, still something to study rather than to do. Symptomatic of the long-range thinking was the commitment to improved human resources and addressing key technological problems.

Apart from the white paper, the Chinese space programme operates within the context of the national five-year plans introduced by the communists. The current 5-year plan is the 10th national five year plan, covering the years 2001–5. This is frequently quoted as a reference point in Chinese statements and its key feature was a commitment to undefined but much increased spending on spaceflight. It has a space subsection called the *10 5 Civil space development programme*. This described the goals and blueprint of the civil space programme in the period 2001–5, but subordinate to the white paper. The two most eye-catching objectives of the period were the commitment to a manned flight by 2005 and the launch, during the period, of the first unmanned moon probe. The programme called for preliminary study of lunar exploration and identification of scientific objectives of lunar missions. An ambitious total of 30 spacecraft was promised during the period, almost half the total launched by China altogether up to that point.

A third reference point is the white paper on defence (2000). This outlined a 20% spending increase, largely in response to what it considered to be a deteriorating international situation, specifically a growing level of American hostility, hegemonism and interventionism.[3] This may have funded military-related space projects.

Having reviewed the broad context, this chapter now reviews the new projects planned, considered or rumoured for the early years of this century.

CHINESE SPACE SHUTTLE

A space shuttle was not mentioned in the white paper. Despite that, there have been intermittent reports over the years of a Chinese interest in building a space shuttle. The Chinese have never made any secret of their interest in shuttle designs. In the late 1970s, a design appeared – attributed to Tsien Hsue Shen – of a Chinese spaceplane on a rocket like the Titan 3, looking very much like the American Dynasoar project of 1960. Some western intelligence experts wondered if they had stolen the Dynasoar design. Obviously they forget that Tsien Hsue Shen had done the preliminary design work that led to the Dynasoar in the first place.

The first recent Chinese studies of a space shuttle were presented at the 1983 International Astronautical Federation congress in Budapest, Hungary when Zu Huang of the Chinese Academy of Sciences presented *Fully reusable launch vehicle with airbreathing booster*, the design of a two-stage ramjet-powered shuttle along the lines first proposed by the Austrian scientists Eugen Sänger and Irene Bredt in the early 1940s. At the 1990 conference of the International Astronautical Federation, Wang Shusheng and Zhang Kexun presented another feasibility study of a Chinese space shuttle on behalf of the Third Institute of Research in Beijing. The bottom stage was 45 m long, with a 198 tonne reusable carrier vehicle using tri-propellants (liquid oxygen, hydrogen and methane) and ramjets. At mach 6.5, the 132 tonne, 35 m long, 15 m wingspan orbiter was released to continue its path toward orbit while the mother craft returned to Earth like a jet plane. The shuttle design was much smaller than the American orbiter or the Soviet *Buran*, and was closer to two then current European concepts – Germany's *Sänger* and the French *Hermes*. The Chinese design could fly three astronauts or bring six tonnes of cargo to an altitude of 500 km. Several years later still, pictures were published of a Chinese spaceplane – one showing the shuttle atop a Long March 2E, another coming in for reentry.[4]

Despite these teasing pictures and presentations, a Chinese spaceplane or shuttle appears to be a distant prospect.[5] The former president of CASC, Liu Jiyuan, issued a statement in December 1999 to say that China had no active shuttle programme and no such programme had started and this remains the formal position.[6] At the 2000 International Astronautical Federation, the paper *A prospect over the development of Long March vehicles in the next decade* indicated that any shuttle programme was a 20-year undertaking and that much preliminary work had to be done first in the areas of propulsion systems, aerodynamics, heat-resistant materials

A model of a Chinese space shuttle. (Courtesy Mark Wade.)

and landing techniques. Limited practical work has been undertaken, however, on some distinct aspects of space shuttle design. News came out in 1998 of the construction of two advanced wind tunnels by the China Aerodynamics Research and Development Centre in Chengdu. The specific aim was to test rockets, missiles and space shuttles.[7] As both the United States and Soviet Union were to find out, the practical work of testing the heat-resistant materials and the aerodynamics involved was expensive and time-consuming. The American and Soviet shuttles which flew in the 1980s could not have flown without practical research dating to the late 1950s. A timescale of several tens of years may also be expected in China. They seem to realise this. In 2002, the Chinese were reported to have begun an examination of the main technology and materials needed to build a shuttle, especially the superlight materials that are needed if a shuttle is to resist the temperatures and shock of reentry. In 2003, CASC president Zhang Qingwei stated that China would eventually seek to develop reusable space vehicles and single-stage-to-orbit technologies. In the meantime, the key technologies involved would be investigated.[8] These would in effect be pre-studies before real work on a shuttle design could actually begin.

ONTO THE LONG MARCH 5

The white paper referred to a new generation of launch vehicles. The decision to proceed with a manned space programme in 1992 was linked a new fleet of launchers. In 1992, at the International Astronautical Federation congress, Xiandong Bao had, in *A modular space transportation system,* outlined a new launcher system able, in different variants, to lift a range of payloads up to 20 tonnes at the top end. His paper marked China's move away from its current toxic fuels to the use of liquid oxygen with kerosene for the bottom stage and hydrogen for the upper stages. The concept of modularization may be traced to the Russian Angara series of rockets, in which a common core is designed to be used in a number of different combinations and sizes (there were seven versions of Angara, from light through to medium and heavy lifters).

In 1996, the *National long and medium-term programme for science and technology development, 2000–2020,* restated the need for a new launcher capable of putting 20 tonnes into orbit. In 1998, Xu Dazhe, then Vice-President of CALT, confirmed the need for an entirely new, more powerful rocket able to lift 20 tonnes.[9] In the west, this has traditionally been called the Long March 5 concept and that will be the term used for convenience here, though it has not yet formally received such a designation by the Chinese themselves. According to one of the CALT Long March designers, Shen Xinsun, the government allocated project 863 funding to develop some of the critical technologies that would be necessary for the Long March 5 to be a success. The project 863 funding was focussed on the lower and upper stage engines, cost containment, the achievement of reliability and preventing toxicity in the engines. During the mid-1990s shopping visits to Moscow, the Chinese had been unable to persuade the Russians to part with the designs of the huge RD-170 engine used on

Long March 5 design. (Courtesy Mark Wade.)

their Energiya rocket, though that they were allowed to buy a RD-0120 engine. The first funds were allocated to the project for new rockets in the 2001-5 five-year plan.[10] We know that the Chinese later bought sophisticated computer-aided-design software to help them refine the engine design.[11]

At the 2000 International Astronautical Federation, China confirmed its interest in building a heavy launcher capable of matching Europe's Ariane 5 and Russia's Proton. The latest thinking was outlined that year by Wu Yansheng and Wang Xiaojun in their presentation called *A prospect over the development of Long March vehicles in the next decade*. A 55 m tall rocket would use liquid hydrogen and liquid oxygen, be flanked by four large strap-ons, weigh 800 tonnes and able to place 23 tonnes in low Earth orbit or send 11 tonnes to geostationary orbit. The new launcher bore some similarities to the European Ariane 5.

Long March 5 climbs to orbit. (Courtesy *Aerospace China*.)

A variation of this idea appeared in the February 2001 *Aerospace magazine* which again presented the concept of a new, modular Angara – style generation of rockets. Here, core stages were proposed with diameters of 2.25 m, 3.35 m and 5 m (the first two will be familiar measurements) with strap-ons of 2.25 m (also familiar), depending on the number of lower stages used and the length of the upper stage. The capacity of the launcher would be 13 tonnes to geosynchronous orbit and now 25 tonnes to low Earth orbit. It would be kerosene – fuelled for the lower stages (120 tonnes thrust) and hydrogen fuelled for the upper (50 tonnes thrust). Once again, the payload target for the Long March 5 seems to have grown slightly. Drawings of these various versions were published on a Chinese internet site in 2001.[12]

By 2003, China was able to firm up details of the proposed Long March 5 launcher, to appear from 2008, now set as its formal date of introduction. The programme had several core concepts, as follows:

- High reliability, of 98–99%.
- Commercial prices 20–30% below the Long March 3 series.
- Launch preparation period of 15 days.
- Non-toxic fuels.

Two new engines were introduced for the series. These were a 120 tonne liquid oxygen and kerosene engine for the first stage and a 50 tonne liquid oxygen and liquid hydrogen engine for the upper stage. The first stage engines were to have a

Long March 5.

thrust of 1,340kN and a specific impulse of 3286 m/sec. The hydrogen engine was to have a thrust of 540kN and a specific impulse of 432 m/sec. Whether or not it appeared on schedule, the intent was clear. This was a family of rockets that China intended to use for many years. It could be the backbone of the Chinese rocket fleet from 2010 to 2050.

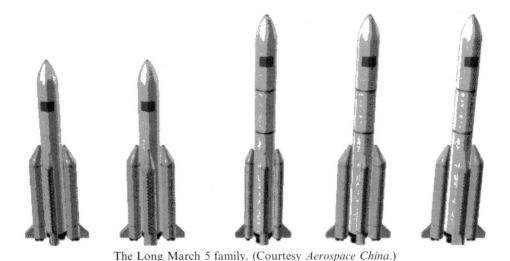

The Long March 5 family. (Courtesy *Aerospace China*.)

SMALL LAUNCHERS

While awaiting the Long March 5 from 2008, many of the most interesting developments in world rocketry have been at the other, small end of the market. There was a huge growth in the requirement for small payloads in low Earth orbit. Typical missions have been small scientific payloads, demonstration flights, imaging satellites and low Earth orbit communications. An American company, Orbital Sciences Corporation, first responded to this market niche by developing the Pegasus, a small rocket lifted to altitude by a Lockheed Tristar aeroplane, whence it was dropped, ignited and flew on to orbit.

The Chinese for their part turned back to the Long March 1, which has not flown since 1971. The modernized version, the Long March 1D, is smaller, lighter and more powerful than the Long March 1. It is 31.28 m tall and 2.25 m in diameter. It uses an improved Long March 1 first stage and a Long March 4 second stage. The Long March 1D has new engines (YF-2A, two YF-40 and GF-36 on the three stages respectively), a slightly lighter weight (81 tonnes) but better performance – it is able to put up a 750 kg satellite. It has improved electronics and guidance systems taken from the CZ-3. The promoters, the China Great Wall Industry Corporation, offer low Earth orbit, sun-synchronous and polar orbiting versions. The rocket has made two sub-orbital flights, on 29th May 1995 and in November 1997 (the first one was not apparently successful). Although the rocket has been available since 1991, it has still not found any customers. The initial price on offer was €14 m. In passing, it is worth noting that another adaptation, the 345-tonne Long March 3C, also awaits customers. This variant has two booster strap-ons, rather than the four of the 3B. It is able to place 3.8 tonnes in geosynchronous transfer orbit.

In the 1990s, Russia was successful in converting a number of solid-fuel cold war missiles (like Topol) into commercial rockets (like Start). Early in the new century, it seemed that China might follow a similar path, with the announcement that development of a small four-stage solid-fuelled truck-based launcher, called Kaitouzhe ('Pioneer' in English) or KT-1, was in progress. Kaituozhe was a product of the Space Solid Fuel Rocket Co and its design was finalized in late 2000. Chief designer was Yang Shucheng. The aim was to place 40 kg to 100 kg into 300 km high polar orbit from a mobile launcher, in the same fashion as Russia's Start. Ultimately, a family of solid-propelled rockets could be constructed from this model, the KT-2, KT-3 and so on.[13] Solid fuel rockets had the advantage that they could be stored for some time and could be prepared for launch quickly.

Models of Kaituozhe went on display at the Shenzhen 4th International High and New Technology Achievements Fair in October 2002. There were unconfirmed reports that an attempt was made to launch Kaituozhe 1 on 15th September 2002, but failed because of a guidance fault. A second launching took place from Taiyuan on 19th September 2003, but little information was available and there were contradictory accounts, one describing it as an orbital mission that failed, another as a successful suborbital flight. Like the Long March 1D, it has still to find its customers. Yang Shucheng was quoted in early 2004 as saying that the rocket would be used to launch small, domesic payloads.

NEW SPACECRAFT PROGRAMMES

The white paper on spaceflight gave commitments to the development of programmes for both applications and space science, although it was reluctant to comment or focus on individual proposals. Here, projects of the Chinese space programme known to be under way are reviewed. Other projects have been mentioned already (e.g. microsatellites) and these are not repeated.

Earth resources

China's first Earth resources satellite, CBERS, was launched in 1999 and the first oceanographic satellite, Haiyang, was launched in May 2002 piggyback on the Long March 4B. Development of a programme of environmental monitoring and radar observation satellites was jointly proposed by CAST, the State Bureau for Environmental Protection and the State Disaster Reduction Committee. The project was approved in 2002, manufacturing got under way in 2003 and launch was set for 2006. These would be variants of Haiyang 1, called Haiyang 1A, 1B and 1C. Haiyang 1A and 1B would be a pair of two 470 kg optical satellites, while Haiyang 1C would be a much larger 690 kg satellite with synthetic imaging radar. It is understood that the satellites will be based on the CAST968 model, much larger than the microsats already in construction for disaster monitoring.[14] Each of the optical satellites will have a pair of wide-field multispectral cameras of 360 km swath and 30 m resolution, supplemented either by infrared sensors or hyperspectral imagers. Radar imaging can produce astonishingly detailed and delineated pictures of the Earth's geology and ocean floors, but requires advanced, energy-gulping radar systems. Radar imaging was developed by Canada and as far back as 1997, China set up a receiving station to pick up data from Canada's Radarsat and has been studying how best to develop a similar satellite of its own. The satellites will, between them, cover the Earth's surface every 12hr, ideal for monitoring floods, drought, forest fires, sandstorms, high tides, environmental disasters and red tides. Another project reported in the pipeline was for an experimental CAST968-type 470 kg multi-mission satellite, with imaging and telecommunications systems developed with Iran and Thailand flown into a sun synchronous polar orbit in 2004. This has also been called the Asian Research Satellite and Mongolia, Korea and Indonesia have also been associated with the project.[15] Staying with the Haiyang bus, China also announced that a version would be adapted for another mission planned after 2004. The aim would be to study the effect of solar activity on the Earth's environment through the use of a far ultraviolet imager, ultraviolet radiation sounder, backscatter radiometer and high-energy electron plasma and wave detectors.[16]

Observing the Sun: Doublestar
In July 2001, China signed an agreement in Paris with the European Space Agency for a mission to follow the European Cluster project for solar observation. Cluster was an ambitious project to send a group of probes into eccentric Earth orbit to study the Sun, but the original probes were blown apart when Europe's Ariane 5 exploded on its maiden mission in 1996. The backup models were eventually flown into orbit on the Russian Soyuz in summer 2000 and proved to be an outstanding success. The Doublestar system comprises two satellites – hence the title 'Doublestar'

Doublestar. (Courtesy *Aerospace China*.)

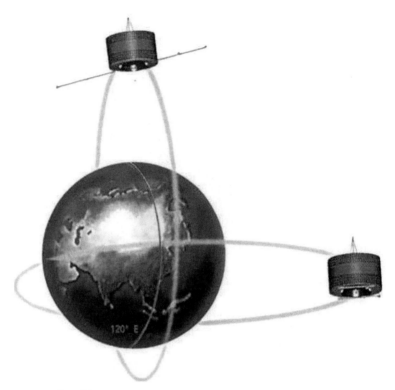

Doublestar's mission. (Courtesy *Aerospace China*.)

– to study the Sun's relationship with the Earth's magnetosphere, their findings to be cross-referenced to those of Cluster. In Chinese, the system is called the Tan Ce, the word for 'explorer' or 'probe'.

Doublestar will carry 18 instruments, of which ten will duplicate those on Cluster and eight will be new purpose – built Chinese instruments. Some are indeed spares or engineering models built during the preparation for the Cluster mission – meaning that they are known to be reliable. Using existing equipment will keep costs down. The European instruments include a flux gate magnetometer, plasma electron experiment, heavy ion detector and particle imaging detector. The European Space Agency contributed a modest €8 m to the mission and will get 4 hr a day of data over the planned 18 months of the missions.

The equatorial satellite was launched first, lifting off from Xi Chang on a Long March 2C. It entered a highly elliptical orbit of 560 – 78,948 km, inclination 28.5°. The equatorial satellite required an adaptation of the Long March 2C called the CTS in order to reach such an altitude, the highest ever achieved by a Chinese satellite. This was the first launch of the Long March 2C from Xi Chang. The polar satellite will fly later from Taiyuan into a 350 by 25,000 km orbit at 90°, circling the Earth every 7.3 hrs. The equatorial satellite concentrates on the Earth's magnetic tail while the polar satellite will check what is going on over the magnetic poles and the

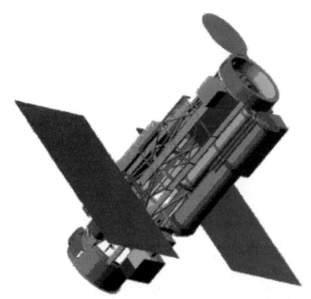

The space telescope. (Courtesy *Aerospace China*.)

resulting aurorae. Hopefully, they will improve scientists' knowledge of magnetic storms which can upset communications, radar and navigation systems on Earth. These are small satellites, about 350 kg weight. There are certain regions of the sky which the Cluster probes cannot reach but which will now be accessible to Double Star, meaning that a three-dimension picture of the effect of solar activity on the Earth can now be built up.

China is also planning a two-tonne solar observatory in cooperation with Germany (the proportions being 80/20). According to Ai Guoxiang, academician of the Chinese Academy of Sciences, this is a project begun in 1992 which involves the launching of a solar observatory with a 1 m telescope. The telescope itself will weigh 2 tonnes and carry five instruments operating for up to 5 years. When launched, it will be the most advanced space telescope in the world, far ahead of Europe's SOHO. The telescope will have a polarising spectrograph, accompanied by four side-mounted telescopes to examine the Sun in wide-band, X-ray and hydrogen rays. The pace of progress of this project has been quite uncertain, but a detailed description is now available.[17] The instruments comprise:

- A 1-m primary optical telescope.
- A 12-cm extreme-ultraviolet imaging telescope.
- A 12-cm white-light telescope.
- A wide-band spectrometer, with soft and hard X-ray spectrometers and a gamma-ray spectrometer.
- A solar and interplanetary radio spectrometer.

The 2,000 kg solar telescope will be launched from Taiyuan on a Long March 4B in 2008 for a three year mission in a circular orbit at 730 km, 99.3 mins, 98.3°. The telescope's observations will be transmitted down to a ground station in Miyun and then relayed 60 km to Beijing by fibre-optic cable.

Lunar and planetary science: project 211

The superpower monopoly on deep space exploration was broken some time ago when Europe sent the Giotto probe to comet Halley and Japan sent its first probes to the Moon in the 1990s. The Chinese would like to fly their first deep space mission early in the 21st century. The white paper was careful in giving a commitment to *studies* of lunar and planetary missions, making it clear that actually flying such a mission was some way off. At a 1995 Chinese Academy of Sciences conference, the director of space research, Prof Jiang Jingshan, told journalists that a pre-study of a lunar satellite was under way following a proposal by Academy of Sciences member Min Guirong.

In 1997, three designers obtained funding under the science programme 863 to research a possible lunar programme. Yang Yiaxi, Wang Dayan and Chen Fangyun published their results under the title of *Recommendations for the development of China's lunar exploration programme*. The following year, an expert group was appointed, issuing a report as *Overall design and key technology elements of a lunar exploratory robot*. This set three objectives for a Chinese moon probe:

- To improve knowledge of the formation of the lunar surface, its gorges and craters.
- To monitor, from the lunar surface, the solar wind, radiation and micrometeoroids.
- To analyse lunar rocks with an onboard laboratory, to detect the presence of helium 3.

Two symposia on lunar probes were held at Tsinghua university in May 2000 and January 2001. In October 2000, Qinghua University completed a study of the robotics involved in a lunar sample return mission modelled on that carried out by Russia's Luna 16, 20 and 24 in 1970-6. An imported Japanese robot was rebuilt so that it could be manipulated, from Earth, to grasp rocks to be lifted up and placed in a recovery capsule for return to Earth. The following year, the university built a model miniature 6-wheel solar-powered lunar rover, not unlike the *Sojourner* rover landed by the Americans on Mars. The Dean of the Department of Computer Science and Technology at Qinghua University, Sun Zengqi told the International Conference on Engineering and Technological Sciences 2000, held in the university in October 2000, that his department had explored a range of robotic technologies that could be used for lunar exploration – in collecting samples, exploring the lunar surface, deploying instruments, sending back television and, ultimately, paving the way for manned landings.[18]

The lunar programme was discussed at the China aerospace forum the following year. The conference, entitled *Policy and perspectives on China aerospace development* and held on 8th–9th October 2001 was told by Xu Dazhe of the Chinese National Space Administration (CNSA) that China was capable of a lunar mission. The following month, on the anniversary of the white paper on space exploration, Liang

Moonward bound? (Courtesy *Aerospace China*.)

Sili of the China Academy of Sciences and Sun Laiyan, vice-director of the CNSA, gave 2005 as the target date for the first Chinese unmanned mission to the Moon.

Evidence that China had progressed in its thinking for its first mission came in January 2002, when *China Space Journal* outlined three prospective lunar missions: an orbiter, soft-lander and sample-return mission. The journal gave indications of the types of payloads that could be carried moonbound:

- Long March 3A, 1,600 kg
- Long March 3C, 2,400 kg
- Long March 3B, 3,300 kg

In a sub-article, the possibilities were explored by four scientists: Zhuang Fenggan and Yu Menglun of the Academy of Sciences; Long Lehao of the Chinese Academy of Engineering and Tang Yihua of the China Academy of Launch Technology. They argued that China could start its lunar programme with relatively sophisticated probes: there was no need to repeat the type of basic missions flown by the United States and Soviet Union in the early years of the Moon race. Indeed, the Long March rocket gave the Chinese just such advanced options. In May 2002, the *Wen Wei Po* reported that chemistry expert and director of the Beijing national observatory Ouyang Ziyuan had been appointed chief scientist in charge of China's Moon exploration project. In 2002, it was learned that among the cooperation areas of

312 The future of China in space

A Chinese lunar robot. (Courtesy *Aerospace China*.)

the Russian – Chinese bilateral space commission were missions to the Moon, Mars and further afield – an indicator of things to come.

The success of the Shenzhou missions gave Chinese scientists greater confidence to talk about lunar projects. In January 2003, Chinese space officials were quoted at a conference in Bangalore, India as saying that China was looking forward to eventually sending astronauts around the Moon. They appeared to be outlining manned missions of the type that Russia had planned with its Zond spacecraft over 1968–70, but which were never actually carried out. Although the reports sounded like windy pronouncements or possibly even mistranslations, an Italian space writer, Paolo Ulivi, took the trouble of checking out Chinese technical papers in the open literature on possible Moon missions. Much to everyone's surprise, he uncovered no less than 67, most written since 1997. These papers covered a remarkably common set of topics, from launching a simple probe to hit the Moon, putting down a soft-lander, how to get a satellite into lunar orbit, tracking and navigation systems, trajectory profiles, robotic systems, lunar roving vehicles and a series of possible missions using existing Long March launchers. These documents explored mission-specific designs to the adaptation of communication satellites for lunar missions (a picture of a Dong Fang Hong 3 comsat, adapted as a lunar orbiter, was even published). Wide-ranging studies were conducted, from trajectories to intelligent robots.[19]

Come 2003, these plans began to harden into something more concrete. Speaking at the 25th Annual Congress of the International Astronomical Union in Sydney in 2003, Ye Shuhua from the Shanghai Astronomical Observatory put forward an outline schedule for Chinese plans for robotic exploration of the moon:

- An orbiter mission in a 200-km polar orbit in 2005.
- Soft-landing in 2010.
- A sample return mission by 2020.

She suggested the orbiter would carry a 130 kg payload comprising the following:

- CCD stereo cameras, taking images of 50 km^2 at oblique angles up to 25°.
- An imaging spectrometer.
- A laser altimeter.
- Gamma-ray and X-ray spectrometers for chemical analysis of the lunar surface.
- A microwave sounder for studying the lunar regolith.
- A high-energy particle detector to research the lunar particle environment and its interaction with the solar wind.
- Two low-energy ion detectors to analyse ions in energy ranges around 12 MeV, with six channels, and protons with energies of 3–300 MeV.[20]

The mission would last a year, with signals relayed to a 50 m dish in Beijing. The launcher will be a Long March 3A. For manoeuvres in lunar orbit, the spacecraft will have a 490N engine.

Confirmation that these plans had solidified into something more definite came in the April/May 2003 edition of *Aerospace China* which reported that an official go-ahead for the Moon probe had been given by the government at its meeting on 28th February. Later, a project name was revealed, project 211, indicating such an approval. The *People's Daily* named the project as the *Chang e* programme, called after a beautiful fairy who took a magic potion, flew to the Moon and became a celestial goddess. In August, the draft design of the Moon probe and its orbiting altitude were settled. Ten key task teams were appointed to progress the project, which had four broad aims:

- Three-dimensional imaging of the Moon to determine its structure, topography, craters, history and structural evolution.
- Determination of the contents and distribution of its chemical elements.
- Measurement of the thickness of its regolith.
- Exploration of the particle and radiation environment around the Moon.

In 2004, Sun Jiadong was appointed chief designer of the project and Ouyang Ziyan the chief scientist. The cost of the endeavour was estimated at ¥1.4 bn (€140 m).

At the same time, India was beginning to talk of preparing its own Moon probe, a development that must not have gone unnoticed in Beijing. After lengthy discussions, the Indian prime minister announced on independence day in August 2003 that India would aim to send a probe to the Moon by 2008. Called Chandrayan Prathim 1 (*Chandrayan* for short), it would be a 525 kg satellite to circle the moon at an altitude of 100 km, carrying cameras and instruments for infrared and x-ray observations. Probes to the planets would follow later. Japan also planned two Moon probes, Lunar A and Selene, both being sophisticated missions. Indeed, in autumn 2003, Hu Wenrui of the Academy of Sciences drew his colleagues' attention to Selene and reminded them that it would be wasteful for China to repeat the research of others.

314 The future of China in space

A Chinese lunar base. (Courtesy Mark Wade.)

Emboldened by the prospects of a Moon probe, Chinese scientists began to talk that summer about a programme for the exploration of Mars to follow. In summer 2003, China Academy of Sciences Centre for Space Science and Applied Research expert Liu Zhenxing reported that Mars had been studied as part of a project 863 planetary exploration project. The first phase in this study had been an examination of the exploration of Mars to date by other countries and the results obtained. This had helped the researchers draw up some initial possible objectives for Mars exploration science and some outline spacecraft designs. The study had identified the lack of a deep space tracking network as a significant problem for China to overcome. Liu Zhenxing ventured the opinion that China should now examine the key technologies for unmanned Mars exploration, the calculation of orbits, appropriate launch systems and a deep space tracking network. He outlined a possible four phase approach by China to Mars exploration:

1 Mission definition.
2 An orbiter mission to probe the martian environment, including its magnetic field, ionosphere and atmosphere, and the physical and chemical properties of its surface.
3 Soft landers with rovers and robonauts.
4 Surface observation stations and the collection of rocks for sending back to Earth.

Again, this suggested an approach similar to the new Moon project: theoretical studies, followed by a debate about the range and scale of possibilities, followed by the hardening of decisions into a concrete project.

Assessment and conclusions 315

Shenzhou orbiting the Moon. (Courtesy Mark Wade.)

Military
It is not known if China plans any military space missions (many argue that Zi Yuan 2 already is a military space programme). According to the United States Department of Defense, China has the capacity to develop, within ten years, a digital imaging satellite (presumably like the Russian Yantar-Neman) and film-based photoreconnaissance satellite.[21] The department has been expansive concerning all the various military capabilities that China *could* acquire by a mixture of overseas purchases and indigenous development, but thin on what China *has* actually done or was concretely known to be planning. Little has come out of China itself about prospective military missions – hardly surprisingly, granted the uneven flow of information about much less sensitive civil projects – but one did which spoke of China developing an anti-satellite system. The Small Satellite Research Institute of the Chinese Academy of Space Technology (CAST) was reported to have ground tested a system of parasitic nanosatellites which would, at a time of tension, attach themselves unnoticed to enemy satellites following a surreptitious rendezvous manoeuvre. Being very small, they would be unnoticed until it was too late. Attached to their hosts, they would await the command to disable them, either by explosion or by electronic interference. Whether this fiendish plan was a paper study, a threat, or a real project, is difficult to tell.[22]

ASSESSMENT AND CONCLUSIONS

What is the ultimate aim of the Chinese space programme? For many Chinese, the development of an indigenous space programme has been a source of pride, or, as one official document put it, inspiring 'lofty thoughts'. They are conscious that they have developed their programme almost entirely on their own, using indigenous human and industrial resources and despite varying levels of American blockade. With the atomic bomb, the hydrogen bomb and now manned spaceflight, China achieved world superpower status. Some foreign affairs commentators see China's space programme as a means of conferring on China the technological leadership – and with it, political leadership of Asia. Manned flight marks China's space programme as a quantum step more developed than that of the other spacefaring powers of the region, Japan and India.[23] Proposals for the new Long March 5 launcher fleet are not only impressive in their own rights, but indicative of serious long-term intent.

In western writings of future space missions, or in what might be called near-term science fiction, China has rarely played any part. An honourable exception is Arthur C Clarke. In his famous novel *2010: Odyssey two*, Arthur C Clarke had a manned Chinese interplanetary spaceship called, appropriately the *Tsien Hsue Shen,* racing to Jupiter and its life-giving moon Europa, ahead of the Americans and the Russians. The Chinese did indeed get there there first, but what happened after that is another tale. The Chinese part of the adventure was, in the event, disappointingly dropped from the film version. What a story it would have made!

A forward glimpse of Chinese aspirations in space may be found in the book *Academicians envisaging the 21st century,* issued to mark the new millennium. Here, Ouyang Ziyuan outlined the building of a lunar base from first landfall in 2010 to a self-sufficient colony by 2020. If China could build bases at Earth's poles, it could do so on the Moon. The chapter described how the lunar colonists would build their own solar power plants, extract minerals from the lunar soil, travel across the Moon in lunar roving vehicles and make astronomical observations of the heavens. At the time the book was published, both the director of the China National Space Administration, Luan Enjie and the chief designer of Chinese rockets, Long Lehao, made futuristic speeches about how, later in the century, China would venture on to the Moon and Mars.[24] When national science week was organized in May 2002 to stimulate children's interest in science, the exhibits included a Chinese base on the red planet, complete with greenhouses and domes and, in an adjoining exhibit, a robotic rover vehicle.

Fantasy? Maybe, but when Zhou En Lai and Tsien Hsue Shen set up the Chinese space programme on the 8th October 1956, who could have imagined that Chinese yuhangyuan would circle the Earth within 50 years? The Chinese space programme has been forged in a hard factory of technological backwardness, political upheaval and international isolation. The imagination, dreams, patience and dogged determination of Tsien Hsue Shen and his colleagues ensured that China could develop a space programme worthy of the country's ancient achievements in science and engineering. Would it be surprising if an interplanetary spaceship called the *Tsien Hsue Shen* one day travelled to that lunar base or to the far ends of the solar system? Granted all that has just happened in recent years, it might be more surprising if one did not.

REFERENCES

1 Craig Covault: China seeks ISS role, accelerates space programme. *Aviation Week & Space Technology*, 12th November 2001; China to develop civil satellite technology base. *Spacedaily*, 24th April 2001. http://www.spacedaily.com/news.
2 Phil Clark: Chinese satellites – status report at the end of 2001. *Journal of the British Interplanetary Society*, vol 55, 2002.
3 David Fulghum: China sees external threats growing. *Aviation Week & Space Technology*, 6th November 2000.
4 See Project 921-3. Mark Wade's *Encyclopaedia Astronautica*. http://www.friends-partners.org/-mwade/articles.
5 Philippe Coué: Les chinois et l'avion spatial. *Air & Cosmos*, 25 septembre 1998.
6 China has no shuttle programme. *Go Taikonauts!* 15th December 1999. http://www.geocities.com,Cape Canaveral/Launchpad/1921/news.
7 Wind tunnels built for aerospace, missile research. BBC report, SWB series, 22nd April 1998.
8 China begins to fund research into space shuttle programme. *SpaceDaily*, 3rd June 2002; China eyes RLV development. *Aerospace China*, vol 4, #3, autumn 2003.
9 China to develop larger launch vehicles for space exploration. BBC reports, SWB series, 25th March 1998.
10 Scientists says national carrier rocket development on a par with USA's. BBC reports, 15th March 2001; China to develop new rockets to launch space station. *Spacedaily*, 14th March 2001, http://www.spacedaily.com/news).
11 Craig Covault: China seeks ISS role, accelerates space programme. *Aviation Week & Space Technology*, 12th November 2001.
12 Chen Lan, internet posting to Friends and Partners in Space, 6th February 2001; Théo Pirard: Chine: Simulators et lanceurs. *Air & Cosmos*, 22 juin 2001.
13 China Space News, 7th June 2001, *Go Taikonauts!* website; Phillip S Clark: *Chinese space activity, 1996 2000*. Molniya Space Consultancy, 2001; Théo Pirard: L'empire du Milieu prépare de nouveaux lanceurs. *Air & Cosmos*, 1875, 31 janvier 2003.
14 Michael A Taverna: India, China to expand Earth-observing nets. *Aviation Week & Space Technology*, 29th October 2001.
15 Craig Covault: China, Iran pursue imaging spacecraft. *Aviation Week & Space Technology*, 1st October 2001. Christian Lardier: 30 à 35 satellites chinois d'ici 2005. *Air & Cosmos*, 5 octobre 2001.
16 Michael A Taverna: Chinese, Italian moves highlight growing scientific collaboration. *Aviation Week & Space Technology*, 29th October 2001.
17 For a description of Double Stark, see Michael A Taverna: Chinese, Italian moves highlight growing scientific collaboration. *Aviation Week & Space Technology*, 29th October 2001 and Zhang Huiting: DSP birds will be launch on Long March 2C. *Aerospace China*, vol 4, #3, autumn 2003; For a description of the space telescope project, see Lu Chang, Geng Lihong, Ai Guoxiang & Jin Shengzhen: Satellite and ground support equipment for space solar telescope. *Aerospace China*, vol 4, #3, autumn 2003.

18 China builds space robot to explore the Moon. BBC reports, 20th October 2000; Wei Long: Chinese robots to land on moon before Yuhangyuan. *Spacedaily*, 18th October 2000, http://www.spacedaily.com/news
19 Paolo Ulivi: The Chinese planetary programme. Posting 3rd November 2002 at http://utenti.lycos.it/paoloulivi/chinamoo.html. Subsequently updated 16th June 2003; see also Feng Jian-xian and others: *Kamado – a lunar robot and its telescience and intelligentization system architecture*. Paper presented to International Astronautical Federation congress, Bremen, Germany, October 2003.
20 China's first lunar steps outlined in Sydney IAU presentation. *Space Daily*, 22nd July 2003.; Sun Qing: China striding to Moon. *Aerospace China,* vol 4, #3, autumn 2003. For the most recent accounts, see Ziyuan Ouyang, Yongliao Zou & Chunlai Li: *The scientific objectives of the first Chinese lunar exploration project.* Paper presented to the 2003 lunar exploration conference in Hawaii, November 2003; and Dayi Wang: *Attitude and orbit control for China lunar satellite*. Paper presented to the 2003 lunar exploration conference in Hawaii, November 2003.
21 Defense Department details Chinese military space capabilities and plans. *Spacedaily*, 28th June 2000. http://spacedaily.com/news.
22 Cheng Ho: China eyes anti-satellite system. *Spacedaily*, 8th January 2000, http://www.spacedaily.com/news
23 Susan Lawrence & Sadanand Dhume: Houston, we have company. China leads Asia toward the final frontier – but is prestige worth the cost? *Far Eastern Economic Review*, 3rd August 2000.
24 Wei Long: Chinese scientists envisages Moon city in early 21st century. *Spacedaily*, 23rd October 2000. http://www.spacedaily.com/news

Appendix 1

LIST OF CHINESE SATELLITE LAUNCHINGS

	Name	Date of launch	Weight (kg)	Inc. (°)	Period (min)	Altitude (km)	Launcher	Launch site
1	Dong Fang Hong	24 Apr 70	173	68.5	114	440-2,386	CZ-1	Jiuquan
2	Shi Jian 1	3 Mar 71	221	69.9	106	267-1,830	CZ-1	Jiuquan
3	Ji Shu Shiyan Weixing 1	26 Jul 75	1,107	69.9	91	183-460	FB-1	Jiuquan
4	Fanhui Shi Weixing 0-1	26 Nov 75	1,790	63	91	179-479	CZ-2C	Jiuquan
5	Ji Shu Shiyan Weixing 2	16 Dec 75	1,109	69	90.2	186-387	FB-1	Jiuquan
6	Ji Shu Shiyan Weixing 3	30 Aug 76	1,108	69.1	108	198-2,145	FB-1	Jiuquan
7	Fanhui Shi Weixing 0-2	7 Dec 76	1,790	59.1	91.1	171-480	CZ-2C	Jiuquan
8	Fanhui Shi Weixing 0-3	26 Jan 78	1,810	57	90.9	161-479	CZ-2C	Jiuquan
9	Shi Jian 2	19 Sep 81	257	59.5	103.4	390-1,600	FB-1	Jiuquan
	Shi Jian 2A		483		103.5	235-1,615		
	Shi Jian 2B		28		103.2	232-1,595		
10	Fanhui Shi Weixing 0-4	9 Sep 82	1,780	62.9	90.2	174-393	CZ-2C	Jiuquan
11	Fanhui Shi Weixing 0-5	19 Aug 93	1,840	63.3	90.1	172-389	CZ-2C	Jiuquan
12	Shiyan Weixing	29 Jan 84	461	31	92	290-460	CZ-3	Xi Chang
13	Shiyan Tongbu Tongxin Weixing	8 Apr 84	461	0.72	1,444	35521-36383	CZ-3	Xi Chang
14	Fanhui Shi Weixing 0–6	12 Sep 84	1,810	67.9	90.3	174-400	CZ-2C	Jiuquan
15	Fanhui Shi Weixing 0-7	21 Oct 85	1,810	62.9	90.2	171-393	CZ-2C	Jiuquan
16	Shiyong Tongbu Tongxin Weixing 1	1 Feb 86	433	0.17	1,450	35895-36225	CZ-3	Xi Chang
17	Fanhui Shi Weixing 0-8	6 Oct 86	1,770	57	90	173-385	CZ-2C	Jiuquan

	Name	Date of launch	Weight (kg)	Inc. (°)	Period (min)	Altitude (km)	Launcher	Launch site
18	Fanhui Shi Weixing 0-9	5 Aug 87	1,810	62.9	90.2	172-400	CZ-2C	Jiuquan
19	Fanhui Shi Weixing 1-1	9 Sep 87	2,070	63	89.7	206-310	CZ-2C	Jiuquan
20	Shiyong Tongbu Tongxin Weixing 2	7 Mar 88	441	0.54	1,455	35784-36612	CZ-3	Xi Chang
21	Fanhui Shi Weixing 1-2	5 Aug 88	2,130	63	89.7	206-310	CZ-2C	Jiuquan
22	Feng Yun 1-1	6 Sep 88	757	99.1	102.3	881-905	CZ-4A	Taiyuan
23	Shiyong Tongbu Tongxin Weixing 3	22 Dec 88	441	0.53	1471	35756-33718	CZ-3	Xi Chang
24	Shiyong Tongbu Tongxin Weixing 4	4 Feb 90	441	0.45	1,472	35780-37199	CZ-3	Xi Chang
25	Asiasat 1	7 Apr 90	1,250	0.26	1,460	35791-36744	CZ-3	Xi Chang
26	Badr	16 Jul 90	52	28.5	96.6	205-983	CZ-2E	Xi Chang
	Aussat demo. model		7,353	Did not orbit				
27	Feng Yun 1-2	3 Sep 90	889	98.9	102.8	885-900	CZ-4A	Taiyuan
	Qi Qui Weixing 1		8	98.9	102..8	884-900		
	Qi Qui Weixing 2		8	98.9	102.6	862-902		
28	Fanhui Shi Weixing 1-3	5 Oct 90	2,080	57	89.7	208-311	CZ-2C	Jiuquan
29	Shiyong Tongbu Tongxin Weixing 5	28 Dec 91	1024	31.1	112.4	219-2,451	CZ-3	Xi Chang
30	Fanhui Shi Weixing 2-1	9 Aug 92	2,590	63.17	89.5	172-330	CZ-2D	Jiuquan
31	Optus B-1	13 Aug 92	1,582	0.33	1,472	35657-37330	CZ-2E	Xi Chang
32	Fanhui Shi Weixing 1-4	6 Oct 92	2,060	63	89.8	214-312	CZ-2C	Jiuquan
	Freja		259	63	109	596-1,763		
33	Fanhui Shi Weixing 1-5	8 Oct 93	2,100	57	89.6	209-300	CZ-2C	Jiuquan
34	Shi Jian 4	8 Feb 94	400	28.5	638	209-36,118	CZ-3A	Xi Chang
	KF-1		1,600	28.5	636	210-36,054		
35	Fanhui Shi Weixing 2-2	3 Jul 94	2,760	63	89.7	174-343	CZ-2D	Jiuquan
36	Apstar 1	21 Jul 94	1,368	0.3	1,579	35269-41820	CZ-3	Xi Chang
37	Optus B-3	27 Aug 94	1700	1.56	1,389	30778-38978	CZ-2E	Xi Chang
38	Zhongxin 6	29 Nov 94	2,232	0.26	1,426	35181-35993	CZ-3A	Xi Chang
39	Asiasat 2	28 Nov 95	1,400	0.6	1,404	34375-932	CZ-2E	Xi Chang
40	Echostar 1	28 Dec 95	3,288	5	981	17572-35072	CZ-2E	Xi Chang
41	Apstar 1A	4 Jul 96	1,400	1	1,553	34089-42010	CZ-3	Xi Chang
42	Zhongxing 7	18 Aug 96	1,200	27	307.5	200-17,229	CZ-3	Xi Chang
43	Fanhui Shi Weixing 2-3	20 Oct 96	2,970	63	89.6	171-342	CZ-2D	Jiuquan
44	Dong Fang Hong 3-2	8 May 97	2,232	0.32	1,436	35778-35788	CZ-3A	Xi Chang
45	Feng Yun 2-1	10 Jun 97	1,380	0.3	1,436	35780-35792	CZ-3	Xi Chang
46	Agila	20 Aug 97	3,770	0.3	1,436	35777-35791	CZ-3B	Xi Chang
47	SD test	1 Sep 97	650	87	97.3	623-632	CZ2CSD	Taiyuan
			650	86.3	97.3	623-633		

List of Chinese satellite launchings   321

	Name	Date of launch	Weight (kg)	Inc. (°)	Period (min)	Altitude (km)	Launcher	Launch site
48	Apstar 2R	16 Oct 97	3,747	0.1	1442	35818-36018	CZ-3B	Xi Chang
49	Iridium 42,44	8 Dec 97	650	86.3	97.3	623-632	CZ2CSD	Taiyuan
			650	86.3	97.2	6233-633		
50	Iridium 51, 61	25 Mar 98	650	86.6	97.4	624-641	CZ2CSD	Taiyuan
			650	86.6	97.4	625-641		
51	Iridium 69,71	2 May 98	650	86.4	97.4	625-641	CZ2CSD	Taiyuan
			650	86.4	97.4	626-641		
52	Chinastar 1	30 May 98	2,984	1.3	0.1	33423-38149	CZ-3B	Xi Chang
53	Sinosat 1	18 Jul 98	2,840	1.9	1436	35785-35790	CZ-3B	Xi Chang
54	Iridium 3, 76	19 Aug 98	657	86.4	97.2	616-630	CZ2CSD	Taiyuan
			657	86.4	97.2	617-630		
55	Iridium 11A, 20A	20 Dec 98	657	86.4	97.5	623-654	CZ2CSD	Taiyuan
			657	86.4	97.5	623-654		
56	Feng Yun 1-3	10 May 99	958	98.8	102	849-868	CZ-4B	Taiyuan
	Shi Jian 5		298	98.8	102	846-865		
57	Iridium 14A,21A	11 Jun 99	657	86.4	97.5	626-647	CZ2CSD	Taiyuan
			657	96.5	99.8	746-752		
58	CBERS 1	14 Oct 99	1,450	98.5	99.6	733-746	CZ-4B	Taiyuan
	SAC 1		60	98.6	99.6	732-747		
59	Shenzhou 1	19 Nov 99	7,600	42.6	89.7	197-325	CZ-2F	Jiuquan
60	Feng Huo 1	16 Jan 2000	2,300	0.75	1,436	35781-35788	CZ-3A	Xi Chang
61	Feng Yun 2-2	25 Jun 00	1,400	1.08	1,436	35778-35792	CZ-3	Xi Chang
62	Zi Yuan 2A	1 Sep 00	1,500	97.4	94.3	474-493	CZ-4B	Taiyuan
63	Beidou 1	31 Oct 00	2,300	24.98	753	1432-35685*	CZ-3A	Xi Chang
64	Beidou 2	21 Dec 00	2,300	0.11	1,436	35768-35809	CZ-3A	Xi Chang
65	Shenzhou 2	10 Jan 01	7,600	42.6	89.8	197-336	CZ-2F	Jiuquan
66	Shenzhou 3	25 Mar 02	7,600	41.4	89.84	195 – 336	CZ-2F	Jiuquan
67	Feng Yun 1-4	15 May 02	950	98.8	102.2	851 – 873	CZ-4B	Taiyuan
	Haiyang		360	98.8	100.7	792-795		
68	Zi Yuan 2B	27 Oct 02	1,500	97.4	94.3	473 – 490	CZ-4B	Taiyuan
69	Shenzhou 4	30 Dec 02	7,600	42.4	94.1	471-483	CZ-2F	Jiuquan
70	Beidou 3	24 May 03	2,300	0.2	1,436	35751-35821	CZ-3A	Xi Chang
71	Shenzhou 5	15 Oct 03	7,970	42.4	91.2	197-328	CZ-2F	Jiuquan
72	CBERS 2	21 Oct 03	1,550	98.5	100.3	773-774	CZ-4B	Taiyuan
	Chuangxin		100	98.5	99.6	732-750		
73	FSW 3-1 (Jian bing 4)	3 Nov 03	3,000	63	89.7	194-325	CZ-2D	Jiuquan
74	Zhongxing 20	15 Nov 2003	2,300	0.31	1,436	35771-35797	CZ-3A	Xi Chang
75	Tan Ce 1	29 Dec 2003	350	28.2	1,643	560-78,948	CZ-2C	Xi Chang

* *transfer orbit*

Launch failures

18 Sep 1973	Feng Bao 1	Ji Shu Shiyan Weixing	Second stage failure
14 Jul 1974	Feng Bao 1	Ji Shu Shiyan Weixing	Second stage failure
5 Nov 1974	Long March 2A	Fanhui Shi Weixing	First stage failure
10 Nov 1975	Fang Bao 1	Ji Shu Shiyan Weixing	Second stage failure
28 Jul 1979	Feng Bao 1	Shi Jian	Second stage failure
21 Dec 1992	Long March 2E	Optus B-2	Launch failure
25 Jan 1995	Long March 2E	Apstar 2	Exploded 70 sec
14 Feb 1996	Long March 3B	Intelsat 708	Failed at 2 sec.

Appendix 2

PRINCIPAL MILESTONES IN THE DEVELOPMENT OF THE CHINESE SPACE PROGRAMME

790	Invention of the rocket in China by Feng Jishen
1083	Rockets used by the Song dynasty to fight Xixia
1275	Kublai Khan uses rockets to drive out the Japanese
16th cent.	Wan Hu develops a two stage rocket kite to ascend into heaven, assisted by kites
	Naval rocket, the Fiery dragon
1911	Birth of Tsien Hsue Shen in Hangzhou
1926	(Liquid fuel rocket fired in United States (Goddard))
1933	(Liquid fuel rocket fired in Soviet Union (Korolev))
1935	Tsien Hsue Shen goes to study in the United States
1937	Tsien Hsue Shen's first writings on modern rocketry
1942 Oct 3	(First modern rocket, the A-4 (V-2) launched, Peenemünde, Germany)
1945	Tsien in Germany to survey German wartime achievements in rocketry
	(Arthur C Clarke proposes the establishment of satellites in 24hr orbit to provide global communications)
1947	(Soviet Union fires German wartime rockets)
1949	(Revolution in China)
1955	Tsien Hsue Shen repatriated to China
1956 Jan	Establishment of Scientific Planning Commission in China
	Adoption by China of *Long-range planning essentials for scientific and technological development, 1956-67*, committing China to the development of rocket and jet technologies
Apr	Central Committee invites Tsien Hsue Shen to outline the potential of guided missiles and rockets
	Appointment of State Aeronautics Industry Commission
Sep	Visit by Nie Rongzhen to Soviet Union

Appendix 2

	Oct 8	Central Committee of the Communist Party of China, presided by Mao Zedong, establishes the Fifth Academy to develop the space effort – the founding date of the Chinese space programme
		Arrival of R-1 missiles from Soviet Union
1957	Jul	Nie Rongzhen goes to Moscow a second time to ask for more advanced missiles
	Aug 20	*New Defence Technical Accord 1957-87* signed with Soviet Union
	Oct 4	(First satellite launched into Earth orbit by the Soviet Union (Sputnik 1)). Chinese Academy of Sciences sets up seven observing stations
1958	Jan	Fifth institute adopts the *Essentials of a ten year plan for jet and rocket technology, 1958-67*. Adoption of project 1059 – to copy and fire a Russian R-2
		Great leap forward
		Russian R-2 arrives in China
	Apr	20th corps leaves for Jiuquan to begin construction of China's first launch site
	May 15	(Soviet Union launches large scientific satellite (Sputnik 3))
	May 17	Mao Zedong declares: 'We too must launch artificial satellites'
	Aug	State Scientific Planning Commission endorses proposal for China to launch an Earth satellite. Proposal codenamed project 581
	Nov	Task of construction of satellite given to Shanghai Institute of Machine and Electricity Design (SIMED)
1959	Jan	Plan to launch an Earth satellite shelved by Deng Xiao Ping
1960	Feb 19	Launch from Laogang of T-7 prototype sounding rocket
	Aug	The great split: Soviet scientists ordered home
	Sep	Flight of T-7M sounding rocket
	Sep	China fires Soviet-made R-2
	Nov 5	China launches the first modern rocket manufactured in China, a copied R-2 (Dong Feng 1)
1961	Apr 12	(First man in space, Yuri Gagarin (Soviet Union))
	Aug 1	Establishment of Shanghai Academy of Space Technology (SAST)
1962	Mar 21	Dong Feng 2 missile crashes at Jiuquan
		SIMED team goes to Beijing to study in detail the idea of an Earth satellite
1963	Jan	SIMED made part of the Fifth Academy
	May	Start of interplanetary flight symposia
	Dec	Flight of T-7A sounding rocket
		Tsien Hsue Shen publishes *An introduction to interplanetary flight*
1964	May	Academy of Sciences begins feasibility study into Earth satellite (concluded a year later)
	Jun 29	First successful flight of Dong Feng 2 missile
	Jul 19	Flight of T-7A biological sounding rocket
	Nov 23	Fifth Academy replaced by the Seventh Ministry
1965	Jan	Tsien proposes an Earth satellite to the Central Committee

Principal milestones in the development of the Chinese space programme

	Mar	Preliminary work begins on liquid hydrogen powered rocket at Liquid Fuel Rocket Engine Research Institute; start of work on the YF-73 rocket engine
	Apr	Tsien's proposal receives support from the party's Defence, Science and Technical Commission, recommending a satellite launch in the 1970-1 time period
	Jul	Academy of Sciences proposes *A proposal on the plan and programme of development work of our artificial satellites*
	Aug 10	Approval of the plan at a meeting chaired by Zhou Enlai
	Aug	Allocation of project to institute 651 in Beijing
	Aug	Decision to disperse production to third-line areas
		Start of design of Dong Feng 4 and Long March 1
	Oct 20	Start of 64-day conference on the design of the artificial Earth satellite, project 651
		Start of construction of rocket engine testing site in Beijing
1966	Mar	Cultural revolution begins
		Closed conference on manned spaceflight project
	May	Earth satellite named Dong Fang Hong, design revised
		Introduction of the chief designer system
	Jul 14	First Chinese space dog, Xiao Bao, flies into space on T-7A-S2 sounding rocket (second flight on 28 July, carrying Shan Shan)
	Oct 27	Dong Feng 2A used for live nuclear rocket test
	Dec 26	First flight of the Dong Feng 3 missile
		Start of conceptual studies for recoverable satellite programme
		Start of construction of China Satellite Launch and Tracking General Control in Beijing
1967		First flight of He Ping 2 sounding rocket from Heilongjiang
	Mar 17	Zhou Enlai orders protection of key space workers from the cultural revolution, puts space programme under military authority
	Jun 23	Setting up of mission control centre, Xian
	Sep 11	Three-day conference on recoverable satellite programme (project 911)
	Nov	Establishment of the Chinese Academy for Launcher Technology (CALT), replacing the Beijing Wan Yuang Industrial Corporation
1968	Feb 20	Establishment of Chinese Academy for Satellite Technology (CAST)
	Apr 1	Setting up of Research Centre into Physiological Reactions in Space
	Aug 8	Testing of solid rocket motor GF-01 on sounding rocket (again on 20 Aug)
1969	Jul 20	(United States lands two men on the Moon)
1970	Jan 30	First flight of the Dong Feng 4 missile
	Apr 24	China launches its first Earth satellite, the Dong Fang Hong
	May	Conference held to prepare the launch of China's first scientific satellite, the Shi Jian

	Aug	Military Commission, under Lin Biao, adopted five year plan for the launching of 14 satellites and eight new launchers over 1971-6 First project conference on a communications satellite
	Oct 5	Start of selection of first squad of yuhangyuan
1971		First flight of He Ping 6 sounding rocket from Jiuquan
	Mar 3	China's first scientific satellite, the Shi Jian 1, which works for eight years
	Mar 15	Enlistment of first yuhangyuan (project 714)
	Jun	Yuhangyauns report for duty
	Sep 10	First flight of the Dong Feng 5 intercontinental ballistic missile from Jiuquan
	Sep 13	Failure of the Lin Bao plot, his plane crashes while fleeing to USSR
	Nov	Start of training of yuhangyuan
1972	May 13	Closure of project 714
	Aug 10	Suborbital test of the Feng Bao rocket Visit of President Nixon to China
1973	Sep 18	Unsuccessful attempt to launch Feng Bao
1974	Nov 5	First attempt to launch Long March 2 (unsuccessful)
	Nov 25	Report on developing satellite communications in China is commissioned by the Central Committee
1975	Feb 17	Report *Concerning the question of development this country's satellite communications* is presented
	Apr	Mao Zedong gives go-ahead for communications satellite proposal (project 331)
	Jul 26	Launch of Ji Shu Shiyan Weixing 1, first of series of three satellites, probably military (project 701)
	Nov 26	Launch of first recoverable satellite, the Fanhui Shi Weixing, recovered after three days. China is the third space power to recover satellites, after United States and Soviet Union
1976	Apr	Death of Zhou Enlai
	Aug	Decision to use liquid hydrogen for third stage of new rocket to launch communications satellites
	Sep	Death of Mao Zedong
	Oct	Military coup – crushing of the Gang of Four
1977	Mar	Stations for communications satellites in geosynchronous orbit requested from International Telecommunications Union
	Sep	Commissioning of comships *Yuan Wang 1* and *2*
	Oct	Project conference on new communications satellite Third plenum of 11th Central Committee adopts new plan for the exploration of space, involving Dong Feng 5, submarine launched missile and communications satellites Start of policy of cooperation with other countries
1978	Aug	Deng Xiao Ping says China will not take part in a space race but will devote its efforts to the practical application of satellite technology

Principal milestones in the development of the Chinese space programme

	Oct	Policy of the four modernizations and rectification. Restoration of the chief designer system
		Start of construction of new launch site, Xi Chang
1980	May 18	First operational test of the Dong Feng 5 intercontinental ballistic missile, impacting in the south Pacific (second test, 21 May)
		China joins International Astronautical Federation
1981	Sep 20	China launches three scientific satellites – Shi Jian 2, 2A and 2B, the third country after the Soviet Union and United States to launch three satellites on one rocket
1982	Apr 9	Establishment of the Space Ministry, replacing the Seventh Ministry
	Oct 12	First firing of submarine launched intercontinental ballistic missile
1984	Jan 29	First launch of the Long March 3. China becomes, after the United States and Europe, the third country to master liquid hydrogen fuels. The Dong Fang Hong 2 communications satellite was put, incorrectly, in low Earth orbit. First launch from new Xi Chang launch site.
	Apr 8	China successfully puts a 24 hr communications satellite into orbit, the Shiyan Tongbu Tongxin Weixing (experimental geostationary communications satellite)
1985	Oct	China puts its rockets on the world launcher market
1986	Feb 1	First operational geostationary communications satellite: Shiyong Tongbu Tongxin Weixing
		Start of construction of Haikou launch site on Hainan Is, for sounding rockets
		Building of Miyun Earth station to receive Landsat data.
1987	Sep 9	Introduction of FSW 1 series of recoverable satellites. Heavier, able to stay in orbit < 10 days. The series introduced microbiology experiments in addition to remote sensing tasks.
1988	Sep 6	First Chinese meteorological satellite, Feng Yun 1. First launch of the Long March 4, first launch from new Taiyuan launch site.
	Feb 1	First of the Dong Fang Hong 2A communications satellites
	Dec 19	Launch of Weaver Girl 1 sounding rocket from Haikou
1990	Apr 7	First commercial launch by Chinese Long March 3 rocket (Asiasat, for a Hong Kong company)
	Jul 16	First launch of new Long March 2E launcher, putting Pakistan satellite into orbit (Badr)
	Oct 5	China flies animals into orbit on FSW 1-3, the third country to do so, after Soviet Union and United States
1991	Jan 22	First launch of Weaver Girl 3 sounding rocket
1992	Aug 9	Introduction of FSW 2 series of recoverable satellites. Heavier, manoeuvrable, able to stay in orbit < 18 days.
		Approval of manned spaceflight project, project 921. Wang Yongzhi appointed as director, Qi Faren as spaceship designer; start of work on new fleet of launchers, including Long March 5

1993		Establishment of Chinese National Space Administration (CNSA) Chief of staff of the People's Liberation Army, Chi Haotian, visits Star Town, the cosmonaut training centre in Moscow – marking renewal of contact broken in 1960
1994	Feb 8	First launch of Long March 3A, carrying Shi Jian 4 scientific satellite
1995	Mar	Commissioning of comship *Yuan Wang 3*
1996	Feb 14	Long March 3B rocket crashes on its first flight – the St Valentine day's massacre
	Mar	FSW 1-5 crashes into the south Atlantic. Launched in October 1993, it had failed to return to Earth and had blasted into an unstable orbit
	Aug 20	Arrival of Chinese astronaut trainers and other specialists in Star town, Moscow
	Oct	47th International Astronautical Federation Congress held in Beijing
1997	May 8	First successful launch of the Dong Fang Hong 3 series of communications satellites
	Jun 10	First geostationary meteorological satellite is launched, the Feng Yun 2
	Jun 18	Congressional resolution to set up the Cox committee
1999	May	Release of unclassified version of Cox report, alleging Chinese espionage of American space secrets
	Nov 19	First flight of the Shenzhou, prototype manned spacecraft (14 orbits)
2000	Oct 31	Launch of Beidou 1, first of two part satellite navigation system
	Nov 22	Publication of government white paper on Chinese space development
		First tests of engines for new rocket fleet
2001	Jan 9-16	Second flight of Shenzhou, this time making 108 orbits
2002	Mar 25 –Apr 1	Third flight of the Shenzhou
	Dec 30 –5 Jan	Fourth flight of the Shenzhou
2003	Apr	Chinese moon probe, *Chang e* is approved (project 211)
	Oct 15	China's first astronaut, Yang Liwei, circles the Earth 14 times on board Shenzhou 5 and is successfully recovered.

Glossaries

The terminology of the Chinese space programme may not be well known in the west. These glossaries contain a listing of terms associated with Chinese satellites, rockets and space equipment; the main institutional bodies and academies associated with the space programme; an introduction to the key personalities of the programme and China's political leadership; and a glossary of the main political terms and developments of the period which had an impact on the space programme. A listing of some technical terms of space terminology is briefly given.

GLOSSARY OF CHINESE SATELLITES, ROCKETS AND SPACE EQUIPMENT AND RELATED TERMS

Beidou	'The plough' or 'big dipper' – the name of the constellation around the pole star – which came to be the name for China's series of navigation satellites.
Chang Zhen	'Long March', the name of China's main family of rocket launchers.
Dong Fang Hong	'The east is red' – the name of China's first satellite. Much later, it was also the name of China's communications satellite. These were called the Dong Fang Hong 2, 2A and 3 series.
Dong Feng	'East wind', the name of China's missiles (Dong Feng 1, 2 etc); also the Jiuquan (q.v.) launch site, also known as the 'east wind site'
Fanhui Shi Weixing	(FSW) Recoverable experimental satellite, series beginning in 1975.
Feng Bao	'Storm', the name of China's launcher developed in Shanghai.
Fen Huo	Beacons lit on the Great Wall, to warn of invading armies – and the name of a new series of Chinese communications satellites from 2000.

Haiyang	Oceanographic satellite launched 2002.
He Ping	'Peace', the name of a series of Chinese sounding rockets (He Ping 2 and He Ping 6).
Ji Shu Shiyan Weixing	Technical experimental satellite (series of three satellite, probably military, launched 1975-6)
Jian Bing	Meaning 'Vanguard', the term was applied to unspecified photography flights within the FSW series. The FSW cabin launched in November 2003 was called Jian Bing 4. The term may have military connotations.
Jiuquan	China's first launch site, in the Gobi desert.
Qi Qi Weixing	Atmospheric satellite (balloons carried into orbit on Feng Yun 1-2 in 1990).
Shi Jian	Name for China's series of scientific satellites. In Chinese, the word Shi Jian means 'practice' and 'construction'.
Shenzhou	'Divine heavenly [or magical] vessel', the name of China's manned spaceship
Shiyan Weixing	'Experimental satellite' (communications satellite launched 29 January 1984)
Shiyan Tongbu Tongxin Weixing	'*Experimental* geostationary communications satellite' (launched 8 April 1984)
Shiyong Tongbu Tongxin Weixing	'*Operational* geostationary communications satellite' (first flown 1988)
Shuguang	'Dawn', China's first manned spaceflight programme, 1970–3
Taiyuan	China's third launch site, used for satellites flying into polar, sun-synchronous orbit (qv).
Tianwen Weixing	'Astronomical satellite', a project devised over 1976-84, but not flown.
Xi Chang	China's second launch site, in south-western China, used for geosynchronous satellites (qv).
Yeti Fadong	Liquid-fuel rocket engine. Chinese rocket engines are designated YF-1 and so on. 'GF' is the acronym used for solid fuel rocket engines.
Zhinui	'Weaver girl', the name of a Chinese sounding rocket programme from 1988.
Zhongxing	'The star of China'. Zhongxing 5 was the name given to an American satellite already in orbit but bought by the Chinese in 1992. Zhongxing 6 was an abandoned comsat launched by the Chinese in 1994. Zhongxing 7 was an American satellite bought in 1996 and launched by China, but which entered the wrong orbit and was abandoned. Zhongxing 1 to 4 were the earlier Shiyong Tongbu Tongxin Weixing satellites 1-4.

GLOSSARY OF INSTITUTIONS AND ACADEMIES ASSOCIATED WITH THE CHINESE SPACE PROGRAMME

ARMT	Chinese Academy for Solid Rocket Motors.
CASC	China Aero Space Corporation.
CASET	Chinese Academy for Space Electronics.
CALT	Chinese Academy for Launcher Technology, formerly Beijing Wan Yuang Industrial Corporation.
CAST	China Academy for Space Technology.
CCF	Chinese Academy of Mechanical and Electrical Engineering.
CHETA	Chinese Electromechanic Academy.
CLTC	China Satellite Launch and Tracking General Control.
Fifth Academy	The main organizational body responsible for the Chinese space programme from 1956 to 1964.
Gt Wall Industries Corp	The main company promoting Chinese launchers and commercial space products in the west.
SAST	Shanghai Academy for Spaceflight and Technology, founded 1961.
Seventh Ministry	The main organizational body responsible for the Chinese space programme from 1964 to 1982.
SIMED	Shanghai Institute of Machine and Electrical Design

GLOSSARY OF KEY PERSONALITIES OF THE PERIOD

Deng Xiao Ping	Leader of China from 1978 to 1997. He restored order in China after the death of Mao, opening the economy.
Jiang Zemin	President of China in the 1990s, associated with the manned spaceflight project
Lin Biao	Leader of a leftist group in the Chinese leadership, supportive of the cultural revolution, influential in the Military Commission in 1970. The following year, Lin Biao fled China for the Soviet Union, but his plane was shot down in flames before he reached the border.
Mao Zedong	Leader of the Chinese revolution (1949). He died in September 1976.
Nie Rongzhen	The military leader of the Chinese space programme, a marshal.
Qi Faren	Chief designer of the Shenzhou.
Tsien Hsue Shen	Born, 1911, father of the Chinese space programme, China's greatest scientist
Zhou Enlai	Foreign, then prime minister under Mao Zedong, he is considered to be a moderating influence during the cultural revolution. He died shortly before Mao, in April 1976. He was the leading political patron of the space programme.

332   **Glossaries**

Hu Jintao	China's president after Jian Zemin
Wen Jiabao	China's prime minister

GLOSSARY OF KEY POLITICAL EVENTS AND TERMS OF THE PERIOD

Cultural revolution	This was launched by Chairman Mao Zedong in March 1966. He encouraged young people, who formed the red guards, to reinforce the communist revolution by seeking out counter-revolutionary elements and thought. This violent campaign of radical egalitarianism lasted ten years and was disruptive of the economy and science.
Gang of Four	Four people, led by Mao's wife, Jian Qing, who tried to maintain the revolutionary path of the cultural revolution. They seized power after Mao's death but were overthrown by a military coup a month later (October 1976).
Four modernizations	The modernization of China announced by Deng Xiao Ping in 1978. The four modernizations were science, agriculture, industry and defence.
Great leap forward	A national campaign launched by Chairman Mao Zedong in 1958 whereby China would rapidly increase its agriculture and industrial production. Steel was requisitioned from every home, with small furnaces set up in every street. A campaign was announced against pests – mosquitoes, flies, rats and sparrows. Within a year, the country was in chaos, people were starving and the campaign was called off.
Long March	The time in 1932 when Mao Zedong led his communist army 8,000km out of a nationalist government encirclement to the north of the country where they subsequently regrouped.
Rectification	The period from 1978 when Deng Xiao Ping restored order to the economy and society that had been upset by the period of the cultural revolution.
Third line regions	Inland regions of China, away from the coastal regions and border with the Soviet Union, where industries were dispersed in the 1960s because of the threat of war.

GLOSSARY OF SPACE TERMINOLOGY

Apogee	The furthest point from Earth in a satellite's orbit.
Geosynchronous orbit	An orbit 36,000km high in which a satellite orbits the Earth once every 24hrs over the equator, thus appearing to hover over the same point continually. It is a perfect altitude for

Glossary of space terminology 333

	satellites designed for global communications. This is also called a 24hr orbit or a geostationary orbit.
Supersynchronous orbit	A manoeuvre, developed by the Long March 3B, to fly out to 44,000km to save propellant on an orbital change manoeuvre, leading ultimately to a normal synchronous orbit.
Inclination	The angle in degrees (°) at which a satellite crosses the equator while orbiting the Earth. This also defines the parts of Earth over which the satellite orbits. Thus, a satellite circling at 58° will fly directly over the land and sea mass of Earth between latitude 58°N and 58°S.
Perigee	The lowest point to Earth in a satellite's orbit.
Period	The time which a satellite takes to orbit the Earth, generally given in minutes.
Sun-synchronous orbit	An orbit, generally polar, which passes over the same point on the Earth at the same time each day, ensuring a constant sun angle on the target or weather being photographed

Bibliography on the Chinese space programme (in English)

BOOKS, MONOGRAPHS AND ARTICLES

Gerald L Borrowman: China's long march to orbit. *Spaceflight*, vol 25, #5, May 1983.
China Space News: World Space Week, Hong Kong, 4-11th October 2000, commemorative publication (in Chinese and English)
Iris Chang: *The thread of the silkworm*. Basic Books, New York, 1995.
Heyi Chen: *Into outer space*. China Pictorial Publications, Beijing, 1989.
Jingwu Bai & Feng Li: *Footprints of China's launch vehicles and their further evolution*. Papers of the 54th International Astronautical Federation, Bremen, Germany, October 2003.
Yanping Chen: China's space commercialization effort – organization, planning and strategies. *Space policy*, #1, February 1993.
Robert Christy: Chinese puzzle no more. *Space flight news*, #35, November 1988.
Phillip S Clark: The Chinese space year, 1984. *Journal of the British Interplanetary Society*, vol 39, #1, January 1986.
Phillip S Clark: China – in business and advancing fast. *Spaceflight*, vol 29, #2, February 1987.
Phillip S Clark: *Chinese space activity, 1987-8*. Astro Info Services, London, 1989.
Phillip S Clark: *China's recoverable satellite programme*. Molniya Space Consultancy, Heston, 1994.
Phillip S Clark: *Chinese launch vehicles*. Molniya Space Consultancy, Heston, 1996.
Phillip S Clark: *The Chinese space programme – an overview*. Molniya Space Consultancy, Heston, 1996.
Phillip S Clark: China speeds up in the space race. *Jane's Intelligence Review*, vol 9, #4, April 1997.
Phillip S Clark: Review of the Chinese space programme. *Journal of the British Interplanetary Society*. 1998.
Phillip S Clark: China's DFH-2 and DFH-2A communications satellites programme. *Journal of the British Interplanetary Society,* Vol 54, 2001.

Phillip S Clark: *Chinese space activity, 1996-2000*. Molniya Space Consultancy, Hastings, 2001.
Phillip S Clark: The Feng Bao I launch vehicle programme. *Journal of the British Interplanetary Society*, vol 55, #7/8, 2002.
Phillip S Clark: Orbital manoeuvres of China's Zi Yuan satellites. *Journal of the British Interplanetary Society*, vol 55, #7/8, 2002.
Phillip S Clark: Chinese satellites – status report at the end of 2001. *Journal of the British Interplanetary Society*, vol 55, #7/8, 2002
Phillip S Clark: *The development of China's piloted space programme – from sounding rockets to Shenzhou 5*. Molniya Space Consultancy, 2003.
Phillip S Clark: *The first flights of China's Shenzhou spacecraft*. Monograph, November 2003.
Chunling Lu & Zhengxi Qin: *The payload system design of China ocean colour satellite HY-1*. Papers of the 54th International Astronautical Federation, Bremen, Germany, October 2003.
Feng Jian-xiang, Zhang, Guo-li, Zhang, Zhe, Han Cun-bing; Li Qiang, Gong Jun-min, Tang Bin & Gong Bo: *Kamado – a lunar robot and its telescience and intelligentsianization system architecture*. Papers of the 54th International Astronautical Federation, Bremen, Germany, October 2003.
Kenneth Gatland: *Missiles & rockets*. Blandford Press, London, 1975.
Kenneth Gatland (Ed): *Illustrated encyclopaedia of space technology*. Salamander, London & New York, 1989 (second edition)
Great Wall Industry Corporation: *Long March 2C – launcher manual*. Beijing, 2000.
Jim Harford & Wilbur Pritchard: *China space report*. New York, American Institute of Aeronautics & Astronautics, 1980.
Vincent Kohler: China's new long march. *Analog*, 1987.
Hormuz P Mama: China's Long March family of launch vehicles. *Spaceflight*, vol 37, #9, September 1995.
Frank Marble: Tsien revisited. *Caltech News*, volume 36, 2002.
G Lynwood May: China advances in space. *Spaceflight*, vol 30, #11, November 1988.
James Oberg: Year of the rocket. IEEE *Spectrum*, May 2001.
James Oberg: Taikonauts prepare for liftoff. IEEE *Spectrum*, December 2001.
Théo Pirard: Chinese secrets orbiting the Earth. *Spaceflight*, vol 19, #10, October 1977.
Lawrence Stern & Jack High: America takes a long march into space. *Spaceflight*, vol 32, #4, April 1980.
Jiaqi Song & others: *China defence, research & development*. Hong Kong, China Promotion Ltd, Hong, undated.
Sun, HYX & Chen, XM: *Payload data management system for Chinese unmanned spacecraft Shenzhou*. Papers of the 54th International Astronautical Federation, Bremen, Germany, October 2003.
Sun Jiadong: Lifting off with flying colours. *China in focus*, #21.
Tangming Cheng, Xiaojun Wang & Dong Li: *The new generation launch vehicles of Long March family*. Papers of the 54th International Astronautical Federation, Bremen, Germany, October 2003.

336 Bibliography on the Chinese space programme (in English)

Reginald Turnill: *The observer's book of unmanned spaceflight.* Frederick Warne & Co Ltd, London & New York, 1976

Paolo Ulivi: *Earth and science payload of the Shenzhou orbital modules.* Sent for publication, 2004.

Zhu Yilin: *An introduction to Chinese space endeavour* (collected works). China Academy of Space Technology, Beijing, 1995.

Zhu Yilin: Space microgravity scientific experiments in China. *Spaceflight*, vol 35, #10, October 1993.

Zhu Yilin: Development of Chinese Earth satellites under Prof. Tsien. *Journal of the British Interplanetary Society*, vol 50, #5, 1997.

Zhu Yilin: Fast track development of space technology in China. *Space policy*, May 1996.

Zhu Yilin: Applications of remote sensing satellites in China. *Earth space review.* Vol 5, #3, 1996.

Zhu Yilin, with Xu Fuxiang: Status & prospects of China's communication broadcast satellites. *Space policy.* Vol 13, #1, February 1997.

Zhu Yilin: Development of small satellites and their operation in the Asia-pacific region. Monograph.

Zhu Yilin: Some results of Chinese space microgravity life science experiments. Monograph.

Zhang Yun (Ed): *The Chinese space industry today.* China Social Sciences Publishing Co, Beijing, 1986 (in four volumes).

PERIODICALS

Aerospace China
Aviation Week and Space Technology
China Today

Air & Cosmos
China Pictorial
Flight International

The British Interplanetary Society not only covers Chinese spaceflight development in its magazine *Spaceflight* and in its *Journal of the British Interplanetary Society* but organizes fora on the Chinese space programme for its members. Contact British Interplanetary Society, 27 South Lambeth Road, London SW8 1SZ, www.bis-spaceflight.com

WEBSITES RECOMMENDED FOR FURTHER INFORMATION

Go Taikonauts!
One of the best known websites on the Chinese space programme is Chen Lan's *Go Taikonauts.!* This has regular news bulletins, frequently asked questions about the Chinese space programme, with lists of satellites, launchings, rockets.
http://www.geocities.com/CapeCanaveral/Launchpad/1921

Websites recommended for further information

Sven's place
Swedish space engineer Sven Grahn runs a website devoted to aspects of the Soviet and Chinese space programmes, many of which illuminate obscure events through his own insights into signalling systems. The side has run several stories on aspects of the Chinese programme, including mission profiles. Downloadable recordings of signals from Chinese satellites are available. Strongly recommended.
http://www.svengrahn.pp.se

Spacedaily
Spacedaily is a daily internet news services on all aspects of spaceflight, generally provided in 1, 2 or 3 page stories. A subsection called *Dragon Space* covers the Chinese space programme, where back issues are available. It is an easy site to navigate and attractively presented.
http://spacedaily.com/news/china

Mark Wade's *Encyclopaedia Astronautica*
Mark Wade's *Encyclopaedia Astronautica* is a truly mammoth internet site, with a host of articles, technical pages, histories and photographs cataloguing the Russian, Chinese and other space programmes. One can spend hours there in fascination.
www.astronautix.com

China Daily
China Daily is the daily, English language newspaper of Beijing, available both in paper format and on the internet at:
http://www.chinadaily.com.cn

Xinhua
Xinhua is China's news agency. Stories and colour photographs are posted on its internet site.
http://www.xinhuanet.com

China Space News
Wednesday and Saturdays, produced by China Space Sciences & Technology Group Company
www.spacechina.com/news

China Great Wall Industry Corporation
Outlines the products of the Great Wall Industry Corporation, focussing on the Long March launchers, with sections for news, launch centres, satellites and gallery.
www.cgwic.com

People's Daily
The Beijing *People's Daily* provides regular reports on China's space programme. It is available both directly on the world wide web and on e-mail subscription.
http://english.peopledaily.com.cn/

Index

Academy of Sciences
 1965 plan, 75
 and Shenzhou 2, 268
 and sounding rockets, 228–9
 Applied Geophysics Research Institute, 228
 first artificial satellite, 49–50
 FSW series, 152–3
 Institute of Automation, 78
 microsatellites, 161
 role in space programme, 181
 seeds experiment, 149
Accidents, 54, 100, 119–122, 142
Aerodynamics Research Institute, Beijing, 178
America: see United States
Apstar 1, satellite, 120
Apstar 1A, satellite, 122
Apstar 2, satellite, 121
Argentina, cooperation with, 191
ARMT: see Chinese Academy for Solid Rocket Motors
Asia Pacific Conference on Space Cooperation, 191
Asia Pacific Mobile Telecommunications Co, 127
Asia Pacific Multilateral Cooperation in Space Technology and Applications Programme, 163
Asia Research Satellite, 307
Asiasat 1, satellite, 117, 125
Asiasat 2, satellite, 121
Asiasat, company 117

Astrium, European company, 162
Australia, and commercial launches, 119–121

Badr, satellite, 117
Bei Shizhang, 239
Beidou, programme, 157–160
Beijing Aerospace College, 169
Beijing Aerospace Command and Control Centre, 185, 252–3
Beijing Centre for Payloads and Applications, 181
Beijing Control Engineering Research Institute, 77
Beijing Environmental Engineering Institute, 97
Beijing Industrial College, 169
Beijing Institute for Spacecraft Systems Engineering, 49, 88, 179
Beijing Rocket Engine Testing Station, 177–8
Beijing Satellite Manufacturing Plant, 173
Beijing University of Aeronautics & Astronautics, 175
Beijing University, 169
Beijing Wan Yuan Industrial Corporation, 49, 171
Boeing Corporation, 129
Brazil, cooperation with, 91, 190–1, 195
 CBERS project, 153–7
Britain, cooperation with, 191
Budget of Chinese space programme, 293–4
Bush, President George W, 130
Bush, President George, 117

340 Index

Cai Qiao, 180, 239
Cao Gangchuan, 9, 261, 269
Carlucci, Frank, 117
CAST968 bus, 91, 140, 306–7
CBERS, satellites, 153–7
Central Meteorological Bureau, 135
Centre for Space Research & Applied Sciences, 181
Centre for Space Science & Applied Research (Academy of Sciences), 268, 314
Chandrayan Prathim, Indian Moon probe, 313
Chang e
    fairy, 14
    Moon probe, 313
Chang Fangyan, 98, 148, 157
Chang Kong, satellite series, 70
Chang Zheng rocket: see Long March rocket
Chang, Iris, 18
    assessment of Tsien Hsue Shen, 63–4
Changchun Applied Chemistry Research Institute, 54
Changchun Institute of Optics and Fine Systems, 78
Chen Fangyun, 310
Chen Nengkuan, 39
Chen Quan, yuhangyuan, 251
Chen Zugui, 253
Cheng Kaija, 39
Chi Haotian, 248
Chien Wi Zhang, 51
China Academy for Solid Rocket Motors (ARMT), 174
China Academy for Space Electronic Technology (CASET), 174–5
China Academy of Aerospace Navigation Technology, 175
China Academy of Launch Technology (CALT)
    and Shenzhou, 262
    communications satellite, 96–7
    description, 171–2
    Long March 2, 76, 79–82
    Long March 3, 98
    role in developing launchers, 210
    test facilities, 176–180
China Aerodynamics Research & Development Centre, Chengdu, 301

China Aerospace Corporation (CASC), 142, 171–5
China Assets Insurance Co, 123
China Children & Teenagers Fund, 163
China National Space Administration (CNSA), 170–1, 248
China Resources Satellite Application Centre, 182
China Satellite Launch & Tracking General Control, 185
China Space & Technology Group, 175
China Space Machinery & Electronics Group, 174, 176
China State Seismological Bureau, 161
Chinasat 7, 123
Chinasat 8, 124, 127
Chinastar 1, 128
Chinastar 22, 111
Chinese Academy of Space Technology (CAST)
    and Shenzhou, 262
    communications satellite, 96
    description, 172–3
    environmental satellites, 306
    Founded, 49
    FSW series, 76
    Long March 1, 52
    parasitic nanosatellites, 315
Chinese Electronic Technology Academy (CHETA), 174
Chinese Society of Aeronautics and Astronautics, 172
Chinese Society of Astronautics, 172, 189
Chinese Telecommunications & Broadcasting Satellite Corporation, 110, 123
Chuang xin, microsatellite, 155, 161–2
Ciu Yajie, 104
Clark, Phillip S, 121
Clarke, Arthur C, 96, 316
Clinton, President Bill, 125–7
Communications satellites, benefits of, 113–115
COPUOS committee, 192
COSPAS/SARSAT system, 191
COSTIND: see State Commission for Science, Technology & National Defence
Cox, Christopher, Commission, 62–3, 125–6
Cultural revolution, 50–2, 78, 93

Da Qi, 138
Daimler Benz, German company, 110
Deng Jiaxian, 38
Deng Qingming, yuhangyuan, 252
Deng Xiao Ping, Chinese leader
    accused by Tsien, 64
    and cultural revolution, 50
    and Dong Feng launch, 42
    and manned spaceflight programme, 245
    and project 863, 148
    and sounding rocket, 30
    cancelation of project 581, 28
    international cooperation, 189
    modernization, 189
    political ascendancy, 64, 68, 93
    rectification, 168
Dingxin, airport, 203
Discoverer, American satellite programme, 75, 85
Dong Fang Hong, first Chinese satellite
    comparison to Shi Jian 1, 61
    design, 49–50, 53
    enters orbit, 58
    launched, 58
    reaction to launch, 59
    still circling the Earth, 58
Dong Fang Hong 2, communications satellite series, 102–7
    and Beidou programme, 158
Dong Fang Hong 2A, communications satellite series, 107–9
Dong Fang Hong 3, communications satellite series, 109–111
    adapted as lunar orbiter, 312
Dong Feng 1, missile, 34, 37
Dong Feng 2, missile, 37–8
Dong Feng 2A, missile, 38
Dong Feng 3, missile, 37, 39
Dong Feng 31, missile, 42, 195
Dong Feng 4, missile, 37, 39, 42, 56, 195
Dong Feng 41, missile, 42, 195
Dong Feng 5, missile, 37, 39–40, 42, 72, 195
Dong Feng, missile, launches from Taiyuan, 206–7
Dong Xiaohai, yuhangyuan, 242
Dorbod Xi, landing site, 9–13
Doublestar, project, 190, 307–9

Echostar, satellite, 121

Environmental Engineering Space Laboratory, Harbin, 180
Environmental Simulation Engineering Test Station, 179
EPKM, motor, 121, 232
European Space Agency, cooperation with, 189–191

Fang Guojun, yuhangyuan, 242
Fang Rongchu, 222
Fang Wi, 245
Feng Bao, rocket:
    construction, 72
    description, 226–7
    design, 71
    first launch, 73
    JSSW series, 73–4
    origins, 40
    Shi Jian 2 series, 88–90
    tests, 72–3
Feng Huo, satellite, 111, 113
Feng Jishen, 16
Feng Shizhang, 169
Feng Yun, satellite series:
    accident, 142
    Feng Yun 1–1, 136, 138
    Feng Yun 1–2 138–9
    Feng Yun 1–3, 91, 139
    Feng Yun 1–4, 140
    Feng Yun 2 series, 141–3
    Feng Yun 2–1, 142
    Feng Yun 2 2, 143
    Feng Yun 3 series, 141
    Feng Yun 4 series, 142
Fifth Academy, 22–4, 54, 168, 170
Four modernizations, 68
Fourth Academy, 245
France, cooperation with, 189–191
Freja, Swedish satellite, 118
Fanhui Shi Weixing (FSW), satellite series:
    first, FSW 0, series reviewed, 92
    and Shuguang project, 239, 243
    challenges of 76–9
    first launch, 82–3
    first successful mission (FSW 0–1), 83–4
    fourth mission (FSW 0–4), 86
    origin, 75–6
    rest of FSW 0 series, 87
    results of programme, 147–153, 188

342 Index

Fanhui Shi Weixing (FSW), satellite series: cont.
  second mission (FSW 0–2), 85–6
  third mission (FSW 0–3), 86
  FSW 1 series, 143–5
  FSW 1–1, 144
  FSW 1–2, 144
  FSW 1–3, 144
  FSW 1–4, 118
  FSW 1–5, 144–5
  FSW 2 series, 145–7
  FSW 2–1, 145
  FSW 2–2, 145
  FSW 2–3, 146–7
  FSW 3 series, 147
  FSW 3–1, 146

Galileo programme, 158–160
Gang of Four, 64, 67, 85, 90, 93, 98, 244
Germany, cooperation with, 190–2
  solar observatory, 309–310
GF series, description, 232
GF–1, solid rocket engine, 55
Grahn, Sven, 200–1
Great leap forward, 27
Great split (Sino-Soviet), 34
Great Wall Industries Corporation, 109, 122, 131–2, 175, 189, 195, 305
Gu Yidong, 280
Gua Guantan, 158
Guo Baozhu, 294
Guo Rumao, 241–2
Guo Yonghuai, 22

H–8, upper stage, 98
Hainan, launch site, description, 207
Haiyang, oceanographic satellite, 91, 140–1, 306–7
Hang Wen, 112
Hao Yan, 99
Harbin Industrial College, 169
Harbin Industrial University, 180
Harbin Military Engineering College, 169
Harbin Polytechnical Institute, 191
Harbin Polytechnical University, 180, 287
Harbin University of Technology, 162–3
Harbin, military launch site, 195
Haxi Chemical & Machinery Co, 232
He Ping, sounding rockets, 228

Heat shield, for FSW, 77
Hebei Semiconductor Research Institute, 152
Hong Chunhui, 7
Hu Jintao, President of China, 2, 3, 261
Hu Wenrui, 313
Hu Yaobang, 42, 106
Hua Guofeng, 68
Huang Chan, 168
Huang Chunping, 218
Huang Weilu, 23, 42, 53, 69
Huayin Machinery Plant, Shanghai, 179
Hughes Corporation, 124–130
Huxi Xincun range control centre, 200

India, cooperation with 191
Indonesia, cooperation with, 195
INMARSAT, 192
Institute for Automation (Academy of Sciences), 77
Institute for Aviation & Space Medicine, 180
Institute for Military Operations Research & Analysis, 64
Institute for Plant Physiology & Econology (Shanghai Institute of Biological Sciences), 279
Institute for Special Equipment, Beijing, 185
Institute for Technical Physics, Shanghai, 142
Institute for Tracking, Command & Control, Luoyong, 185
Institute of Control Engineering, Beijing, 173
Institute of Environment Test Engineering, 76
Institute of Genetics, 148
Institute of Medical Space Engineering, 267
Institute of Space Engineering, Beijing, 262
Institute of Space Medicine, 240–1
Institute of Space Physics, Xian, 182
Institute of Space Technology, Xian, 182
Intelsat 708 satellite, 122, 125
Interagency Space Debris Coordination Committee, 139
International Astronautics Federation, 69, 172, 192, 246, 253, 300–2, 312
International Organization for Standardization, 192
International Space Station, 191, 284, 286
International Telecommunications Union, 69, 97, 191–2

Iridium, communications satellite, 126, 131–3
Italy, cooperation with, 97, 189, 191

Japan, cooperation with, 189
Jen Hsinmin, 245
Ji Shu Shiyan Weixing, satellite series, 70–4, 92–3
Jia Shugung, 240
Jian Bing, series, 76, 147, 156
Jian Qing, 48
Jiang Jingshan, 310
Jiang Zemin, President of China, 63, 192, 218, 247–8, 254, 259, 261, 269, 274–5, 285
Jing Haipen, yuhangyuan, 252
Jingxi, hotel conference, 239
Jingyun: see Harbin, military launch centre
Jiuquan, launch centre:
  American overflights, 32–3
  area 2, 56
  decided as launch site, 31
  description, 196–204
  flight of Yang Liwei, 1
  launch of Feng Bao, 73
  missile launches from, 34–41
  sounding rocket launches, 45

Kaitouzhe, solid fuel rocket, 306
Karshi, ground station, 183, 278
Kiribati, ground station, South Tarawa atoll, 186
Kompsat, satellite, 132
Korea, cooperation with, 191
Kua Fu, satellite, 91

Laboratory for the Detection of Microwaves, Technology & Information, 181
Landsat, ground station, 69, 185
Lanzhou Institute of Physics, 76, 178, 182
Lei Fanpei, 175
Li Benzhen 185
Li Biyong, 169
Li Boyong, 80, 231
Li Chunqi, 109
Li Jinai, 7, 14
Li Jiyuang, 300
Li Naiji, 54, 169
Li Peng, prime minister, 64, 108, 192, 215

Li Tsinlong, yuhangyuan, 248–9
Li Xue, 115
Li Zaishen, 32
Li Zhankui, 82
Li Zhengjun, 242
Liang Shoupan, 29, 169
Liang Sili, 81, 169, 310
Lin Bao plot 67, 242
Lin Bao revolutionaries, 60
Lin Bao, 61
Lin Huabao, 45
Liquid Fuel Rocket Engine Research Institute, 98
Liquid Fuel Rocket Engine Testing Station, 104
Liquid Rocket Engine Company, 210
Liu Buoming, yuhangyuan, 251
Liu Chengjun, 7
Liu Chuanru, 100
Liu Chuanshi, 96
Liu Hong, 287
Liu Jiyuan, 215
Liu Mui, 277
Liu Shaoqi, President of China:
  and reorganization of space programme, 48
  and rocket programme, 38
  and sounding rocket, 30
  and Taiyuan launch site, 207
Liu Shuzhi, 242
Liu Wang, yuhangyuan, 251
Liu Yongdong, 268, 270
Liu Zhenxing, 3145
Liu Zhongying, 174
Liu Zhusheng, 280
Lockheed Martin Corporation, 121, 129
Long Lehao, 74, 311, 316
Long March, rocket:
  Long March 1 series:
  1 M version, 211
  1D version, 305, 211
  description of series, 211
  design, 52–6
  1C version, 211
  origins, 39
  Long March 2 series:
  2A and 2C, description, 212–3
  2B, 212
  2C launcher manual, 196

Index 343

344 Index

Long March, rocket: *cont.*
  2C version adapted for Iridium, 131–2
  2D description, 214–5
  2D replaces 2C model, 145
  2E and Cox report, 125
  2E commercial missions, 119–121
  2E first mission, 117
  2E, description, 215–6
  2EA, description, 216–7
  2F and flight of Shenzhou 5, 3–7
  2F facilities at Jiuquan, 202
  2F first rollout, 257
  2F, description, 218–220
  compared to Long March 1, 92
  CTS version, 308
  description of series, 212–230
  design, 79–82
  first mission, 82–3
  origins, 50, 76
Long March 3 series:
  3, description, 220–1
  3A and lunar mission, 311
  3A, description, 221–2
  3B and Cox report, 125
  3B and Moon mission, 311
  3B, description, 223–4
  3B, first mission, 122
  3B, subsequent missions, 122–4, 133
  3C, 305, 311
  description of series, 220–4
  design, 97–100
  first launch, 103
  origins, 96
  second launch, 104–5
Long March 4 series:
  4A, description, 225
  4B, description, 225
  4B–8S, description, 225–6
  crash of upper stages, 157
  description of series, 225–6
  first launch, 138
  origins, 137
  upper stage explosions, 139, 155
  Long March 5 series, 301–4
Long Xuehao, 224
Loral Corporation, 124–130
Lu Sicheng, 197
Lu Xiangxiao, yuhangyuan, 242
Luan Enjie, 130, 170, 284, 294, 316

Ma Zuozin, 52
Malindi, Kenya, tracking station, 187
Mao Zedong, Chinese leader:
  and communications satellite, 96
  and Feng Bao, 227
  and Shanghai Academy of Space Technology, 174
  and sounding rocket, 30
  and Taiyuan launch site, 207
  approval of Earth satellite project, 25
  commends launch, 60
  death, 67
  ending of Shuguang programme, 243
  foundation of Chinese space programme, 22
  FSW mission, 84
  launched cultural revolution, 50
  medallions on FSW 1–5, 145
  preparation of first satellite, 58
Marble, Frank, 63
Marembi Corporation, 146
Mars probe, Chinese plans for, 314
Matra, French company, 87
Meng Zhizhong, 136, 174
Messerschmitt Bolkow Blohm, Germany company, 109
Micro Electronics Research Institute, 81
Military satellites, plans for, 315
Min Guirong, 69, 181, 310
Ministry of Machine Building, 96
Ministry of the Space Industry, 170, 189
Mir, space station, 191, 284–5
Moon base, Chinese plans for, 316
Moon probe, Chinese plans for, 310–314
Motorola Corporation, 131–3

Namibia: *see* Swakopmund
National Aerospace High Technology Space Robotic Engineering Research Centre, 160
National Defence Committee, 239
National Defence Science & Technology University, 169
National Remote Sensing Centre, 182
National Research Centre for Small Satellites & Related Applications, 163
National Satellite Meteorology Centre, 182
Netherlands, cooperation with 191
New Decade Institute, 124

Ni Zhongliang, 104
Nie Haisheng, yuhangyuan, 3, 251–2
Nie Rongzhen, Marshal of China:
   and cultural revolution, 51
   and sounding rocket, 30
   and tracking system, 182
   early Chinese space programme, 22
   first Chinese missile, 34
   project 581, 27
   role in space programme, 167–8
   visits Moscow, 24
Nixon, President Richard, visit to China, 95
Northwest China Industrial College, 169
Northwest Institute for Electronics, Xian, 182
Northwest Polytechnical Institute, Xian, 262

O'Keefe, Sean, 130
Olympiad microsatellite, 163
Optus 1, satellite, 119
Optus B2, satellite, 120
Optus B3, satellite, 120
Ouyang Ziyuan, 311, 313, 316

Pakistan, 117–8, 187
Pan Zhanchun, yuhangyuan, 252
PKM motor, 232
Projects:
   Project 1059 (first rocket), 28–9
   Project 211 (Moon probe), 311
   Project 331 (communications satellite), 96
   Project 581 (Earth satellite), 27
   Project 651 (Earth satellite) 47–50
   Project 701 (JSSW series), 70
   Project 711 (Shuguang), 239–254
   Project 863 (horizontal technology programme), 148, 160, 180, 246, 287, 301, 310, 314
   Project 911 (recoverable Earth satellite), 76 (see also: Fanhui Shi Weixing)
   Project 921 (manned spaceflight) 247, 253
Purple Mountain Observatory, 91, 182

Qi Faren, designer, 4, 57, 237, 253, 261, 280
Qi Qui Weixing, satellite balloons, 138
Qi Zing, 135
Qian Ji, 75, 87
Qian Qi, 240
Qin Wenbo, 249
Qinghua 1, satellite, 160–1

Qinghua Satellite Technology Company, 160, 163, 176
Qinghua Tongfang Company, 160, 176
Qinghua University Enterprise, 160, 176
Qinghua University, 160–1, 163, 169, 310

Reagan, President Ronald, 117, 245
Remote manipulator systems, 286–7
Ren Xinmin, 52, 56, 59, 69, 89, 98, 169, 211
Rockets, early Chinese, 15–17
Russia, cooperation with, 191
   in connexion with manned spaceflight programmes, 248
   in connexion with space stations, 284–6
   in connexion with rocket engines, 301–2
   see also Soviet Union

SAC–1, Brazilian satellite, 154–5
Sardenberg, Ronaldo, 154
Satellite Assembly Plant, Beijing, 173
Satellite Maritime Tracking & Control, 185
Satellite Oceanic Application Centre, 182
Satellite Survey Department, 184
Scientific Instrument Factory and Institute of Technical Physics, Shanghai, 182
Seeds, experiments with, 148–9
Seventh Ministry, 48, 52, 90, 96, 170
Shaanxi Liquid Rocket Engine Company, 182
Shaanxi Space Dynamics High Technology Company, 175
Shan Shan, dog, 44
Shang Yushe, 182
Shanghai #2 Bureau of Machine and Electrical Equipment, 71
Shanghai Academy of Space Technology (SAST):
   Chuang xin microsatellite, 161
   description, 174
   Feng Yun series, 136
   founded, 49
   role in development of launchers, 210
Shanghai Aerospace Technology Research Institute, 142
Shanghai Astronomical Observatory, 312
Shanghai Electronic Equipment Factory, 182
Shanghai Huayin Machinery Plant, 74
Shanghai Institute for Satellite Engineering, 174

346 Index

Shanghai Institute for Technical Physics, 146, 174
Shanghai Institute of Machine & Electrical Design, 27, 29, 46–7, 98
Shanghai Research Institute of Satellite Engineering, 181
Shanghai Scientific Instruments Plant, 61
Shanghai Telecom, 161
Shansi Electronic Equipment Plant, 182
Shen Jungjun, 248
Shen Qizhen, 239
Shen Weigou, 222
Shen Xinsun, 301
Shen Yuang, 245
Shen Zhongbao, 228
Shenjian: *see* Long March 2F
Shenzhou, manned spaceship:
  and SAST, 174
  description, 262–6
  tracking arrangements, 187
  Shenzhou 1 mission, 259–262
  Shenzhou 2 mission, 266–272
  Shenzhou 3 mission, 273–8
  Shenzhou 4 mission, 278–282
  Shenzhou 5 mission, 3–14
  Shenzhou 6 mission (prospective), 283
  Shenzhou 7 mission (prospective), 283
Shi Jian, satellite series:
  Shi Jian 1, 61–2
  Shi Jian 2 (including 2A, 2B) 87–90, 92
  Shi Jian 3 (cancelled), 91, 154
  Shi Jian 4, 91–2
  Shi Jian 5, 91–2
Shi Jinmaio, 71, 87, 211
Shijiendu, launch site, 30
Shin Jingmiao, 253
Shiyan Tongbu Tongxin Weixing, satellite, 105–7
Shiyan Weixing, satellite, 103
Shiyong Tongbu Tongxin Weixing series:
  Shiyong Tongbu Tongxin Weixing 1, satellite, 107
  Shiyong Tongbu Tongxin Weixing 2, satellite, 108
  Shiyong Tongbu Tongxin Weixing 3, satellite, 108
  Shiyong Tongbu Tongxin Weixing 4, satellite, 108
  Shiyong Tongbu Tongxin Weixing 5, satellite, 108–9
Shiyong Tongbu Tongxin Weixing, lifetimes, 112
Shuguang, manned spaceship precursor project, 239–254
Sinosat 1, satellite, 128
Sinosat, satellite, 111
Sirio, satellite, 97, 189
Small Satellite Research Institute (CAST), 315
Solar observatory, 309–310
Solid Fuel Engine Research Academy, 232
Song Jian, 99, 169
Sounding rockets, 29–30, 43–5, 228–9
South West Institute of Electronics Technology, Shijiazhuang, 182
Soviet Union, Chinese relationships with, 24–5, 34
Space Leading Group of the State Council, 170
Space Mechanical & Electrical Research Institute, Beijing, 155
Space Physics Research Institute, 228
Space Research & Technology Centre (Academy of Sciences), 181
Space shuttle (American), Chinese experiments on, 191
Space shuttle, Chinese plans for, 299–301
Space Solid Rocket Fuel Rocket Carrier Company, 174
Space station, Chinese plans for, 283–8
Space Vegetable Foundation, 149
Spacenet 1, 109
Spacenet 1, satellite, 109
Spar Communications Group, 113
Sputnik 1, Chinese observations of, 25–6
St Valentine's Day Massacre, 122
Star 63 rocket motor, 120–2
State Bureau for Environmental Protection, 306
State Commission for Science & Technology, 170
State Commission for Science, Technology & National Defence (COSTIND), 170–1, 185
State Disaster Reduction Committee, 306
State Meteorology Administration, 174
State Oceanic Administration Science & Technology Department, 140

Su Shuangning, 260, 280
Submarine launched missile, 42
Sui Qisheng, 252
Sun Baosheng, 280
Sun Jiadong, 49, 61, 69, 78, 86, 96, 313
Sun Jingliang, 137, 211
Sun Jixian, 31
Sun Laiyan, 311
Sun Zengqi, 310
Surrey Satellite Technology Ltd, 160-2, 176
Swakopmund, ground tracking centre, 10, 187
Sweden, cooperation with, 187 (*see also*: Freja)
Swedish Board for Space Activities, 118

T-5, sounding rocket, 29
T-7, sounding rocket, 29-30, 43-5, 228
T-7A, sounding rocket, 43-4, 77-8
T-7AS, sounding rocket, 44
T-7M, sounding rocket, 29-30
Taiyuan launch centre:
    description, 206-8
    selected, 137
Tan Ce, satellite (Doublestar), 308-9
Tang Jiaxuan, 156
Tang Xianming, 207
Tang Yihua, 311
Tansuo, microsatellite, 162
Tao Feng, 278
Tao Hong, 78
Tao Jiaqi, 175
Third Institute of Research, Beijing, 300
THNS, microsatellite, 163
Tian Xing Ecoagriculture Company, 277
Tianwen Weixing, 91, 272
Tracking Data and Relay System (TDRS), 160
Tsien Hsue Shen, founder of the Chinese space programme:
    and cultural revolution, 50-1
    and first Chinese missile, 34
    and sounding rocket, 30
    arrested, jailed, expelled, 19-21
    assessment, 62-4
    at Jiuquan, 32
    book 'Introduction to interplanetary flight', 46
    campaign to resume Earth satellite project, 46-9
    celebrating first satellite, 59
    cofounded Jet Propulsion Laboratory, 19
    Cox report, 127-8
    early space programme, 168-9
    first FSW mission, 83
    first satellite, 57
    foundation of Chinese space programme, 22
    in Germany, 19
    International Astronautical Federation, 69
    later life, 64
    legacy, 316
    origins and childhood, 17-18
    preparations for first satellite, 56-7
    rediscovery by Americans, 38-9
    reorganization of space programme, 48
    returned to China, 21
    shuttle designs, 299
    spaceship Tsien Hsue Shen, 316
    study of manned flight, 239, 254
    with model of spaceship Shenzhou, 63
    work in the United States, 18-20
Tu Shancheng, 173
Tu Shoue, 29, 40, 69, 78, 82, 169, 211

Ukraine, cooperation with 191
Ulivi, Paolo, 91, 312
United Nations, 191
United States:
    cooperation with, 189, 191
    poor relations with China over space programme, 124-131, 133-4
    restrictions on Chinese space activities, 115-117
    visit of experts to China, 97

Wan Hu, 14, 17
Wang Bingzhang, 48, 67
Wang Daheng, 78, 148
Wang Dayan, 310
Wang Dechen, 82
Wang Enmiao, 105-6
Wang Gangchan, 39, 148
Wang Heng, 100, 222
Wang Jintang, 78
Wang Liheng, 253
Wang Lin, 42

Wang Shuanglui, 246
Wang Shusheng, 300
Wang Xiaojun, 302
Wang Xiji, 27, 30, 43, 75, 78
Wang Yongzhi, 82, 237, 253, 282
Wang Yusheng, 146
Wang Zheng, 81
Wang Zhenyin, 88
Wang Zhiren, 99–100, 177
Wang Zhiyue, yuhangyuan, 242
Wang Zhuanshuan, 245
Weaver Girl, sounding rockets: *see* Zhinui
Wei Jiqing, 25
Weinberger, Casper, 117
Wen Ho Lee, 127, 129
Wen Jiabao, Chinese prime minister, 11
White paper on space programmes, 296–9
Wu Bangguo, 275
Wu Dehu, 169
Wu Faxian, 242
Wu Tse, yuhangyuan, 248–250
Wu Yansheng, 171, 302

Xi Change launch site:
    description, 204–6
    development, 100–1
    on tourist route, 191
Xian Institute of Radio Technology (XIRT), 181–2
Xian tracking system:
    and FSW programme, 76, 83
    beginning, 50
    description, 182–5
Xiandong Bao, 246–7, 301
Xiangyuanghong tracking ships, 185–6
Xiao Bao, dog, 44–5
Xiao Gan, 54
Xie Gyuangxian, 69, 98, 210
Xin Zhong Hua factory, 220
Xinyue Mechanical Electronics Plant, Shanghai, 182
Xu Dazhe, 301, 310
Xu Fuxiang, 173, 261
Xu Guozhi, 22
Xu Qijiang, 175
Xu Xin, 161
Xue Lun, 242

Yang Anning, 277

Yang Guoyu, 242
Yang Liwei, first yuhangyuan:
    criteria for being a yuhangyuan, 249–250
    space mission, 2–14
Yang Nansheng, 27, 30, 53, 232
Yang Shucheng, 306
Yang Yiachi, 25, 69, 77, 83, 148, 168
Yang Yiaxi, 310
Yao Tongbin, 168
Ye Dinghou, 174
Ye Shuhua, 312–3
Yeti Fadong (rocket engines):
    YF engines for LM–1D, 305
    YF–1 series, 230
    YF–2, 52
    YF–20 series, 80–1, 231–2
    YF–40 series, 231
    YF–73, 98–100, 203, 231–2
    YF–75, 232
Yu Guiliu, yuhangyuan, 242
Yu Menglun, 81–2, 311
Yu Qiulu, 24203
Yuan Jiajun, 179
Yuan Jiajun, 247, 274
Yuang Wang, tracking ships:
    description, 185–7
    first missile tests, 40–1
    flight of Shenzhou 5, 4–10
    following Shenzhou 1, 259
    return to port after Shenzhou 1, 261
    Second Long March 3, 104
    Shenzhou 2 deployments, 266, 269
    Shenzhou 3 deployments, 274, 276

Zeng Guang Shuang, 222
Zhai Zhigang, yuhangyuan, 3, 251–2
Zhang Aiping, 34, 40–1, 64, 83, 103, 105–6, 168
Zhang Fengxian, 185
Zhang Guitian, 52
Zhang Heqi, 268
Zhang Jianqi, 197
Zhang Kexun, 300
Zhang Lianfu, 41
Zhang Qingwei, 301
Zhang Tong, 175
Zhang Wannian, 261
Zhang Wenzhong, 174
Zhang Xiaomin, 163

Zhang Xinxia, 175
Zhang Yinchun, 163
Zhang Yumei, 9, 14
Zhao Jiyzhang, 50, 75, 240
Zhao Ziuzhang, 25
Zheng Jiwen, 177
Zheng Songhui, 86
Zheng Yuanxi, 81
Zhinui (weaver girl), sounding rockets, 208, 229
Zhong Chuandong, yuhangyuan, 251
Zhong Guofo, 77
Zhongwei 1, 128
Zhongxing 20, satellite, 113
Zhongxing 22, satellite, 111
Zhongxing 6, satellite, 110
Zhongxing 6B, satellite, 111
Zhongxing 7, satellite, 110
Zhongxing 8, satellite, 111
Zhongxing, satellites 1–5, 109
Zhou Enlai, Chinese leader:
    announcing launch of first satellite, 58, 291
    at Jiuquan, 39
    building of Jiuquan, 32
    choice of Taiyuan, 207
    choice of Xi Chang, 206
    cultural revolution, 51–2, 55, 168
    foundation of Chinese space programme, 22, 168
    inspects Long March 2, 81
    preparations to launch first satellite, 56–7
    project for communications satellite, 96
    protecting space programme, 60, 93
    reorganization of space programme, 48
    resumption of satellite project, 48
    tracking system, 182
    welcomes Tsien Hsue Shen back to China, 21–2
Zhou Jianping, 3, 280
Zhou Jung, 280
Zhou Sumin, 161
Zhou Xiaofei, 283
Zhu Gangrui, 173
Zhu Guangya, 38
Zhu Juingren, 169
Zhu Lilan, 149
Zhu Senyan, 100
Zhu Yilin, 30, 46, 78, 88, 112, 173
Zhu Zheng, 169
Zhuang Fenggan, 77, 168–9, 274–5, 311
Zhung Jun, 170
Zi Yuan series, 156–7
Zijnshan Astronomical observatory, 268
Zu Huang, 300